PRINCIPLES OF
BIOLOGICAL MICROTECHNIQUE

PRINCIPLES OF BIOLOGICAL MICROTECHNIQUE

A Study of Fixation and Dyeing

by

JOHN R. BAKER

D.Sc.(Oxon.), F.R.S., *Emeritus Reader in Cytology, Oxford University*

METHUEN & CO LTD

11 NEW FETTER LANE, E.C.4

First published 15th May 1958

Reprinted Twice

Reprinted 1968

1.4

S.B.N. 416 60080 8

Printed in Great Britain
by Richard Clay (The Chaucer Press), Ltd,
Bungay, Suffolk

Distributed in the U.S.A.
by Barnes & Noble Inc.

DEDICATED

TO THE MEMORY OF

PAUL EHRLICH

GENIUS

IN CHEMOTHERAPY

IMMUNOLOGY

MICROTECHNIQUE

Preface

The principles of biological microtechnique may perhaps be reduced to one—the principle that when we make a microscopical preparation of any sort, we ought to try to understand what we are doing, for otherwise we shall examine an unknown object that has been treated in an unknown way. A scientific outlook has been introduced into certain branches of our subject, particularly histochemistry, but there are others in which rule of thumb rules indeed. One thinks at once of the most ordinary processes of the histological or cytological laboratory: of fixation, embedding, dyeing, and mounting. Here the empirical outlook is often manifest, and some workers are content to follow the recipe-book blindly, as though a scientific result could be obtained by unscientific means.

A very long book—and a very learned author—would be necessary if the attempt were made to illustrate the principles of microtechnique by a full consideration of all its branches. It seems best to concentrate on the most familiar processes, so that the principles may find most frequent application in practice. In fixation and in dyeing the tissues are *responsive*: they react to what we do to them. In embedding and mounting they are more passive, allowing us to surround them with what we will. I therefore choose fixation and dyeing as being even more interesting than the other familiar branches of microtechnique. I hope that a study of the principles that should guide us in these branches may engender an outlook towards microtechnique that will find application in a wider field. This outlook is at least as necessary when tissues are prepared for the electron microscope as in our more homely endeavours within the realm of light.

The book is addressed to research-workers, teachers, and students in the fields of pathology, histology, cytology, zoology, and botany. The primary intention has been to make it as useful and attractive as possible to the consecutive reader, but certain features have been introduced to help the casual inquirer. Thus there are numerous cross-references and rather a lot of repetitions. Three of the chapters (5, 6, and 9) can be used as though they were parts of a work of reference. A full index is provided.

This is in no sense a text-book. There is, I hope, nothing dogmatic in it. Its purpose is as much to show the gaps in knowledge as to knit together what is surely known. The possibilities for research in microtechnique seem endless, and every effort has been made to point out as many of them as possible.

The book contains a good deal of new material of three sorts. First, there are new contributions to the theory of fixation and dyeing. Secondly, there are many factual observations that have not been published previously. Thirdly, the Appendix contains full descriptions of new experiments illustrating the principles underlying the processes of fixation and dyeing. Most of these can be carried out in practical classes.

There is in the whole book no practical instruction on how to make a microscopical preparation. For this the reader should turn to one of the excellent guides to the subject, such as Langeron's *Précis de microscopie*,[282] or Pantin's *Notes on microscopical technique for zoologists*.[383]

Apart from my own little book on *Cytological technique*,[34] I know of only three that cover more or less the same field as the present work. These are Fischer's *Fixierung, Färbung und Bau des Protoplasmas* (1899),[169] Mann's *Physiological Histology* (1902),[322] and Zeiger's *Physikochemische Grundlagen der histologischen Methodik* (1938).[553] There cannot be many equally important fields of science to which so few books have been devoted—and those few of such merit as these three.

In the historical parts of the book I have adopted the usual convention of giving as the date of a discovery the year in which it was first made known in print.

Acknowledgements. It was Mr Frank Sherlock, Head Technician of the Department of Zoology, Oxford, who first gave me instruction in microtechnique, and I have always owed a debt of gratitude to him. Dr H. M. Carleton was generous with good advice throughout our association of thirty-five years. Prof. A. C. Hardy, F.R.S., has encouraged cytological studies in his Department and I am thankful for all that he has done to help me. Dr M. Wolman generously sent me the proof of his article on fixation[541] before it was published. One learns best, perhaps, by continual association with lively young minds, and it has been my good fortune for many years to have a splendid succession of research-pupils from many lands, to whom I owe much. I have always been lucky in my assistants and must particularly mention Mrs B. M. Jordan-Luke, whose skill in microtechnique has been of great benefit in my research. She has given much practical help in connexion with the experiments described in the Appendix. Mrs J. A. Spokes has helped me with this book more than anyone else, by acting as my secretary with uniform accuracy and good nature. Several of the illustrations have been copied photographically from old books and journals. Mr P. L. Small and Mr J. S. Haywood have taken a lot of trouble to produce as good copies as possible. Dr W. G. B. Casselman has most kindly made several readings of the oxidation-potentials of fixatives specially for this book.

I take the opportunity of mentioning the benefit I have derived during the last eleven years from my association with Dr C. F. A. Pantin, F.R.S., in joint editorship of the *Quarterly Journal of Microscopical Science.*

JOHN R. BAKER

Cytological Laboratory,
Department of Zoology,
University Museum,
Oxford.

Postscript.—The reprinting of the book has enabled me to correct a few small errors and to delete three short passages, but there has been no major change.

J. R. B.

November 1959.

A second reprinting has enabled me to report on p. 221 some recent work on the differentiation of mordant-dyes by acids. A slip in the structural formula for iodo-nitro-tetrazolium has been corrected (p. 293).

J. R. B.

September 1962.

Preface to the reprint of 1968

Since this book was reprinted in 1963, the emphasis of research in microtechnique has shifted. Interest now centres on the preparation of tissues for study by electron microscopy. As a result, the amount of new information relating to the fixation and dyeing of tissues for study by optical microscopy has scarcely been sufficient to warrant the preparation of a new edition of this book. The fifth edition of my smaller book on 'Cytological Technique', published by Messrs. Methuen & Co., Ltd, in 1966, incorporates a good deal of such recent information as exists. Readers of the present work may care to look at the chapter (5) in its smaller counterpart that deals with practical fixative solutions, because it introduces, at the end, a new classification of fixative mixtures that may be thought to have some value. Chapter 6 in the same little book also contains an account of newer work on the mode of action of mordants and a description of the dye 'haematal 8', which may perhaps have some claim to be regarded as a particularly convenient aluminium-haematein dye for routine use. In view of the continued importance of aluminium-haematein dyes in histology and micro-anatomy, it seems legitimate to mention my paper on their mode of action (*Quarterly Journal of Microscopical Science*, **103**, pp. 493–517 (1962)). I still have some copies of this paper, which I should be happy to send to applicants.

A paper of special importance to those interested in the direct attachment of dyes to tissue constituents was published by D. J. Goldstein in 1963 (*Quarterly Journal of Microscopical Science*, **104**, 413–439).

JOHN R. BAKER

Department of Zoology,
Parks Road,
Oxford.

Contents

APPENDIX

Illustrations

PART ONE

FIXATION

B

CHAPTER I

Introduction to Fixation

Every cytological investigation should start, if possible, with the study of the living cell. A lot of useless controversy would have been avoided if this guiding principle had been obeyed, for too much reliance has been placed on the study of dead cells. It may be queried, then, why we do not restrict ourselves to the study of living material. There are four main reasons.

(1) When tissues are cut in thin sections, it is very easy to determine the relations of the cells to one another and to the intercellular material; the structure of the cells themselves is often very clearly revealed. Some kinds of cells cannot be isolated for vital study, and these are best examined in sections. There are usually practical difficulties in cutting sections of living tissues thin and uniform enough for convenient microscopical study, and anyhow living cells are necessarily damaged by being divided. It is easy to cut thin, uniform sections of dead tissues that have been treated in particular ways.

(2) Although vital colouring gives very important information, yet it is not a method of general application, for many tissue-constituents are not revealed by it. Almost all the constituents of dead tissues can be dyed in brilliant, contrasting colours.

(3) Few histochemical tests are applicable to living cells.

(4) It is convenient to have permanent preparations.

We need an understanding of the kinds of changes that protoplasm undergoes when the processes of microtechnique are applied to it: armed with that, we can profit greatly from the study of dead material. Such understanding presupposes familiarity with living material. In this book we are not directly concerned with the techniques used in the study of living material, except those of vital colouring, but a general knowledge of the structure of living cells will be assumed throughout. We shall consider what happens to the living cell when it is fixed, and to the fixed cell when it is dyed.

It is necessary at the outset to distinguish between preservation and fixation.

A small piece of tissue that has been cut out of an organism will generally cease to retain its structure at the microscopical level quite soon, unless special precautions are taken to keep the cells alive. The main potential causes of damage are evaporation, osmotic swelling or shrinkage, attack by bacteria or moulds, and autolysis.

Autolysis is the self-digestion of cells by enzymes that are always present within them and presumably synthesize protoplasmic proteins during life. On the cessation of normal vital activity their action is reversed, at any rate in the sum-total of its effect. These enzymes are known collectively as kathepsin. Two of them are proteinases, similar to pepsin and trypsin: these shorten the protein chains to peptides. Two others are an aminopeptidase and a carboxypeptidase, capable of pulling the terminal amino-acids off the newly-exposed ends of the peptide molecules. As a result there is a general dissolution of the proteins, which are rendered no longer capable of coagulation by heat or chemical agents. The rate of autolysis can be measured by finding what proportion of the protein has become non-coagulable in a given time. The process is slow at 6° C but still occurs even at 0°.[162]

To preserve a piece of tissue, then, it is necessary to place it in a fluid that will neither shrink nor swell it, nor dissolve or distort its constituent parts; will kill bacteria and moulds; and will render kathepsin inactive. A fluid that will do this is a *preservative.*

It would be very convenient to have fluids in which separate cells or teased fragments of tissues might be indefinitely preserved and in which they could be examined microscopically at any time. Although it has been customary for centuries to preserve organisms and their parts in suitable fluids for subsequent study with the naked eye or hand-lens, not many cytologists or histologists have sought to elaborate fluids of this kind that would serve their needs. The best-known of the few such fluids that exist are Petit's (usually known as the fluid of Ripart and Petit [107]) and Amann's,[6, 7] especially the latter's lactophenol. Amann called his fluids *Beobachtungsmedien,* to indicate that they not only prevented the decay of tissues, but were media in which cells might remain during microscopical examination. It is desirable that research should be undertaken to find new, improved preservatives, but the subject does not fall within the scope of this book.

A *fixative* must do everything that a preservative does, and something else as well. We do not say that we fix a door open when we merely open it: fixing implies that we take action to ensure that it will retain its position when other forces subsequently act upon it. Similarly, the essence of fixation is that the various tissue-constituents are modified in such a way that they retain their form as nearly as possible when the tissue is subjected to treatment that would have damaged them in their initial state. It follows that fixation is a forward-looking process: it exists only in relation to subsequent events.

The subsequent event in which the early microscopists were chiefly interested was the cutting of sections with a hand-razor. They wanted above all to make the tissues hard, and they called the process 'hardening'. Some of their hardening fluids are used as fixatives to the present day. When it was discovered that the necessary support could be given to the tissues by embedding them in collodion or other media, less emphasis was placed on hardening, and the term 'fixation' came into general use in the early eighteen-eighties.[478]

In modern microtechnique the processes against which it is especially important that tissues should be protected are embedding, sectioning, and mounting. The first- and last-named often involve dehydration, which has a strong tendency towards distortion; embedding often requires a high temperature; sectioning can cause mechanical damage (especially cracking and crumbling). A fixative is a fluid that stabilizes the tissue-constituents as far as possible against these and other potentially damaging processes.

Although this is the primary function of fixatives, yet there are others, scarcely less important. In particular, most fixatives make the tissues much more easily colourable by dyes than when they were alive, and colourable in particularly informative ways. After suitable fixation almost every part of the cell and of the intercellular material can be dyed, often with great selectivity, so that neighbouring parts show up brilliantly in different colours. Chromatin, which is scarcely colourable during life, is one of the most easily dyed of all tissue-constituents after fixation. The wealth of our knowledge of cytogenetics is to a large extent due to this fact.

Living cytoplasm commonly has a refractive index (r.i.) in the neighbourhood of 1·353; [429, 430] that is to say, not very much higher than that of the saline solutions in which cells are commonly immersed for vital study. An aqueous solution of sodium chloride at 0·9% has an r.i. of 1·335.[431] When cells are examined alive in such media, a water-immersion objective will give almost as good resolution as a first-rate oil-immersion objective, for the high numerical aperture of the latter will be partly wasted. One may surround living cells with innocuous media of the same r.i. as the cytoplasm [44] and thus obtain slightly higher resolution (as well as gaining other advantages), but the difference will not be great. As soon, however, as a fixative acts, a profound change occurs. The evidence suggests [44] that the protoplasm is now represented by interlacing sub-microscopic fibres having the r.i. of dry protein (about 1·54). These fibres lie in water, if the fixative is aqueous; this can be replaced by media of any desired refractive index. If a medium of r.i. close to that of dry protein is used (Canada balsam, for instance), two results ensue: almost perfect transparency is obtained (which may be modified as desired by the use of dyes), and oil-immersion objectives can be used at their full aperture. This fact should not, however, be too strongly stressed, for the making of a permanent microscopical preparation involves considerable shrinkage of the tissues of organisms (p. 76), and this reduces or nullifies the advantage of higher microscopical resolution. The advantage can be fully secured only if the object to be examined happens to be unshrinkable by the processes involved in making a permanent preparation. The valves of diatoms provide an example.

Some of the constituent parts of organisms do not require fixation, because they are not subject to autolysis and are resistant to bacteria and moulds and to most of the reagents ordinarily used in microtechnique. Examples are chitin, cellulose, scleroproteins, certain inorganic crystals, and amorphous silica. Most such substances not only do not need fixation but are not acted upon by fixatives; or, if acted upon, may be dissolved (for instance, spicules of calcium carbonate by fixatives containing acid). Many fixatives leave droplets of triglyceride untouched; these do not require to be fixed if no lipid-solvent will be used subsequently.

Apart from such substances as these, it is the main purpose of fixation to alter the tissue-constituents in such a way as to render them no longer subject to autolysis, to decay through the action of bacteria or moulds, or to distortion by subsequent treatment. The framework of the cell is of protein, and disintegration would occur instantaneously if this constituent were to disappear. Neither lipid nor carbohydrate is essential for the cohesion of protoplasm. For this reason it is always necessary to fix protein, whether other substances are stabilized by a particular fixative or not. Fixation is therefore primarily the stabilization of protein.

Fixation can be achieved either by chemical means or by the application of heat. The latter method involves the coagulation of proteins. It tends to cause distortion and does not commend itself, as a general rule, in purely morphological studies. It was especially recommended by Ehrlich, however, for the fixation of blood-smears. He advised short treatment ($\frac{1}{2}$ to 2 min.) at 110° C.[158] In a blood-smear the shape of the cells is anyhow distorted, but the cell-contents may react with special clarity to dyes after heat-fixation. The method could probably be used with advantage in many histochemical studies. Its possibilities have been somewhat overlooked.

Fixation by the use of reactive substances may be called chemical fixation for short, without prejudgement of the question whether chemical fixatives necessarily participate in all cases in chemical reactions with tissue-constituents. Most of the substances used for fixation are solids, used in aqueous solution; some are liquids that can be used without the addition of water.

The number of substances that are really useful in chemical fixation is very small. The substances may be divided into two major groups, according to their effects on proteins. The members of the two groups can easily be distinguished by testing their effect on a solution of albumin (p. 32). Some of them act, like heat, by coagulating the protein; others do not. The coagulant fixatives usually transform protoplasm into a network, while the non-coagulant do not. The principal fixative substances are listed on p. 24. The ones not marked 'absolute' are used in aqueous solution. (For the method of expressing percentage concentrations, see Appendix, p. 313.)

The names used in this list are those of the undissolved substances. These names will be used throughout this book. Thus, fixation by mercuric chloride or potassium dichromate must be understood to mean fixation by some or all of the various ions that are produced when mercuric chloride or potassium dichromate is placed in water (pp. 99 and 126). This usage will be adopted for two reasons. First, for simplicity: it would be tedious to have to mention all the products of solution every time. Secondly, for accuracy in the statement of concentrations. If, for instance, one adds distilled water to 0·5 g of chromium trioxide to make 100 ml of solution, the latter should not be called a 0·5% solution of 'chromic acid'; for the oxide takes up water in ionizing, and the acid produced (cations plus anions) is present at more than 0·5%.

	Suitable concentrations for fixation
Coagulant	
methanol	100% (absolute)
ethanol	100% (absolute)
acetone	100% (absolute)
nitric acid	0·5N
hydrochloric acid	0·5N
trichloracetic acid	2% w/v
picric acid	Saturated
mercuric chloride	Saturated
chloroplatinic acid	0·75% w/v
chromium trioxide	0·5% w/v
Non-coagulant	
formaldehyde	4% w/v
osmium tetroxide	1% w/v
potassium dichromate	1·5% w/v
acetic acid	5% v/v

In the experiments described in the succeeding chapters, the substances were always used at the concentrations given in the list, except where the contrary is distinctly stated. It will therefore be possible to omit mention of concentration and thus avoid unnecessary repetition.

Non-fixative or 'indifferent' substances, such as sodium chloride or sodium sulphate, are often added to solutions of these fixatives. The effects of this will be considered in a later chapter (p. 80).

None of the listed substances has all the qualities of a perfect fixative, and mixtures are therefore made with the intention of combining the virtues of the ingredients. Most mixtures contain two or more of the listed substances, and often an indifferent substance as well. The principles that should guide the design of fixative mixtures will be considered in chapter 7 (p. 139).

Mixtures are generally called by the names of those who introduced them. It is usual to use the surname alone: thus 'Bouin' means Bouin's fluid.

Chemical fixation is generally achieved by putting an organism or part of it in a fluid. In micro-anatomy and gross histology it is sometimes necessary to fix a large volume of tissue in a single piece. The fixative may be injected into a blood-vessel and thereby reach all depths in the piece at almost exactly the same moment. This results in an evenness of fixation that cannot be achieved in any other way with large pieces. The total volume of fixative that can be held in the blood-vessels is, however, small in comparison with that of the tissue; the fluid that has been injected has to diffuse through their walls before it can exert any effect on cells other than those of the blood-vessels; and many cells lie a long way (on the microscopical scale) from any blood-vessel, even a capillary. Perfusion should therefore not be used as a general rule. In cytology it is seldom necessary or desirable. The single cell that it is desired to study can easily be obtained by cutting out a piece of tissue a millimetre or so in diameter. If this be simply thrown into a much larger volume of fixative, the latter will reach all parts without much delay. The rate of penetration of fixatives is discussed in detail below (pp. 37, 67, 150).

Vapours are occasionally used instead of fluids, but most fixatives are not volatile, and the method is applicable only to very minute pieces. The advantage gained is that nothing is dissolved out of the tissue.

It is possible to fix in two stages, by taking preliminary action to stop autolysis and then applying a fixative fluid. Autolysis may be stopped almost instantaneously by placing tissues in some harmless, non-fixative fluid maintained at a very low temperature. Isopentane chilled to below −160° C is suitable. Ice-crystals form in the tissues, but if the temperature is low enough they are very small, and surprisingly little damage is done. In the process of 'freezing-thawing', the tissue that has been chilled is simply allowed to thaw in an ordinary fixative at room-temperature or thereabouts.[471] In 'freezing-substitution' an organic fixative fluid replaces the ice directly, without any melting of the ice-crystals. Ethanol [471] or methanol [141] may be used. The alcohol is chilled to −40° C or lower before the tissue is transferred to it from the isopentane, and then allowed gradually to warm up. It acts on the protein in the total absence of fluid water. These methods have not been sufficiently used to allow their value to be judged with confidence, but high claims have been made for the faithful stabilization of cellular structure by freezing-substitution.[141] The necessity to use a non-aqueous fixative limits choice severely.

A piece of tissue may be suddenly chilled and the ice completely evaporated off from it while the low temperature is maintained. This process of 'freezing-drying' is an old one.[4] It is often carried out in a very elaborate way, but quite simple apparatus is nowadays available.[273] After warming to room-temperature the dry tissue may be fixed as desired. The main advantage of the method, however, is that the tissue can be obtained in an unfixed but fairly stable state, the stability being due to the absence of water. This state is suitable for certain chemical and enzymological studies. The subject does not fall within the scope of this book, because neither fixation nor dyeing is directly involved.

Although it is the purpose of fixatives, as a general rule, to leave the structure of the tissue so far as possible unchanged, yet this cannot apply to its chemical composition. Indeed, it may be said that the whole purpose of fixation is to alter the chemical composition of certain tissue-constituents. If the chemical composition of the tissue were unaffected, the fixative would be without effect: the cells, in fact, would be still alive or like recently dead cells.

It might be thought that the need to alter chemical composition would render fixation inapplicable in histochemical studies, but this is far from being true. Indeed, the service of fixation to histochemistry is as important as its service to dyeing. Tests that would smash a living cell are applicable when it has been fixed. Many fixatives affect the proteins only, leaving DNA, RNA, lipids, carbohydrates, and various inorganic constituents, unchanged; and the proteins themselves, though radically altered, generally retain reactive groups that respond to histochemical tests. The retention of the original structure in the fixed tissue is nearly as important in histochemical as in purely morphological studies, for the purpose of histochemistry is to disclose the distribution of particular substances in space.

Although it is the purpose of fixation to stabilize the form of tissue-constituents, yet it is not always desirable to preserve all of them. In studies of chromosomes, for instance, it is generally best to use a fixative that will either destroy mitochondria or allow them to be destroyed by subsequent treatment, for otherwise they will obscure the view.

With such exceptions as this it is the purpose of fixatives to make the structural resemblance between the final preparation and the living tissue as close as possible.

The appearance given in a fixed microscopical preparation is nearly always to some extent artificial. The artifacts (p. 329) are of two main kinds: intrinsic, or caused by distortion of the original tissue-constituents; and extrinsic, or caused by the deposition within the tissue of extraneous objects (such as the mercury precipitate mentioned on p. 103). Intrinsic artifacts are of two kinds: primary, or produced by the fixative itself and therefore visible while the tissue still lies in the fixative; and secondary, or produced by subsequent treatment.

Intrinsic artifacts are much more frequent than extrinsic. On the whole they have a more serious effect on cells than on intercellular matter, though connective tissue fibres are often wrinkled and intercellular spaces distorted. Cells are subject to several different kinds of artificial changes.[404] Objects that are present in the living cell may simply disappear. Lipid globules, for instance, may be dissolved away. Protoplasm may be coagulated in the form of networks or lumps. So frequent is the network appearance in microscopical preparations that it was formerly supposed to be a character of the living substance; but Hardy [213] showed in 1899 that fixatives produced just such a microscopical network when they coagulated a solution of albumin, and that some fixatives made a coarse and others a fine net. Another frequent artifact is the thickening of membranes. Protein is often coagulated on the inner side of the nuclear and cell-membranes. Retractions also occur. The nuclear membrane is frequently pulled away from the surrounding cytoplasm, and both this and the cell-membrane are frequently wrinkled. Blebs of cytoplasm are often extruded from the surfaces of cells directly exposed to the fixative. The technical term for this process is clasmatosis, but it is often called bubbling. The latter is a misleading term, for no gas is formed.

Only by reference to the living cell can the quality of a fixative be judged. This fact can scarcely be too strongly emphasized. If, however, a particular fixative gives faithful stabilization of structure with certain chosen cells that have been carefully studied in the living state, we have reason to trust it with others that cannot be isolated for vital study.

Fixation for electron microscopy presents us with special problems. Electron micrographs that show objects of microscopic size enable us to judge the quality of fixation at the level of resolution of the light-microscope. Thus, an object of microscopic size, known to be smooth in outline in life, may appear irregular in an electron micrograph, and the irregularities may be of such a size that they would have been visible with the light-microscope in life, had they existed. We have here clear evidence of faulty fixation for electron microscopy; and if the fixation is faulty at the microscopical level, we cannot trust it at the submicroscopical. So far as submicroscopical objects are concerned, we have no direct means of knowing whether a fixative is reliable or not. It is customary to rely upon a fixative that is known to give faithful stabilization of form at the microscopical level. It would be safest not to rely on the submicroscopical detail shown in electron micrographs unless similar pictures are given after fixation by several different fixatives, all known to be reliable at the microscopical level.

Many cytologists have examined separate living cells with the light-microscope and noticed the changes produced in them when a fixative was run below the coverslip. The classical work of this kind was done by Strangeways and Canti (1927) with cultured cells, by dark-ground microscopy. The invention of the phase-contrast microscope led to a whole series of similar studies. Such work is of great interest and value, but it is of course an incomplete way of studying fixation, for only the primary artifacts are usually studied, and no evidence is obtained about the ability of the fixative to stabilize the form of the cellular components against subsequent treatment. In fact, fixatives are not tested as such in these studies, but only as preservatives.

A defect of the method described in the last paragraph is that the cells studied are immediately exposed to the action of the fixative. Now when a piece of tissue is fixed in the ordinary way, the cells that are thus exposed and not protected by any special membrane (such as the free border of the intestinal epithelium of vertebrates) are often seriously distorted, while those below, protected as they are by overlying cells or membranes, are well fixed. Even in a piece of tissue only a millimetre or so in diameter, the great majority of the cells are not superficial. It therefore follows that one might reject a fixative wrongly on the basis of observations made on separate cells.

It is realistic to test fixatives on small pieces of tissue that will be embedded and sectioned, with subsequent dyeing and mounting of the sections. Care should be taken to chose delicate tissues as test-objects, for some are so robust that they withstand the action of indifferent fixatives. It would be proper to take several bits of material fixed in each of the fluids that it is desired to compare, and embed them in a variety of different media; for a fixative that gives good results with one embedding medium may fail badly with another. A full investigation of this sort has never been undertaken, but details of a test involving paraffin embedding will be given on a later page (p. 72). Paraffin is probably the most drastic medium.

The minimum length of time during which a fixative should act depends partly on its rate of penetration (see pp. 37, 67) and partly on its mode of action. In general, the coagulant fixatives have achieved their full effect at any particular depth in the tissue as soon as they have penetrated to that depth at the concentration necessary to cause coagulation; but some fixatives, especially formaldehyde (p. 60), may continue to act progressively on the tissues after the latter has been fully permeated. Unless a fixative produces an extrinsic artifact, prolonged action is seldom harmful. Tissues may be left at least three years in Bouin's fluid without harm.[38] It is convenient as a general rule to fix overnight (about 18 hours). Very much shorter periods are sufficient for the tiny pieces, only about a millimetre thick, that are often used in cytological work; but overnight fixation ordinarily causes no damage. If the proteins of the interior of a piece of tissue have not been fully acted upon in 18 hours, it is unlikely that satisfactory fixation will be achieved by prolonging the period. For this reason it is not sensible to try to fix pieces of tissue several centimetres thick.

Careful instructions are often given as to the exact length of time during which particular fixatives should act, but these may often be disregarded. The truth of this may be established by getting someone else to fix three similar bits of tissue in the same fixative for 8, 16, and 32 hours. When these have been sectioned, dyed, and mounted, it will be found difficult or impossible to guess which is which, unless there happens to have been progressive deposition of an extrinsic artifact and this has not been removed.

It is sometimes desirable to 'postchrome' or 'postosmicate' a piece of tissue; that is to say, to leave it for a long time in a solution of potassium dichromate or osmium tetroxide after initial fixation. (See pp. 129 and 126.)

After fixatives have acted, they must be washed out of the tissues so as to prevent the formation of extrinsic artifacts by their incompatibility with fluids in which the tissue must subsequently be placed. In some cases the fluids that must be used for other purposes (for instance, the alcohols used for dehydration) themselves act as solvents for a fixative; in other cases special methods of washing out must be adopted. Since the methods of elimination are specific to each fixative, they will be considered separately in chapters 5 and 6.

When a tissue has been fixed, one may sometimes wish to preserve it indefinitely before embedding. This applies especially on scientific expeditions, when there are usually no facilities for embedding. The fixative itself may be unsuitable for indefinite preservation, either because it is volatile or because it will eventually produce artifacts. In such cases a post-fixation preservative may be used. Solutions of *p*-hydroxybenzoic acid and its esters are suitable.[379] These fluids are of an entirely different nature from the preservatives mentioned at the beginning of this chapter, for they are quite unsuited to the preservation of fresh tissues.

CHAPTER 2

The Reactions of Fixatives with Proteins. 1. The Visible Effects

For reasons that have already been mentioned (p. 23), fixation is primarily the stabilization of proteins. It will be best to study the reactions between fixatives and proteins before turning to the more complex events involved in the fixation of tissues and cells.

Some of the long chains of amino-acids that make up the proteins of plants and animals are wound into submicroscopic balls, others are extended in long fibres, often interlacing with one another and held together by chemical bonds of various kinds [180] to form elaborate networks in three dimensions. The proportion of globular to fibrous proteins varies widely in different tissues, the globular existing alone in certain body-fluids, the fibrous predominating in certain elements of the connective tissues of animals. In the present state of knowledge it is not possible to assess the relative amounts of each in protoplasm (cytoplasm and nuclear sap), but it is supposed that the fibrous here predominate and form a 'cytoskeleton' in the form of a submicroscopic network.[441]

When fixatives are added to protein sols, there is usually (though not always) a change in visible appearance. Sometimes there is an aggregation of protein molecules into microscopic granules, which are seen as a cloudiness in the fluid. Stronger action produces a flocculus of visible particles, which may remain suspended or fall gradually as a precipitate. Stronger action again may convert the whole of the protein into a single, coherent clot. The appearance of new interfaces between solid protein and surrounding water results in the scattering of light and thus the production of whiteness or colour: transparency is in some degree lost. Cloudiness, flocculation, and clot-formation are grades or stages in a single process of coagulation. The more concentrated the protein sol and the more vigorous the fixative, the greater will be the tendency of the coagulated particles to cohere in a single clot.

Some fixatives act in quite a different way. Instead of coagulating, they make protein sols more viscous or convert them into gels; if the protein is already a gel, they stiffen and stabilize it. In these processes there is no dissociation of protein from water and therefore no production of surfaces that will scatter light. Instead of a coagulum there is a viscous sol or an aqueous gel, transparent unless opaque matter has been deposited from the fixative itself.

To observe the effects of fixatives on proteins, it is convenient to begin with naked-eye observations on what happens when they are added to globular proteins in the form of aqueous sols. Experiments of this kind were made more than half a century ago by Fischer [169] and Mann.[322] A convenient way of carrying out such experiments is described in the Appendix (p. 315). The protein used is egg-albumin.

For the benefit of those who may read this chapter without having read the first, it is necessary to say once more that in all the experiments with fixatives described in this book, the substances were used at the concentrations shown on p. 24, except where the contrary is distinctly stated.

When a coagulant fixative is added to an albumin sol, a coherent coagulum usually appears at once and occupies most of the space filled by the fluid; in some cases (picric acid, mercuric chloride, chromium trioxide) it has fallen somewhat towards the bottom of the tube by the next day. It is either white (ethanol, acetone, trichloracetic acid, mercuric chloride) or coloured (picric and chloroplatinic acids, yellow; chromium trioxide, orange). With methanol and nitric acid it is at first in the form of a fine, white flocculus; the particles eventually cohere, and indeed nitric acid produces in the end rather a firm clot. With hydrochloric acid there is no immediate flocculation, but fine, white particles have begun to form within 5 minutes, and they fill the space occupied by the fluid within an hour; the next day they are coherent and the tube can be held horizontally without loss of water.

When formaldehyde, osmium tetroxide, potassium dichromate, or acetic acid is added to the albumin sol, no coagulum is formed. (Formaldehyde solution usually contains methanol, and very sparse, fine particles are then formed in the fluid.)

It must not be thought that non-coagulant fixatives are necessarily without effect on albumin, simply because they produce no visible result in this experiment. It can easily be shown, for instance, that both formaldehyde and osmium tetroxide can render albumin non-coagulable by ethanol[26] (see Appendix, p. 320). It has been known for many years that osmium tetroxide can set undiluted egg-white into a gel, without coagulation.[26] The ability of osmium tetroxide to make protein sols into gels naturally depends on the concentration of the protein. Thus 12% serum albumin is gelled but 8% is not.[405] Different proteins also differ considerably, serum globulin being more easily gelled than albumin, and fibrinogen than globulin.[405] Such gels are generally opaque. They have a tendency to liquefy eventually.[405, 23]

Although potassium dichromate does not coagulate albumin sols, it does gradually render undiluted egg-white more viscous and eventually transforms it into a weak, semi-transparent gel. (See Seki.[455] Potassium dichromate was used in the form of Müller's fluid,[363, 358] that is, with the addition of sodium sulphate.) Acetic acid has no such effect.

There is unfortunately no substance that will represent the fibrous proteins of the tissues so conveniently in test-tube experiments as albumin will represent the globular ones. Gelatine gel (25% w/W) can be used. It is convenient to cast the material in 15- or 30-grain pessary moulds. Most fixatives will not stabilize these gels in such a way that they retain their form when put in warm water. Gels that have been left for 24 hours in a solution of acetic acid or potassium dichromate dissolve in water at 37° C in a quarter of an hour or a little more. After fixation in methanol, ethanol, picric acid, mercuric chloride, or chromium trioxide, the gels resist complete solution for an hour or more but eventually dissolve. Formaldehyde and osmium tetroxide contrast strongly with these fixatives, for the gel does not dissolve, even if kept in water at 37° C for days. The gel fixed by osmium tetroxide is black and swells slightly in the warm water; that fixed by formaldehyde retains its original appearance, and one would not guess that it had been fixed.

Interesting information can be obtained from experiments with gelatine gels containing an admixture of albumin. Instructions for preparing gelatine/albumin gel are given in the Appendix (p. 314). The gel is cast in pessary moulds. It will serve for certain purposes as a crude model of protoplasm. The refractive index is about 1·365,[431] which is within the range shown by the protoplasm of ordinary cells. Thus the proteins are at about the same concentration as in the living cell, and both globular and fibrous ones are present.

Gelatine/albumin gels are stabilized against warmth by fixatives that will not stabilize gelatine alone. If gels that have stood in fixatives for 18 hours be rinsed and transferred to a large volume of water at 37° C, two hours later they will present the appearances recorded in table 1.

TABLE I

The appearances shown by gelatine/albumin gels soaked in various fixatives for 18 hours and then left for 2 hours in water at 37° C.[38]

	Form of gel
Coagulant fixatives	
methanol	Original form roughly maintained
ethanol	,, very roughly maintained
acetone	,, ,, ,, ,,
nitric acid	Flattened at bottom of vessel
hydrochloric acid	Shapeless mass at bottom of vessel
picric acid	Original form roughly maintained
mercuric chloride	,, ,, ,, ,,
chromium trioxide	,, ,, ,, ,,
Non-coagulant fixatives	
formaldehyde	Original form exactly maintained
osmium tetroxide	,, ,, ,, ,,
potassium dichromate	Completely dissolved
acetic acid	,,

The appearances will be the same after four days. (The gel fixed with osmium tetroxide was not observed beyond one day.)

Certain facts stand out from the experiments with gelatine and gelatine/albumin gels. The most obvious relate to the non-coagulant fixatives. It is clear that two of these (acetic acid and potassium dichromate) do not fix these two proteins, in the circumstances of the experiment; the other two (formaldehyde and osmium tetroxide) give much the best fixation of all the substances tried. The strong mineral acids, though coagulant fixatives, give very feeble stabilization of form. The other coagulant fixatives roughly stabilize the form of the gel.

The reactions of most fixatives with nucleoproteins are very different from those with albumin. For the pioneer work on this subject, see Berg.[61, 62] Convenient experiments of the same nature are described in the Appendix (p. 317). The results of such experiments are recorded in table 2.

The table shows that although picric acid, mercuric chloride, and chromium trioxide are coagulants of both albumin and nucleoprotein, yet these two proteins are not always coagulated by the

TABLE 2

The reactions of fixatives with nucleoprotein solution.[38]

(See Appendix, p. 317.)

	Result	
	Immediate	*Next day*
Coagulant of nucleoprotein		
acetic acid . .	Thick coagulum	Precipitate has fallen and occupies bottom one-third of tube
nitric acid . .	Coarse coagulum	,, ,, ,,
picric acid . .	Fine flocculi, gradually thickening	,, ,, ,,
mercuric chloride .	Fine flocculi	Fine flocculi remain suspended
chromium trioxide .	Coagulum	Precipitate has fallen and occupies bottom half of tube
Intermediate		
hydrochloric acid .	Fluid whitens (becomes slightly opaque); no visible coagulum	A small precipitate (only a few mm deep) has fallen to bottom of tube
formaldehyde . .	No coagulum	Extremely minute flocculi have formed
methanol . .	,,	No coagulum, but the fluid has become much more viscous
Non-coagulant of nucleoprotein		
ethanol . . .	No coagulum	No coagulum (except thin skin on surface)
osmium tetroxide .	,,	No coagulum
potassium dichromate	,,	,,

same substances. Thus acetic acid, a non-coagulant of albumin, is a precipitant of nucleoproteins, a fact that must be correlated with its almost invariable inclusion in fixatives for chromosomes. Conversely, ethanol coagulates albumin but not nucleoprotein. Potassium dichromate coagulates neither.

When gelatine/albumin gels are acted upon by fixatives, they generally either shrink or swell. The results of a series of experiments carried out with gels cast in pessary moulds are shown in fig. 1. The fixatives acted for 18 hours. Those fixatives that are not aqueous solutions are seen to shrink the gels, while the acids, especially acetic acid, swell them strongly; the two saturated aqueous solutions (picric acid and mercuric chloride)

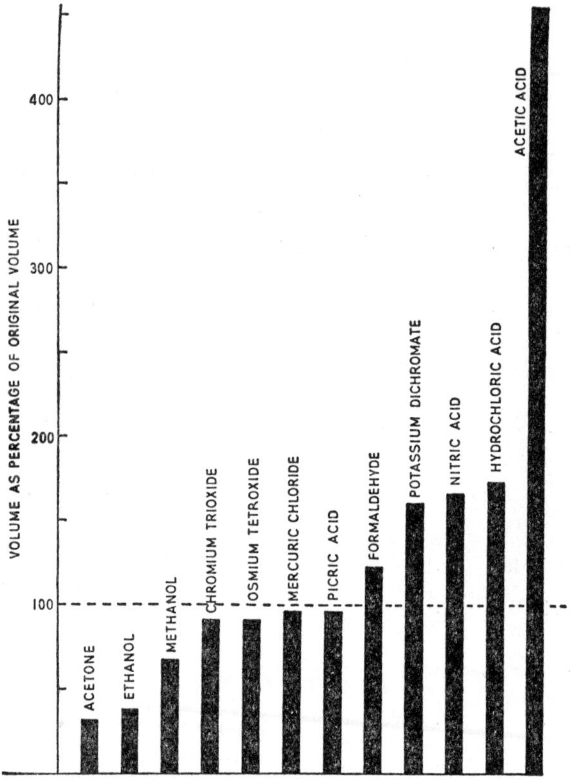

FIG. 1. Graphical representation of the changes in volume undergone by gelatine/albumin gels during 18 hours in various fixatives used at the concentrations shown on p. 24.[38]

change the volume least. The swelling effect of acids is progressive. A simple aqueous gelatine gel (15% w/W) shows the effect well. A gel cast in a pessary mould swelled in acetic acid solution to about 13 times its original volume in a week.

Gelatine/albumin gels placed in distilled water increase in volume by about 40% in 18 hours. If sodium chloride is added to the water at $\frac{1}{2}$, 1, 2, or 4%, the swelling is almost exactly the same as in distilled water.

Different fixatives penetrate into protein gels at different rates. Our understanding of the subject dates from Medawar's investigation [337] of the rate of their penetration into coagulated blood plasma. In his experiments the fresh blood of cocks was cleared of corpuscles by centrifuging and poured into tubes closed at one end; the plasma in the tubes was then coagulated by the addition of a trace of tissue-extract. The tube was placed in a large volume of fixative. The rate of penetration of coagulant fixatives was easily noted, since the plasma became opaque on coagulation. A sharp line divided uncoagulated from coagulated plasma. To test non-coagulant fixatives, Medawar dissolved indicators in the plasma; these showed the progress of the fixative into the plasma by change of colour.

The experiments showed that in penetrating into a protein gel, fixatives obey the laws of diffusion. If d is the distance penetrated, t is time, and K a constant depending on the fixative used, then

$$d = K\sqrt{t}.$$

It is convenient to measure d in millimetres and t in hours. It follows that K represents the distance in mm penetrated in one hour.

When indicators are used to measure the penetration of fixatives, what is being measured is the progress of the fixative at the concentration necessary to give a visible colour-reaction with the particular indicator used. There is no reason to suppose that this will be the same as the concentration necessary to fix the protein. The following experiments were designed to overcome this difficulty.[38] For full practical details, see Appendix (p. 318).

Gelatine/albumin was chosen as the protein gel into which the fixatives were to penetrate. The gelatine gave the necessary solidity, while the albumin marked the progress of coagulant fixatives by becoming opaque. Small tubes were filled with the gel. A rubber pipette-bulb closed one end, while the other was left open (fig. 2). The tube was placed in a large volume of fixative, in which it floated vertically. The penetration of coagulant fixatives could easily be measured from time to time. A diamond-mark near the top of the tube served as reference-point. The distance penetrated, *d*, is a–b in fig. 2. It is to be noted that the distance measured is not necessarily the same as the depth of the fixed protein: that would only be so if the process of fixation caused no swelling or shrinkage.

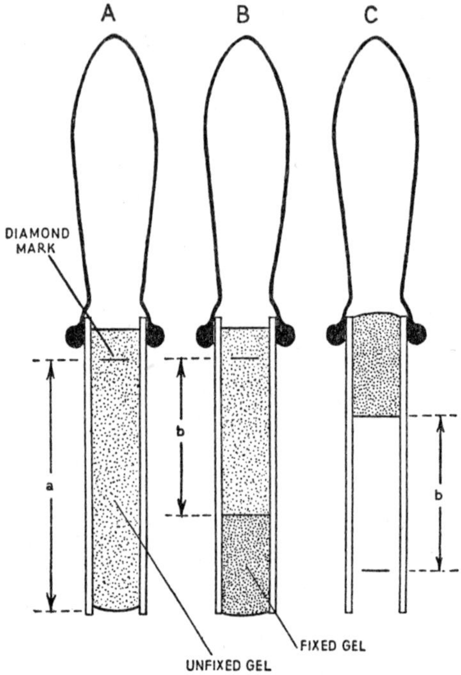

FIG. 2. Pipettes used in the measurement of the rate of penetration of fixatives into gelatine/albumin gel.[38]

A, the pipette filled with gel, before being placed in the fixative. B, the pipette after immersion for a period in a fixative. If the fixative is a protein-coagulant, the line of demarcation between unfixed and fixed gel can be seen. C, the glass tube has been turned upside down and the unfixed gel has run out.

The distance the fixative has penetrated is *a–b*.

The measurement of the rate of penetration of noncoagulant fixatives is more complicated. When the fixative has acted for a particular time (*t*), the tube is lifted out; the pipette-bulb is removed and placed on the other end of the tube. The tube is then floated in water maintained at 37° C. The unfixed gel soon melts and runs out; the fixed part remains in the tube (fig. 2, C).

Acetic acid fixes neither gelatine nor albumin, and its rate of penetration can therefore not be measured with gelatine/albumin gel. If nucleoprotein be substituted for albumin, however, the limit of fixation is clearly seen. For the method of preparing gelatine/nucleoprotein gel, see Appendix, p. 315.

Since potassium dichromate is not a fixative of gelatine, albumin, or nucleoprotein, this method cannot be used to measure its rate of penetration.

It is convenient to make observations at 2^2, 3^2, 4^2, and 5^2 hours, and so on, if desired, up to 12^2 hours. If the distance penetrated is

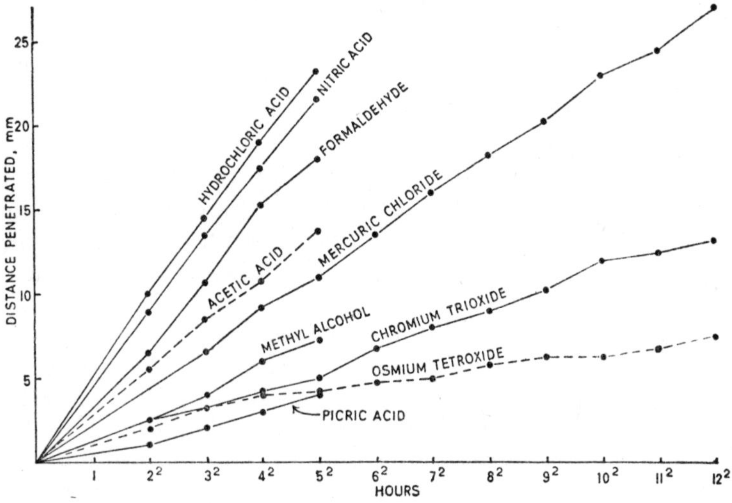

FIG. 3. Graph showing the rate of penetration of fixatives into gelatine/albumin gel (acetic acid into gelatine/nucleoprotein gel). The fixatives were used at the concentrations shown on p. 24.[38]

plotted as ordinate against an abscissa divided into equal parts representing 0 to 1, 1 to 2^2, 2^2 to 3^2, 3^8 to 4^2, and 4^2 to 5^2 hours, etc., the results can be shown as lines that would be straight if the equation $d = K\sqrt{t}$ were exactly obeyed. It will be seen from fig. 3 that the lines are nearly straight. It is probable that they would have been even more nearly so if the temperature of the room had remained constant. The only substance that does not show general obedience to the equation is osmium tetroxide. The fixed gel in this case appears to offer some resistance to penetration, for the curve falls with the passage of time. Up to 4^2 hours, the K-value is fairly constant at 1·0, but from 4^2 to 12^2 hours only 0·31.

The values of K are shown below. The figures are based on

observations at 5^2 hours. Apart from those for osmium tetroxide, they would not have differed much if periods other than 5^2 hours had been chosen instead.

	K
hydrochloric acid	4·65
nitric acid	4·3
formaldehyde	3·6
acetic acid	2·75
mercuric chloride	2·2
methanol	1·45
chromium trioxide	1·0
osmium tetroxide	0·85
picric acid	0·8

It is to be remembered that acetic acid penetrated into a somewhat different gel from the others.

It follows from the equation that a fixative with a K value of 1·0 (chromium trioxide) will penetrate 20μ (the diameter of a large cell) in 1·44 sec.; that is to say, it penetrates that distance at the rate of 50 mm per hour, but the rate falls off so rapidly that in fact it only penetrates 1 mm in an hour, and it takes 100 hours to penetrate 1 cm. As we shall see (p. 68), penetration into tissues is slower than into gelatine/albumin gels, and it is obvious that the internal parts of pieces of tissue several cm thick cannot be effectively fixed even by the most rapidly penetrating fixatives.

The results with gelatine/albumin gel agree in general with Medawar's, so far as coagulant fixatives are concerned, but fixatives penetrate more quickly into coagulated blood-plasma than into the gel used in the experiments described here.

We turn now from naked-eye observations to microscopical study.

The minute structure of protein coagula was first investigated by the German botanist, Berthold,[68] in 1886. He believed that the protoplasmic network seen in preparations of plant cells was a coagulation artifact. He put a drop of egg-white on a slide and noticed the formation of a microscopical network in it on the addition of water (presumably by the coagulation of globulin and ovo-mucoid). He also noticed the formation of separate granules on the addition of an aqueous solution of iodine.

Another German botanist, Schwarz,[451] who was aware of Berthold's findings, published the results of a much fuller investigation in the following year. He added absolute ethanol, picric acid solution, tannic acid solution, and Flemming's fluid to aqueous solutions of dried egg-white at various concentrations, and studied the resulting material under the microscope. He noticed that if the solution of egg-white was dilute, fine granules were seen, while with higher concentrations these joined together to form a network of fibres. He made similar observations with peptone and soluble gelatine (probably metagelatine). His figures of the networks produced in soluble gelatine by the action of tannic acid are reproduced here in fig. 4, A, B. Schwarz considered

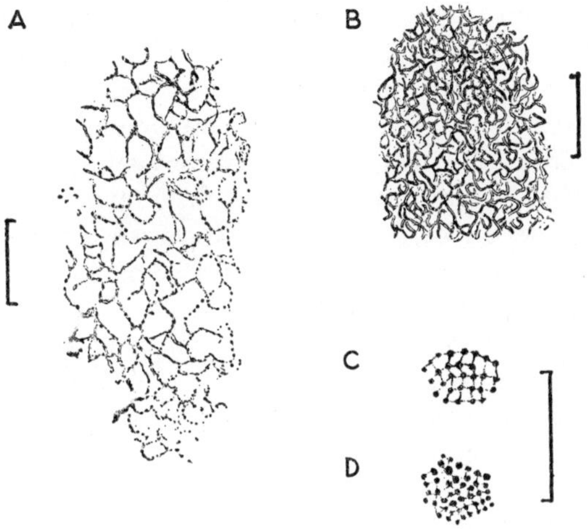

FIG. 4. Protein coagula seen under the microscope. Each scale represents 10μ.

A, dilute solution of soluble gelatine, fixed by 0·5% tannic acid. B, ditto, fixed by 6% tannic acid. C, egg-white fixed by mercuric chloride (saturated solution in 0·6% sodium chloride); paraffin section, 1μ or less. D, ditto, fixed by potassium thiocyanate. (A and B from Schwarz;[451] C and D from Hardy.[313])

that the granules or fibres lay in a continuous *Grundsubstanz*. They were more colorable by picric acid than was this interstitial material.

Schwarz's important work has been very much overshadowed by Hardy's,[313] and indeed the Cambridge physiologist carried the investigation a good deal further. He worked chiefly with egg-white, but also with gelatine. He caught up a drop of protein solution in a loop of silk thread and placed it in an aqueous solution of mercuric chloride or some other fixative; he then usually embedded it in paraffin and cut thin sections (down to $0{\cdot}6\mu$). He also used frozen sections and minute, teased fragments of unembedded material. As fixatives he also used the vapour of osmium tetroxide and the heat of steam.

Two of Hardy's figures are shown in fig. 4, C, D. He found that weak protein solutions generally showed separate microscopical particles after fixatives had acted, while stronger ones resulted in the appearance of a sponge or net. As the figures show, the net was thickened at the nodal points. The size of the meshes varied with the fixative used. The vapour of osmium tetroxide (a non-coagulant fixative, as we have seen) was alone in not producing a microscopically visible structure.

Hardy's main contribution was his discovery that there is no *Grundsubstanz*. The whole of the coagulum consists of granules or spongework, with water in between. This was evident when sufficiently thin sections were cut, for there was nothing in the meshes of the net that could be demonstrated even by saturated solutions of dyes. In thicker sections the out-of-focus appearance of other parts of the net gave the misleading appearance of an interstitial substance, but Hardy realized that in finished microscopical preparations the meshes were occupied only by Canada balsam or other mounting medium. Thus, as he said, the essence of fixation by coagulant fixatives is the separation of fluid from solid, of water from protein; and the solids, if sufficiently abundant, hang together to form a network (or in some cases a honeycomb-like structure). As a result, water can often be squeezed by hand from a fixed protein, though a pressure of 400 lb to the square inch will not separate it from an unfixed protein gel. When water has been separated from protein, it can be replaced by other fluids. While the water is still present, there is usually little change of volume; but when the replacement occurs, there is usually shrinkage. Similarly a silicic acid gel shrinks when its water is replaced by another fluid.

It has already been remarked (p. 22) that the network produced by the action of coagulant fixatives has the same refractive index as dry protein, and that is why mounting media having about the same r.i. as dry protein give such glassy transparency to microscopical preparations of the tissues of organisms.

Hardy considered that osmium tetroxide in aqueous solution produced a network in egg-white, but the effect was more probably due to the water of the solution acting in the absence of salt on globulin.

It might be thought that the structure produced in protein sols and gels by coagulant fixatives would have very deleterious effects in microtechnique. So long, however, as the network is a very fine one, the effect is not wholly damaging, for the spaces produced by coagulation give access to embedding media, especially paraffin, that would otherwise be unable to enter. It is a striking fact that the great majority of the familiar fixative mixtures contain a coagulant (see p. 148).

CHAPTER 3

The Reactions of Fixatives with Proteins. 2. The Chemical Changes

The most familiar fixative substances will be considered one by one, from all points of view connected with their use as fixatives, in chapters 5 and 6 (pp. 89 and 111). Their chemical composition and the ions they form when dissolved in water will be mentioned there. To prevent unnecessary repetition, this information will not be given in the present chapter.

It is important to explain certain terms and conventions that will be used in this chapter and throughout the rest of the book when reference is made to the chemical structure of proteins.

A protein has two constituent elements, which will be called the

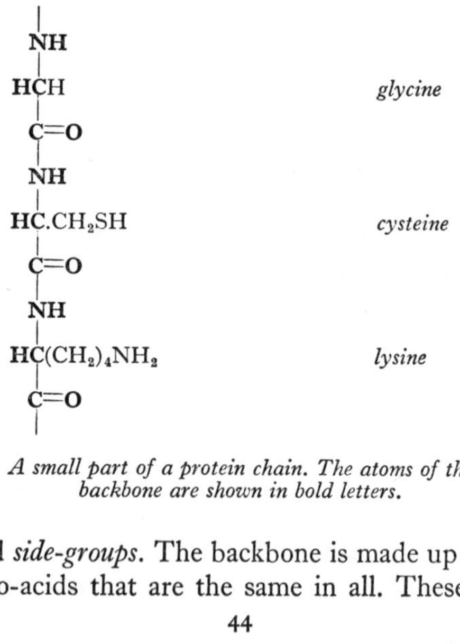

A small part of a protein chain. The atoms of the backbone are shown in bold letters.

backbone and *side-groups*. The backbone is made up of those parts of the amino-acids that are the same in all. These are repeated over and over again, like vertebrae; they articulate through the peptide link, $-\overset{\overset{O}{\|}}{C}-NH-$. The backbone is folded in various ways; this folding will not be represented in the structural formulae used. The side-groups are those parts of the amino-acids that distinguish one from another. These project laterally from the backbone, like ribs. They alternate in direction, but in this book they will all be represented as projecting to the right. This convention will be adopted partly for economy of space, partly because the different side-groups are more easily recognized if always written in the same sequence (just as words are more easily read if the sequence of the letters is not reversed).

A sequence of linked amino-acids (backbone and side-groups) will be called a *chain*. The word *molecule* will not be used loosely. It seems inapplicable when protein chains are linked together by chemical bonding through their side-groups to form net-like structures of indefinite size, perhaps extending from one end of a cell to the other.

A substance cannot act as a fixative if it attacks the peptide link. Far from being proteolytic, fixatives tend to link protein chains together and thus give mechanical stability.

An important feature of the structure of proteins is that a particular side-group may react chemically without necessarily changing the nature of the chain as a whole. Suppose, for instance, that a particular protein containing very few cysteine side-groups is exposed to a substance that reacts with nothing in the chain except –SH groups. Obviously the chain as a whole will be scarcely affected, and no fundamental change in the nature of the protein will have occurred. If a particular side-group is capable of reacting with a particular dye, it will retain this property when a fixative has only blocked some other side-groups.

Despite these facts, the appearance and 'nature' of a protein, especially its solubility in water, are often markedly changed by fixation, and the substance is often said to be 'denatured'. Unfortunately this word has been used vaguely, and with different meanings by different authors. So long as it referred to a loss of solubility, one knew what it meant; but loss of solubility is accompanied by increase of reactivity, and this fact has altered the meaning entirely: for if the increase in reactivity occurs, the process is often called denaturation, even though solubility is actually increased. It is obvious that fixation cannot involve increase in solubility, and there are therefore many 'denaturing agents', such as urea, that could not possibly be fixatives. It is easier to study denaturation, however, if the product be soluble; and for this reason much of our knowledge of the process, admirably summarized in several reviews,[12, 13, 366, 408] is not directly applicable to the problems of microtechnique. For similar reasons most of the study of denaturation has been devoted to proteins in the form of sols, yet gels are more interesting to the biologist, because protoplasm is essentially a soft gel; and gelled proteins can also be denatured.

We may broadly distinguish *additive* from *non-additive* fixation. In the former, the fixative molecule or a considerable part of it adds itself to the protein; in the latter it does not. The word 'denaturation' was formerly held to imply that the change involved was non-additive, but this usage is not quite general today. It has even been said that any chemical change in a protein, not involving disintegration into amino-acids, is denaturation.[366] In this book, however, the word will be used to mean a non-additive change in a protein, causing it to become less capable of remaining in intimate relation with water as a sol or gel, and more reactive. The resulting loss of solubility ordinarily manifests itself in coagulation, if the protein be a sol; a gel is rendered harder and opaque.

Polypeptides, in the sense of short chains of amino-acids, cannot be denatured: the process occurs with the very long chains of amino-acids that constitute proteins.

The chief non-additive or denaturing fixatives are these: methanol, ethanol, acetone, nitric acid, hydrochloric acid.

If a protein sol be mixed with a denaturing fixative, the reaction is usually so quick that coagulation appears to be instantaneous. Careful experiment has shown, however, that there are in fact three stages. First, reactivity is increased; then flocculation follows, but the flocculus is soluble in weak acids or alkalis; finally the flocculus hardens into a coagulum, only soluble by proteolysis. Some so-called denaturing agents, such as urea, only cause the first change; and it is for that reason that they are not usable as fixatives.

Denaturation may be brought about in many different ways; for instance, by subjection to very high pressure, extension in extremely thin films, or exposure to ultrasonic waves or ultra-violet light. Freezing and thawing can cause denaturation, especially of lipoproteins, though freezing-drying does not: indeed, freezing-drying should be used instead of fixation if undenatured protein is needed. The only means of denaturing that are used in routine microtechnique are heat and certain chemical agents; the former is seldom used except for the fixation of blood-films.

We cannot establish for each protein a particular temperature of denaturation. Most proteins are denatured if held in water at 60° C for a long time, though ribonuclease is extremely resistant. If a soluble protein be dried, heated to 100° C, and cooled, it retains its solubility. Thus heat alone does not suffice to denature: it is hot water that causes the reaction.

Methanol, ethanol, and acetone are the chief organic substances used in microtechnique as denaturing fixatives. They ally their effects to that of heat; or, to put it in other words, the reaction has a high thermal coefficient. A very low concentration of ethanol suffices to denature proteins if the temperature is above 60° C; at −8° C a concentrated solution produces a flocculus, but this is soluble in water at room temperature;[214] there is no measurable denaturation of any sort below −15° C. (It must be mentioned that alcohol actually favours the solution of a few proteins, such as zein and gliadin.) The alcohols and acetone dehydrate strongly, but it is clear that more than this is involved in denaturation, since mere drying is ineffective.

Hydrochloric and nitric acids are denaturing fixatives, but they are not very commonly used. Denaturation by heat or other means occurs rather readily at the iso-electric point of each protein; at a slightly more or less acid pH than this the tendency to denature is less; from about pH 2·5 downwards and again above pH 10, denaturation occurs at room-temperature. The strong mineral acids at about 0·5N are quite useful coagulant fixatives. In concentrated solution they disintegrate proteins into amino-acids. Acetic acid does not ionize freely enough to produce a pH sufficiently low to coagulate proteins; in addition, the acetate ion interferes with denaturation (p. 64).

We must picture the proteins in the tissues of the living organism as maintaining a special relation with water through their various water-soluble groups. This is so whether the proteins be globular or fibrous, and if the latter, whether bound into gels or not. Since the protein molecule is very long, the amino and carboxyl groups of the terminal amino-acids are not of much significance in this respect, and indeed if the polypeptide chains are bound together into a gel, there need not be any terminal amino-acids. The chief hydrophil constituents are the carboxyl, hydroxyl, and amino-groups of the side-chains, and the carbonyl of the backbone; the imino-groups of several amino-acids, the amido-group of asparagine, and the sulphydryl of cysteine can also associate with water.

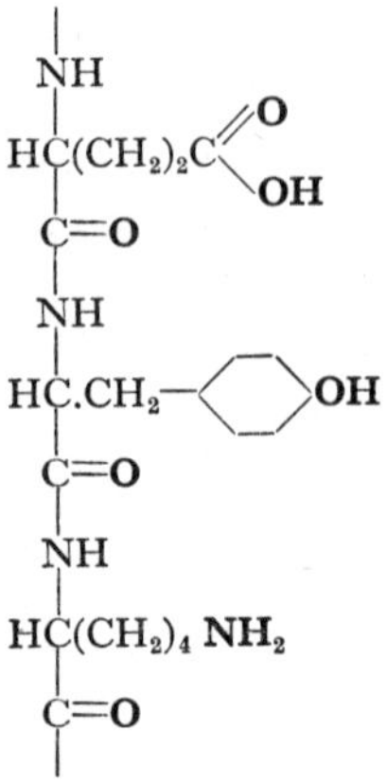

glutamic acid

tyrosine

lysine

Part of a protein chain. The hydrophil groups are shown in bold letters.

The most striking effect of a coagulant fixative on a protein sol or gel is an alteration in the effects of these hydrophil groups, which formerly held water so firmly that it could not be separated by powerful mechanical force, but now can be squeezed out by hand (p. 42). The effect is irreversible, for the protein loses its power to imbibe water and resume its former properties.

To the chemist the most significant result of denaturation is the increase in reactivity. Certain reactive constituents of the protein were previously present, wholly or partly, in latent form: now they exhibit themselves and respond freely to tests for their presence. An increased digestibility by enzymes results, and indeed accounts for the fact that we cook our protein food. Two views have been held about the nature of the latency of the reactive groups before denaturation. It has been suggested that they do not exist in the natural protein, but originate during the process of denaturation (and also in proteolysis). Most students of the subject, however, consider that the reactive groups are present in the original protein, but something hinders the access of test-reagents to them. They may be involved in linkages between protein chains or between different parts of the same chain, or simply inaccessible because the chain is so tightly folded.

The reactive groups liberated by denaturating agents are all ionizing side-chains of amino-acids. Chemists have given most of their attention to those reactive groups that are particularly easily shown by simple tests. It is probably for this reason that they have concentrated so much on the sulphydryl (–SH) of cysteine, the disulphide (–S–S–) of cystine, the phenyl of tyrosine, and the indolyl of tryptophane. It seems that the sulphydryl group, which has been particularly carefully studied in relation to denaturation, does not arise in this process by the splitting of disulphide bonds. Its emergence from latency gives reductive properties to the protein.

The groups that become reactive on denaturing are of great importance to the histochemist, but in general microtechnique we are especially concerned with the amino and carboxyl groups, for these provide the main points of attachment for acid and basic dyes respectively (see p. 167); and since we usually colour cytoplasm and nuclear sap with acid dyes, we are above all interested in the reactivity of the amino-groups of lysine and arginine. Unfortunately, not very much attention has been paid to the release of these particular groups from their latent condition in the natural protein. This is partly due to the fact that their appearance is not so dramatic, since some of them are readily accessible in the protein before denaturation.

If there were a special freeing of basic groups from a latent condition, the protein would become more reactive with acids and acid dyes; and conversely. Such changes would produce a shift in the iso-electric point of the protein. Denaturation by alcohols produces less change in the iso-electric point than fixation in other ways, and that is why these substances are chosen as fixatives when we want to study the basicity or acidity of proteins by the simultaneous use of acidic and basic dye-ions (p. 262). In general, there is a small shift (roughly pH 0·5) of the iso-electric point in the less acid direction on denaturation.

When a conjugated protein (lipoprotein or nucleoprotein) is denatured, the protein constituent commonly separates from the substance with which it was combined, and the latter then reveals its presence more readily.

The specificity of proteins, especially any immunological property, is generally irreversibly lost by denaturation. Trypsin, however, can be reversibly denatured: it loses and regains its power to digest. As a rule enzymatic properties are destroyed, and special fixatives must be chosen if they are to be displayed in microscopical preparations. If the right denaturing agent is chosen, it will coagulate the proteins in general but leave certain enzymes more or less intact. Sections may then be cut and placed in a solution of a suitable substrate. The latter must be carefully chosen, for it must leave microscopically visible evidence of the places in the tissue in which it was attacked by the enzyme: the product of the reaction must be immobile and either visible or capable of being made visible. In this roundabout way the original site of the enzyme can be determined. It is chiefly for work of this sort that acetone deserves to be listed as a fixative. Certain enzymes, such as alkaline phosphatase, are resistant to its action. Acetone distorts tissues seriously and would never be chosen for purely morphological studies. It has, however, certain uses as a fixative in other branches of histochemistry beside enzymology.

An important effect of denaturation on globular proteins is their elongation into fibres. This change can best be witnessed when loss of solubility has not yet occurred. The extension of the protein molecules renders the sol more viscous, and since they tend to arrange themselves parallel with one another in a moving fluid, the formerly isotropic sol now exhibits the birefringence of flow. The protein never extends fully into a straight chain on denaturation, but always remains to some extent folded. Remarkably enough, the fibrous proteins are affected in the opposite way: the chain shortens somewhat by folding. There is thus an approximation of both kinds of proteins towards a similar structure, but this is fibrous, not globular. A coagulum could not be formed if the product of denaturation had not been fibrous. The firmness of the coagulum must depend on the nature of the bonds that tie the fibres together.

A microscopical preparation of denatured protein generally shows a network that appears to consist of interlacing fibres, fused where they touch; there may or may not be swellings at these points. The appearances are similar to those shown in fig. 4 (p. 41). The difference in scale between these microscopically visible fibres on one hand and the polypeptide chains of the proteins on the other is enormous; still, the latter underlie and make possible the former. The visible network is an artifact, but it reminds us of an important truth about the submicroscopical structure of proteins. In protein gels we must imagine the polypeptide chains keeping a small and fairly regular distance from one another, with water everywhere intervening between one and the next except where they actually adhere. When denaturation has taken place, the water is no longer firmly bound, and the polypeptide chains can come close up against one another to form microscopically visible strands, while microscopically visible spaces are formed between one such strand and another. A network of strands is thus formed. If globular proteins are present, these will extend on denaturation and participate in the formation of the strands.

It may be said in brief summary that fixation by denaturing renders globular proteins fibrous; irreversibly alters the relations of all proteins with water, so that this is no longer firmly held; and greatly increases the reactivity of certain side-groups of the constituent amino-acids, so that they respond much more readily to tests for their presence and attach themselves freely to dyes; but the specificity of many proteins is lost. The chemical denaturing agents are non-additive: they achieve their effects while remaining essentially detached.

Many fixatives are additive to proteins. The difference between additive and non-additive fixation is not quite so great as might be supposed. Most substances undergo a profound change if an atom be added to them, but proteins are necessarily different in this respect. A fixative might be capable of adding itself to the side-group of a particular amino-acid that was very scantily represented in a particular protein; the fixative might nevertheless denature the protein as a whole. For this reason one must not draw too sharp a distinction between denaturation by non-additive fixatives and the changes brought about by additive ones. The distinction is partly a matter of the arbitrary use of words. Many of the changes described in the preceding part of this chapter occur also when certain additive fixatives react with proteins.

Nevertheless, a valid distinction does exist in most cases. Further, certain additive fixatives, unlike any non-additive ones, are non-coagulant of albumin. The familiar additive fixatives may be grouped thus:—

coagulant	*non-coagulant*
mercuric chloride	formaldehyde
chromium trioxide	osmium tetroxide
picric acid	

These substances will now be considered in turn.

The reactions of mercuric chloride and other mercuric salts with proteins have been carefully studied by several authors.[205, 244, 245, 196] The metal becomes attached to protein in several different ways at the same time, unless special precautions are taken to isolate the reactions.

Mercuric salts react with the sulphydryl (–SH) group of the cysteine component of proteins. If the amount of mercury be restricted to what will combine with the sulphur present in the protein, no other reaction will occur; this may therefore be called the primary reaction. It is possible to isolate a fraction of serum albumin that contains only one sulphydryl group in each molecule.[245] This may be crystallized in the form of a compound with mercury, which contains one atom of mercury to two molecules of the albumin fraction. The evidence suggests strongly that mercury forms a link between cysteine residues. Each of the latter is potentially *mercurium captans*, a mercury-catcher, ready to form a compound (mercaptide) with the metal.

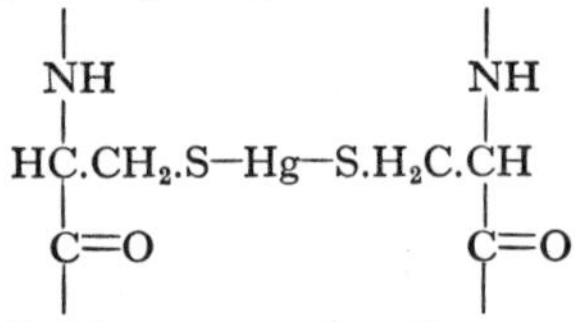

Mercury forming a link between cysteine side-groups in two protein chains

Many such links could be formed between protein chains in which cysteine occurred repeatedly, and many chains could be bound together into a single, polymeric whole. This would tend towards coagulation.

It is unlikely that mercaptide-formation is the main cause of coagulation. When an excess of mercuric salt is present, as in ordinary fixation, much more of the mercury is taken up in other ways, to which we must now turn our attention.

The uptake of mercury by proteins is profoundly affected by acidity and alkalinity. The main features of the process are set out schematically in table 3. A scale of pH is not provided, because each protein would require a different one. The lines dividing the degrees of alkalinity and acidity must not be regarded as separating sharply the several reactions, for in fact these overlap, so that more than one reaction may occur at any particular pH. The table calls attention to the reactions that are dominant in certain broad regions of alkalinity and acidity.

TABLE 3

Diagrammatic representation of the chief reactions of mercuric chloride with proteins, in the presence of excess of mercuric chloride

	Mercury is taken up as	*Nature of bond*	*Strength of bond*	*Coagulation*
STRONG ALKALINITY	Hg^{++}	To amino-groups by secondary valencies; to carboxyl groups by main valencies	Firm	Little tendency to coagulation
WEAK ALKALINITY neutrality— WEAK ACIDITY	$HgCl_2$	To amino-groups by secondary valencies	Very loose	Opposed by sodium chloride
iso-electric point of protein	$[HgCl_4]^=$	To amino-groups by main valencies	Very loose	Promoted by sodium chloride
STRONG ACIDITY	(No reaction except with sulphur of cysteine)			Little tendency to coagulation

The reactions of mercuric chloride with protein are affected by the ability of mercury to form bonds through subsidiary valencies. For instance, mercuric chloride can in certain circumstances react with ammonia to form a diammine.[470] This results from the fact that mercury can accept extra electrons from donor atoms.

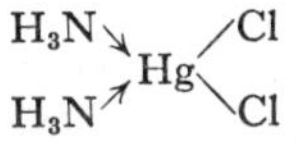

Mercuric chloride combined with ammonia

In strongly alkaline conditions, the mercuric ion is taken up by the amino-groups of proteins (on the side-chains of lysine and arginine), by bonds of this sort.[205] The mercury then reacts as

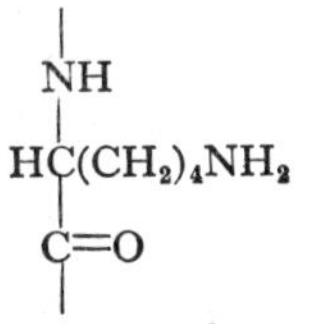

Lysine forming part of a protein chain

though present as mercuric hydroxide. The mercuric ion may perhaps combine in a similar way with the –NH.CO– groups of the protein backbone. Beyond this, the mercuric ions form salt-linkages with the carboxyl side-groups of certain amino-acids.

These reactions appear not to have a strongly coagulative effect, but the mercury is firmly held. This can easily be shown by testing with mercuric potassium iodide, $HgI_2.2KI$, a useful reagent for distinguishing between mercuric ions or loosely held mercury on one hand, and the firmly held metal on the other. The reagent is yellow, but gives the red colour of mercuric iodide in the presence of the ions or loosely held metal.

A different kind of combination is characteristic of weakly alkaline, neutral, or weakly acid conditions down to the iso-electric point of the protein. The mercuric chloride is taken up as a whole molecule through subsidiary valencies, mainly by the amino side-groups of lysine and arginine, and is now held very loosely, so that a red reaction is given with mercuric potassium iodide. The formation of this loose compound is associated with coagulation. Coagulation is opposed by sodium chloride. This is because the chloride ions have a stronger affinity for mercuric chloride than the protein has. The anion $[HgCl_4]^-$ is formed.[205] If albumin is coagulated by mercuric chloride and washed, the clot is reaily dissolved by a saturated solution of sodium chloride (or potassium iodide).

On the acid side of the iso-electric point of the protein, but still within the range of weak acidity, the ion $[HgCl_4]^-$ combines through main valencies with the amino-groups of the protein, which are ionized to some extent on the acid side of the iso-electric point. The compound formed is again very loose, but coagulation is now promoted by the presence of sodium chloride, which increases the amount of the reactive mercuric ion.

In conditions of strong acidity mercuric chloride does not react with proteins, except to form mercaptide; there is little tendency to coagulation.

In practical microtechnique, acids and sodium chloride are often added to solutions of mercuric chloride, apparently without much consideration of the complex consequences. Tissues that have been fixed with mercuric chloride are sometimes placed in Lugol's solution. This is likely not only to decompose the mercaptide with oxidation of the former sulphydryl groups to disulphide,[196] but also to dissolve the coagulum.

If mercuric chloride were to block all $-NH_2$ groups by combining with them, but left the $-COOH$ groups untouched, the protein would become very acidic; conversely, if it were to block all –COOH groups, the protein would become very basic. These changes would control the reactions with dyes, for the coloured ions of the latter associate with the acidic and basic groups of proteins (see p. 192). It is rather a strange fact that most authors who have considered mercuric chloride in this connexion have taken it for granted that this salt simply blocks –COOH groups and thus makes the protein more basic (that is, more attractive to 'acid' dyes (p. 167)).[505, 455, 264] It is stated, however, by the American histochemist Gomori [196] that mercuric chloride *reduces* the attraction of proteins for acid dyes, by blocking $-NH_2$ groups. As we have seen, the reaction of this fixative with proteins is in fact very complicated.

Mercuric chloride is pre-eminent among fixatives for leaving tissues in a condition conducive to brilliant dyeing. It is relevant to consider here the characters of a fixative that will achieve this end. It must not fix proteins in such a tight gel that dyes cannot penetrate, but must somehow make them porous. It must knock DNA off from combination with protein and precipitate it in a form in which it will associate readily with basic dyes. It must leave the protoplasmic proteins in a state in which they will accept acid dyes, so that these parts may be given a colour that will contrast with that given to the DNA. It is evident that mercuric chloride achieves these ends, in the circumstances of ordinary fixation; but we await a full explanation of its superiority over other fixatives in these respects.

Most fixatives leave proteins in a digestible state; denaturation, as we have seen (p. 48), promotes digestibility. In experiments of this sort the excess of fixative must be very carefully washed out: for instance, by washing in running water for 20 days. When this has been done, proteins fixed by additive fixatives generally remain digestible. It has been shown, however, that mercuric chloride slows down slightly the rate of digestion of the proteins of blood plasma by pepsin and trypsin.[338] This may be due to a change in the protein, but it is also possible that some of the bound mercury transfers itself to the enzyme and fixes it.

Picric acid is an alkaloidal reagent: that is to say, it precipitates alkaloidal salts from solution in water. This is a usual property of large, complex anions. Free amino-acids combine with the anion to form picrates; each amino-acid gives a crystal of characteristic form. For a plate showing the crystals of eleven picrates of amino-acids, see Schmidt.[440] Picric acid resembles other alkaloidal reagents in coagulating soluble proteins, but the chemistry of the process has not been worked out. The obvious point of attack would be the amino side-groups of lysine and arginine, but if all these were blocked, the protein would lose most of its affinity for acid dyes. In fact, however, picric acid gives egg-white a strong affinity for acid dyes, but scarcely any for basic ones.[455] It seems probable that acid dyes are able to replace picric acid at its points of association with amino-groups. It is to be recollected that picric acid is not only a fixative but also a dye (p. 185), and that dyes can replace one another in this way. The reduction in affinity for basic dyes has not been explained. Picric acid also forms additive compounds with phenols, and combination with the side-group of tyrosine is not excluded.

Chloroplatinic acid is another alkaloidal reagent. The incorrect name of platinum chloride disguises the fact that the metal forms part of a complex anion, $[PtCl_6]^{=}$.

The mode of action of chromium trioxide on proteins in the process of fixation is not well understood. We have a considerable amount of knowledge about the reactions at high temperatures, because they have been studied by the textile chemists.[192, 410, 182] These reactions must be briefly mentioned here, though their relevance is doubtful.

Compounds of chromium are used in the pre-treatment of wool before the application of dyes. The chemistry of the process of 'mordanting' will be dealt with in chapter 11. In industry, sodium dichromate is commonly used for the purpose, often with the addition of acid. When a dichromate is dissolved in water, the anions produced are essentially the same as those produced by chromium trioxide (see pp. 105 and 126). The chromium is anionic, and it makes little or no difference whether chromium trioxide or acidified dichromate is used. Industrial mordanting is carried out at boiling point, and the fibre is often treated afterwards with a reducing agent.

Anionic chromium appears to associate itself chiefly with those side-groups of the wool proteins that contain $-NH_2$ and $-OH$, and with the $-CO.NH-$ group of the protein backbone. According to the textile authorities, the metal is taken up partly as anions, but some of it also transforms itself into a non-ionic, sexi-covalent form, some of the six links being with the groups just mentioned.[192]

The industrial process results in the firm binding of chromium to protein. When wool is simply steeped in the mordant at room-temperature, however, this is not so.[182] The fibre takes up the mordant and becomes yellow, but the metal is loosely held and can be washed out by a buffer at pH 8.

In microtechnical fixation chromium trioxide is used at room-temperature, yet the metal is firmly bound and cannot be removed even by prolonged washing.[556] Solutions of chromium trioxide are strongly acid (p. 105), and simple denaturation by acidity might be thought to be partly responsible for the results; but proteins are much more violently coagulated than by hydrochloric or nitric acid, and are so altered that they cannot be digested by pepsin or trypsin.[338] Less is known about the chemical changes underlying the action of this fixative on proteins, in the circumstances of the fixation of tissues in microtechnique, than about the changes underlying the action of any other common coagulant fixative. If the reaction were largely with the amino-groups of the proteins, as is supposed to be the case in the mordanting of textile fibres, these groups would presumably be blocked and no longer available to ordinary acid dyes; yet this fixative, perhaps above all others, is favourable to the action of such dyes (pp. 109 and 204). As an oxidizing agent, it is likely to react with the –SH of the cysteine side-group; also with the phenyl of tyrosine, the indolyl of tryptophane, and the iminazol of histidine. It has recently been shown that prolonged fixation in solutions of chromium trioxide does in fact interfere with histochemical tests for the three last-named amino-acids.[111]

A considerable amount of misunderstanding has arisen from the fact that chromium trioxide is used in the tanning of leather. It has been supposed that the process throws light on fixation by chromium trioxide in microtechnique. Unfortunately this is not so, for anionic chromium plays no part in tanning. The chromium trioxide is reduced to a basic salt, and it is cationic chromium that reacts with the proteins of skin to make leather.[544] The reaction of chromic sulphate with proteins has also been investigated chemically,[85] but this again is irrelevant to our subject, for the same reason. Cationic chromium is seldom used deliberately as a fixative in microtechnique, but it may be formed by the reaction of chromium trioxide with the tissues, and may then perhaps itself react with the tissues.

We turn now to the non-coagulant additive fixatives, formaldehyde and osmium tetroxide. These are of particular importance, for the very fact that they are non-coagulant makes them unlikely to distort tissues and cells seriously. It must nevertheless be kept in mind that subsequent treatment, especially embedding in paraffin, often results in gross distortion.

Formaldehyde can be caused to form compounds with various amino-acids,[177] but most of these reactions appear to be irrelevant to microtechnique. The compound with tyrosine, for instance, is only formed by heating for several hours in acid solution.[400]

The reactions of formaldehyde with polypeptides, each consisting of chains of only one particular amino-acid, are extremely

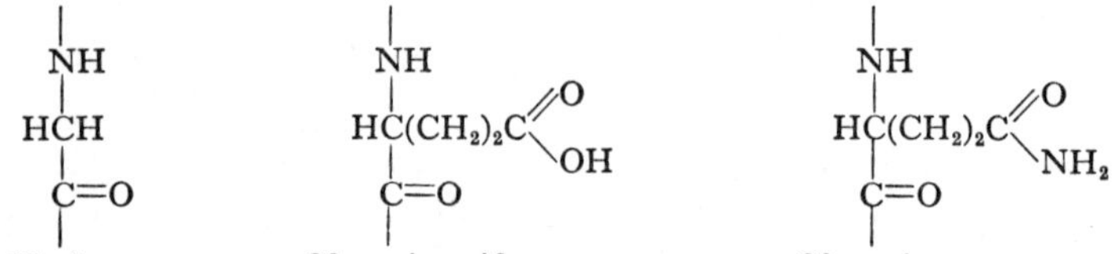

Glycine *Glutamic acid* *Glutamine*

The amino-acids of polyglycine, polyglutamic acid, and polyglutamine

instructive. Experiment shows how much formaldehyde such polypeptides will remove from solution.[174] Polyglycine binds very little formaldehyde; so does polyglutamic acid. Polyglutamine, on the contrary, binds more formaldehyde than any other macromolecule, so far as is known.

Silk-fibroin consists mainly of alanine and tyrosine. Like poly-

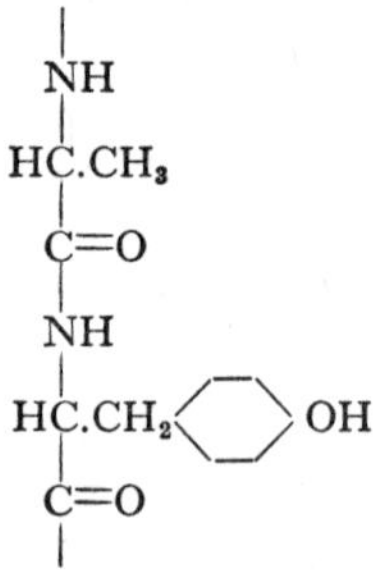

alanine

tyrosine

Part of a molecule of silk-fibroin

glycine and polyglutamic acid, it binds very little formaldehyde: less than one-twentieth as much as polyglutamine.

These facts suggest strongly that formaldehyde reacts with the $-NH_2$ groups of proteins. Lysine is largely involved.[546, 177, 203]

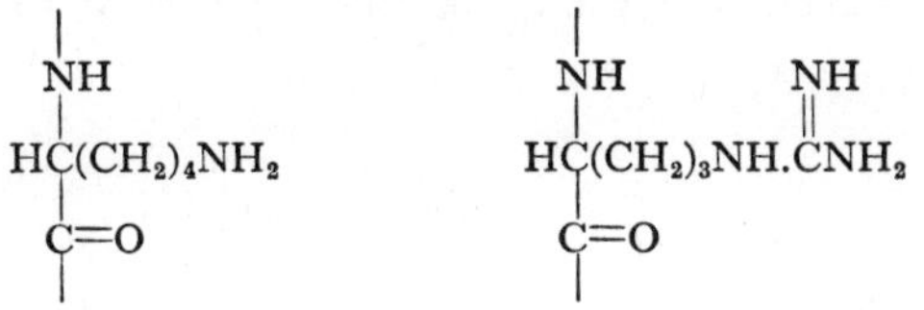

Lysine (left) and arginine (right), as components of proteins

The side-group of arginine also reacts, but only above pH 8, a degree of alkalinity unusual in microtechnical fixation. This reaction is not fixative, for it does not stabilize the protein; indeed, it increases susceptibility to swelling by acids (p. 64) and shrinkage by high temperature.[203]

When proteins are deaminized by the substitution of $-OH$ groups for $-NH_2$, their capacity to bind formaldehyde is much reduced. The work on this subject has been done largely with collagen and casein, because the toughening of these substances by formaldehyde is important in industry. The physical changes associated with tanning by formaldehyde do not occur after deamination.

Formaldehyde could react with lysine by simple addition, with formation of the side-group $-(CH_2)_4NH.CH_2OH$, or alternatively the reaction could be a condensation to form $-(CH_2)_4N{=}CH_2$ and water. The former reaction would be a special case of the general equation $RH + CH_2O{=}RCH_2OH$. The $-OH$ in the additive compound is reactive, and a methylene bridge, $R{-}CH_2{-}R^1$, is thus easily formed. Formaldehyde is commonly thought of as a reducer, but it is to be noted that in this particular reaction it acts on the contrary as an oxidizer. This was pointed out long ago by Kingsbury,[266] who did so much to help to place fixation on a scientific basis.

A methylene bridge between lysine side-groups on two previously separate protein chains at once suggests itself; but the stoichiometric relation between lysine and formaldehyde is nearer 1 : 1 than 2 : 1, and it is therefore probable that the linkage is between the lysine of one protein chain and a different group containing nitrogen in another protein. This could easily be glutamine. A link between lysine and the nitrogen of the peptide (–NH.CO–) link of the main protein chain is also possible. It will be remembered, however, that the peptide link of polypeptides is scarcely reactive, for polyglycine and polyglutamic acid bind very little formaldehyde. Nylon also binds very little formaldehyde,[174] though the –NH.CO– group occurs repeatedly in the molecule.

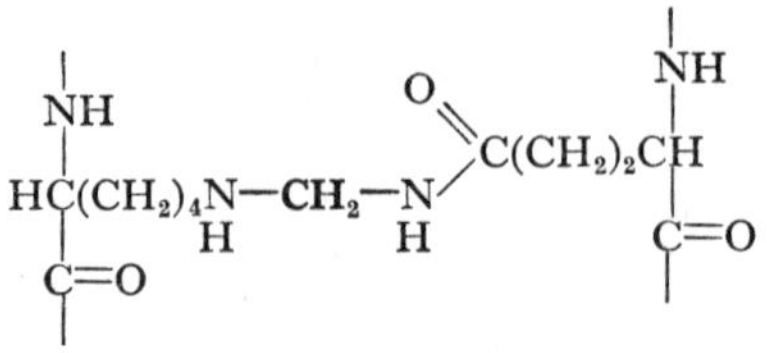

A methylene bridge (bold type) linking two protein chains through lysine (left) and glutamine (right)

Bridges of these kinds would tend towards gel-formation in proteins, or to the strengthening of pre-existent gels; coagulation does not result. In general, cross-linkages of any kind between protein chains result in brittleness, loss of elasticity, reduction in the ability to bind water, and, in the case of soluble proteins, lessened solubility.[77]

Zein is interesting in this connexion. The mechanical strength of fibres and films made of this protein is considerably increased by the action of formaldehyde, yet it contains no lysine (and very little arginine). It must be presumed that links are formed between the nitrogen of an amide (–CO.NH_2) group in one protein chain with a similar group in another, or possibly with the nitrogen of a peptide link.[177]

The reaction between proteins and formaldehyde is remarkably slow and contrasts strongly in this respect with the quick ionic reactions of some fixatives. In experiments with soluble proteins and polypeptides, carried out at 70° C, only one-half of the total amount of formaldehyde that eventually combined was taken up in 8 hours; after the lapse of 24 hours one-tenth of the amount still remained unbound.[174] In practice, fixation by formaldehyde is presumably nearly always incomplete, except when tissues are not only fixed but stored in a solution of formaldehyde. That is why tissues remain capable of taking up a limited amount of acid dyes.

The effect of pH on the reaction between proteins and formaldehyde is complex. The reaction with amide groups is said to be promoted by acidity,[174] but that with simple amino-groups is antagonized; the total effect is a slowing down, and the reaction is very slow below pH 3.[203] The greatest binding of formaldehyde occurs at slight alkalinity (pH 7·5 to 8).[203]

The importance of formaldehyde as a tanning agent for proteins in industry has ensured a careful study of its mode of action. The other important additive, non-coagulant fixative has no industrial applications of this kind, since it is far too expensive for practical use. This is unfortunate, for osmium tetroxide is one of the most valuable fixatives in microtechnique. In electron-microscopy it is pre-eminent on account of the faithfulness of its fixation and the ability of chemically bound osmium atoms to scatter electrons and thus make an image possible. It should be noticed that osmium tetroxide could not fulfil the latter function if it were a non-additive fixative. A mere deposition of unbound osmium would be useless, and indeed as misleading as it often is in light-microscopy.

That osmium reacts with unsaturated lipids (olein and oleic acid) was known long ago to Altmann,[5] but our understanding of its reactions with other tissue-constituents has grown slowly and is still very imperfect. In this chapter we are concerned only with those reactions that throw light on the way in which it fixes proteins.

The first relevant discovery was made in 1920 by Dutch chemists,[83] who showed that osmium tetroxide could be used to convert cyclohexene to cyclohexane-diol. Two points must be noticed

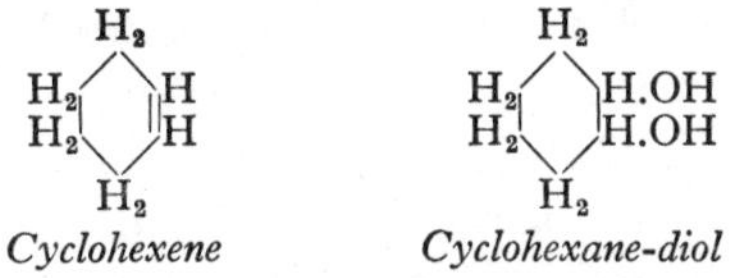

here: the reagent acts at the two ends of a double bond, and the product of reaction is a compound with two adjacent hydroxyl groups (a diol).

Reactions of this sort, whether caused by oxides of manganese

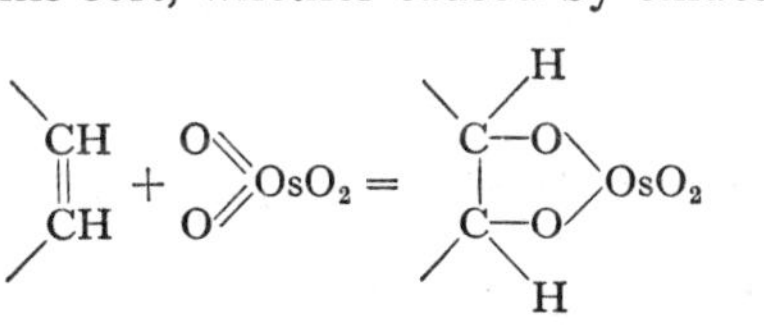

The reaction of osmium tetroxide at the ends of a double bond

or of osmium, may proceed in two stages. First, the double bond disappears and an additive compound is formed; then a second reaction may occur, with production of a diol and osmic acid. (It may be remarked in passing that the expression *osmic acid* is here used correctly. See p. 119.)

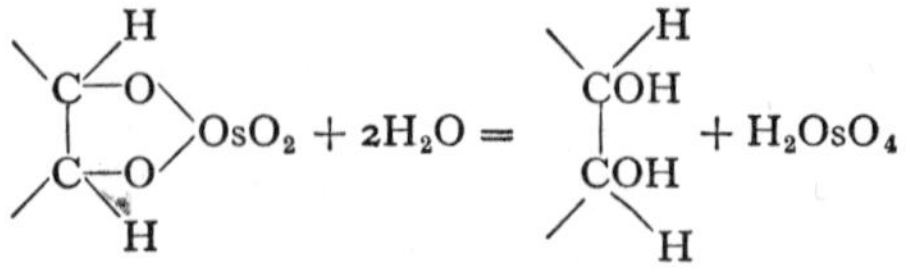

Diol *Osmic acid*

That osmium tetroxide could react in this way with many ring-compounds was shown by the German organic chemist, Criegee, in 1936.[135] One such substance is indene, which is related to indole,

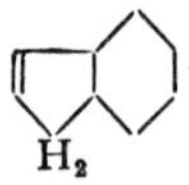

Indene

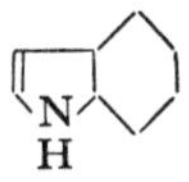

Indole

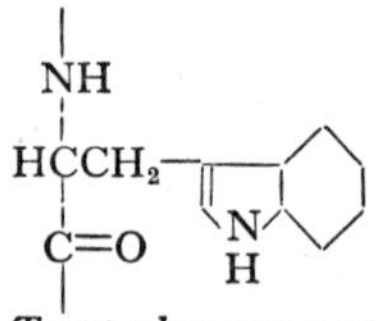

Tryptophane as part of a protein chain

a constituent of tryptophane. It follows from work on related substances [372] that the point of attack of osmium tetroxide would be the two ends of the double bond shown here in the five-membered pyrrol ring of indole and tryptophane.

Osmium tetroxide is capable, in certain circumstances, of joining molecules together. This was shown by Criegee and his associates.[135, 136] They found that osmium tetroxide acted most readily in this way when it had first combined with pyridine, but there are reasons for supposing that other organic substances might take the place of pyridine. It is clear that osmium can form a link between two ring-compounds by joining with both of them

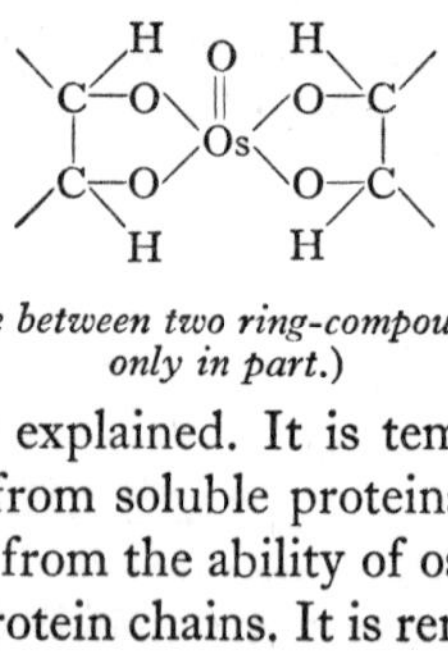

Osmium acting as a bridge between two ring-compounds. (The latter are shown only in part.)

in the way previously explained. It is tempting to suppose that the formation of gels from soluble proteins (or of firm gels from weak ones) may result from the ability of osmium to form bridges of this kind between protein chains. It is remarkable that such gels are often temporary (see p. 33). This may be connected with the fact that osmium tends to free itself from combination with ring-substances, while transforming these into diols.

Many amino-acids react with osmium tetroxide; this can be seen by the development of a dark colloidal reaction product, or sometimes a dark precipitate. It is doubtful, however, whether all amino-acids that react when free necessarily do so when they are constituents of a protein. Thus, the basic amino-acids, lysine and arginine, are reactive when present in proteins; but the protamines, which contain a high proportion of basic amino-acids, scarcely react with osmium tetroxide at room-temperature.[23] Tryptophane, histidine, and cysteine react especially strongly when free, forming dark precipitates.[23] Histidine, like tryptophane, possesses a double bond between carbon atoms; the nature of the reaction with cysteine has not been worked out. Among the various proteins, the most reactive are those that contain a high proportion of these three amino-acids.[23] The higher the amount of tryptophane in a soluble protein, the lower the concentration at which that protein can be gelled by osmium tetroxide.[405]

The anomalous non-coagulant fixatives, potassium dichromate and acetic acid, remain to be discussed. They are anomalous because they appear not to be fixatives of ordinary proteins.

At the end of the last century Fischer [169] showed that potassium dichromate did not coagulate albumin or globulin. We saw in chapter 2 that it does not render gelatine gels insoluble in warm water nor coagulate nucleoprotein. All this is radically altered if the solution be acidified: the reactions are then the same as though we had used chromium trioxide (see p. 106). We are concerned here with unacidified dichromate. The reaction with protein is slow at room-temperature. The unacidified salt is sometimes used in preparing wool for subsequent dyeing, but the process is carried out at a high temperature and the chemistry of it is unlikely to throw much light on fixation. In ordinary microtechnical use unacidified potassium dichromate seems to do little to proteins beyond making them somewhat more basic: that is to say, more capable of being coloured by acid dyes.[455] There is no good evidence that it has a definitely fixative effect on simple proteins, and nucleoproteins it positively dissolves. It is, however, one of the few fixatives that make certain lipids capable of resisting solution during embedding in paraffin. This is almost certainly the cause underlying the continued use of potassium dichromate in microtechnique (see p. 128).

Acetic acid behaves quite differently. The ionization constant being low ($1{\cdot}8 \times 10^{-5}$), the hydronium ion is not present in sufficient abundance to cause denaturation, but is able to exert its swelling effect. Indeed, this effect will be more readily produced if proteins are not coagulated by denaturation. Collagen fibres are swollen most by acids at pH 2·5,[314] and this is near the pH of 5% acetic acid (see p. 134). Acids will tend to break the salt-links between the chains; that is, the links binding together an amino-group in one with a carboxyl group in another. The breakage will expose hydrophil groups of atoms, and it is the entry of water attracted by these groups that is the cause of swelling.[544]

Weak organic acids swell wool much more than mineral acids do at the same pH. This is thought to be due to the action of the undissociated part of the acid, which tends to attach itself to amide groups through hydrogen bonds. The amide groups on

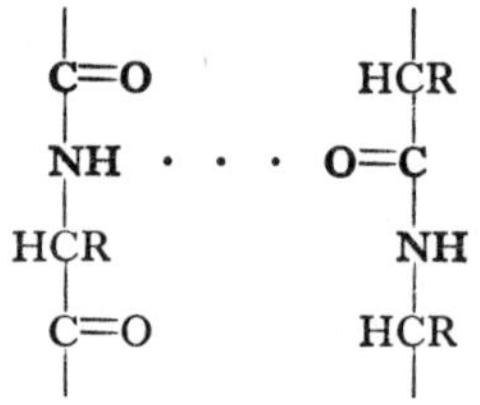

Parts of two protein chains held together by a hydrogen bond (. . .) between two amide groups. The latter are indicated by bold lettering.

neighbouring protein chains are supposed to associate with one another through hydrogen bonds, and the breakage of these by undissociated acid would permit swelling. The fact that acetic acid has much more capacity to swell proteins than mineral acids have is well shown in fig. 1 (p. 36).

In addition to the hydronium ion and undissociated acid, there is the acetate ion to consider. This has a separate effect in fixation, which has been carefully studied by the American botanist Zirkle.[557] He found that all acetates, including hydrogen acetate (acetic acid), tend to act similarly at the same pH (except that the salts of heavy metals show the effects of the cation rather than that of the anion). The acetate ion does not fix ordinary proteins; indeed, it opposes the action of denaturing agents.[408] It can react, however, with nucleoproteins. Whether it does so or not depends on whether the pH is on the less or the more acid side of pH 4·4 (or thereabouts). On the less acid side (for example, the unacidified sodium or ammonium salts), the acetate ion is not fixative, but macerates badly; on the more acid side there is coagulation of nucleoprotein, or separation of DNA from combination with protein with immediate precipitation.

No attempt will be made to give a general account of the fixation of lipids and carbohydrates, because these substances are not stabilized by most fixatives. The reactions of particular fixatives with them will be mentioned in chapters 5 and 6.

CHAPTER 4

The Reactions of Fixatives with Tissues and Cells: Methods of Research

Flemming [172] claimed that osmium tetroxide produced a network in the *Zellsaft* of *Spirogyra.* In making this statement in 1882, he appears to have been the first person to mention the production of a microscopical network in the tissues of an organism through the action of a fixative. This was a strange effect to attribute to osmium tetroxide, the very last fixative one would choose to produce a network. Hardy,[213] however, also described fine-meshed nets in paraffin sections of tissues fixed with solutions of osmium tetroxide; the vapour left the protoplasm homogeneous. As a general rule the non-coagulant fixatives leave tissues more or less transparent, apart from any extrinsic artifact they may themselves deposit. This is because they do not produce a microscopic network. Subsequent dehydration or other treatment may cause opacity.

The coagulant fixatives usually make tissues opaque, and white or yellowish: that is to say, the fixed material reflects light but does not readily transmit it (unless special steps are taken to cause transparency). The origin of the new surfaces responsible for this optical change in the tissues was first disclosed by Berthold,[68] who in 1886 described the coagulation of the protoplasm of plant cells by alcohol and other reagents. He realized clearly that the coagulation of protoplasm was essentially the same as that of egg-white. Similar observations on plant cells were made by Schwarz,[451] who supposed in error that coagulation was the separation of the firmer from the more fluid parts of the proteins of the protoplasm. It was Hardy [213] who showed that the net was the whole of the coagulated protoplasm, the meshes being filled with nothing but the mounting medium in the finished preparation.

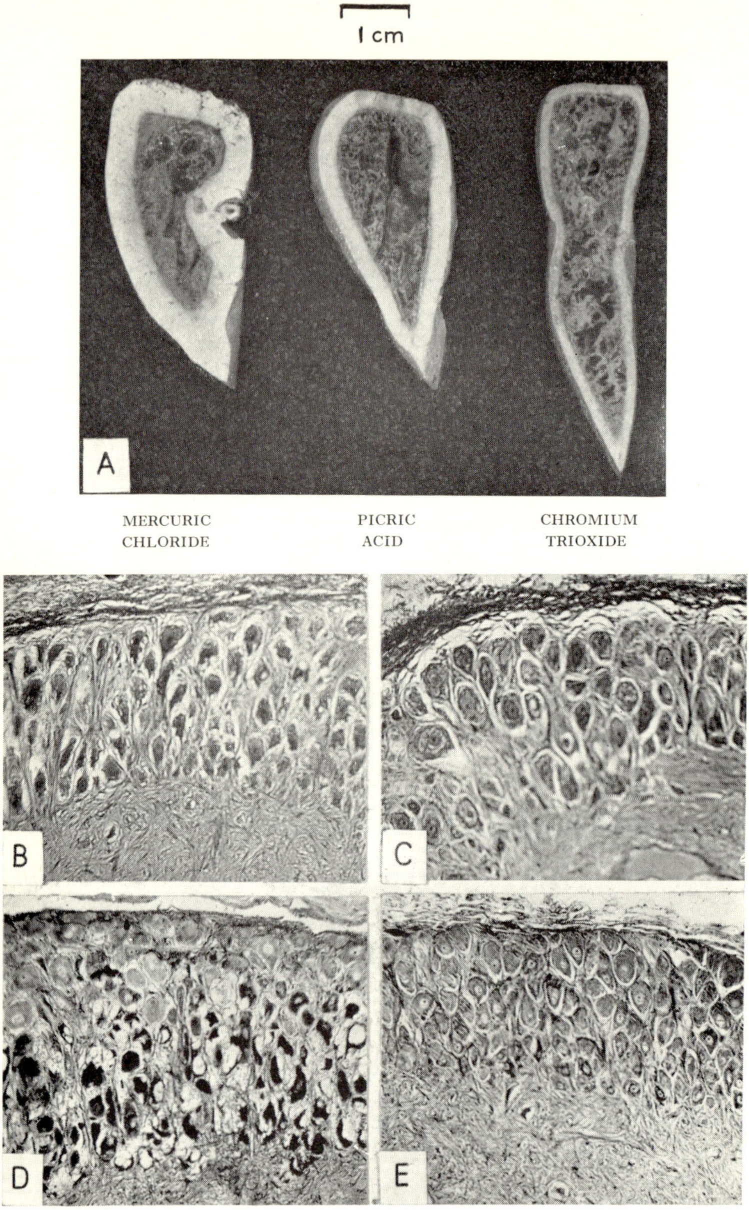

FIG. 5.

A, lobes of the liver of the rabbit left for 24 hours in fixatives (at the concentrations shown on p. 24) and then cut across.[38] B–E, photomicrographs illustrating Young's experiments on the addition of indifferent salts to fixatives. The cells are neurones in the stellate ganglion of *Sepia officinalis*. B, fixed in 4% formaldehyde dissolved in distilled water; C, in the same dissolved in sea-water. D, fixed in Champy's fluid made up with distilled water; E, in the same made up with sea-water.

(B–E are reproduced by kind permission of Prof. J. Z. Young and of the Editors and Publishers of *Nature*.)

Hardy worked with the intestinal epithelium of the woodlouse *Oniscus* (fig. 6) and with various cells of the frog and mammals. 'On the whole', he remarked, 'living cell-substance does react to fixatives just as does solid or fluid soluble colloid.' By solid or fluid colloid he meant protein gels or sols.

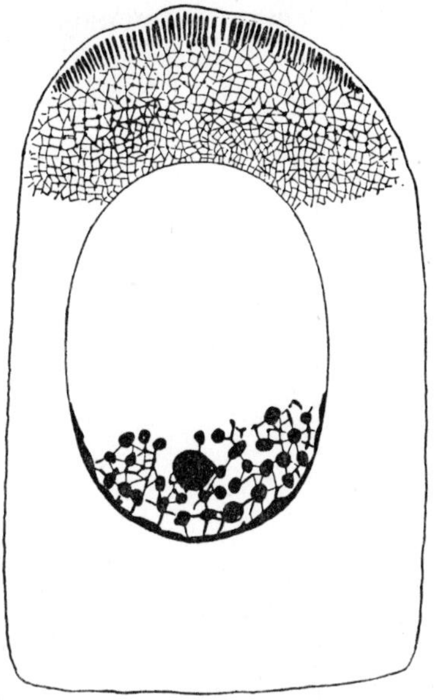

FIG. 6. A cell from the intestine of *Oniscus* (woodlouse), fixed by mercuric chloride (saturated solution in 0·6% sodium chloride); 1·4μ paraffin section, dyed with iron haematein: to show the coagulation of protoplasm. (From Hardy.[213])

Since tissues become opaque when invaded by coagulant fixatives, it is possible to obtain information about their rate of penetration. If mammalian liver or some other rather homogeneous organ be chosen, the line of demarcation between fixed and unfixed tissue is often extraordinarily clear and regular (fig. 5, A), though the fixative runs more quickly along the loose tissue accompanying large blood-vessels than through compact masses of cells. It makes little difference whether excised blocks of liver-tissue be used, or whole lobes; the latter, however, give the clearest results. It is evident that the connective tissue sheath of the organ forms a negligible barrier.

The earlier work on this subject [500, 510, 462] was done before Medawar [337] had disclosed the principles governing the diffusion of fixatives into protein gels (p. 37). Experiments on the penetration of coagulant fixatives into whole liver-lobes of the rabbit prove that the same principles govern penetration into tissues.[38] It is only necessary to suspend liver-lobes by threads in a large volume of the fixative under test, and to remove these at measured intervals and cut them across. The thickness of the coagulated tissue may be measured rather accurately. The figure obtained is not necessarily the same as the distance penetrated into the fresh tissue, because the fixative may change the volume of what it coagulates. Mercuric chloride is convenient in this respect, for it does not greatly alter the volume of mammalian liver (see p. 76). The results with this fixative are shown in fig. 7. The approximation of the graph to a straight line is remarkably close. It is clear that the rate of penetration falls off with time exactly as it does when fixatives enter clotted blood-serum or gelatine/albumin gel. The rate of entry is, however, considerably slower. Thus mercuric

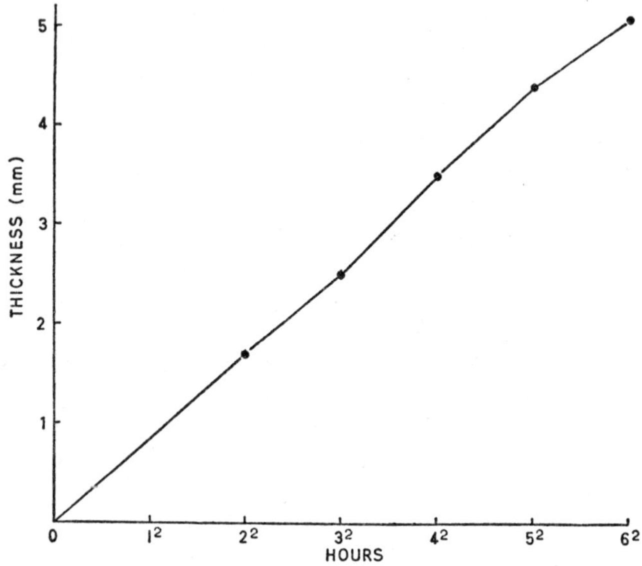

FIG. 7. Graph showing the thickness of rabbit-liver fixed by a saturated aqueous solution of mercuric chloride in various times.[38]

chloride penetrates 4·2 mm into liver in 25 hours (K = 0·84), but 11 mm into gelatine/albumin gel (K = 2·2). A survey of the work done in this field shows that the rate of penetration of coagulant fixatives into gelatine/albumin gels is in general agreement with their rate of penetration into tissues, except that the latter is much slower. Picric acid, however, goes more slowly than chromium trioxide into the gel, but more quickly than the trioxide into liver (fig. 5, A). (The reader will remember that in all such experiments, the concentrations of the fixatives were those listed on p. 24; see p. 32.) It is not obvious why fixatives should go so much more slowly into tissues than into the gel, for the latter contains about the same concentration of protein as ordinary protoplasm (p. 33).

It seems probable that the lipids of the cell-membranes act as barriers.

Fixatives penetrate more quickly into small than into large pieces of tissue. This is of practical importance. It might be thought a matter of indifference whether a large or a small piece were cut out of an organism for fixation: attention could be concentrated on the edge of the large piece, which would be as well fixed as the smaller. This is not so, for the following reason.

Imagine two blocks of tissue, with opposite sides parallel. One of them, block A, is exposed to the fixative on one side only (say the lower side): the other block, B, is exposed on the two opposite sides.

Consider first the passage of the fixative into A. It may be regarded thus. The fixative instantly reaches the upper side at infinitesimal concentration; later at progressively higher concentrations. For a considerable time the concentration at the upper side will be below that at which protoplasm is fixed (that is, either coagulated or changed into a stable gel). Meanwhile the substance will be gradually penetrating at its fixative strength, according to its K-value, and the line separating the fixed from the unfixed tissue will be advancing at continually diminishing rate towards the upper side, in accordance with the equation given on p. 37.

In block B the course of events will be different. On each side the substance would tend to penetrate at its fixative concentration at the same rate as it did on one side of block A, but from the first moment onwards there will be an infinitesimal contribution from the opposite side, and this will continually increase. The concentration at any given place at any given time will therefore be higher than it would have been if only one side had been exposed to the fixative. The thinner the piece, the more evident this effect will be. It follows that a fixative will penetrate more quickly into a small than into a large piece of tissue, and small pieces should therefore be used unless there is some particular reason for not doing so.

It is instructive to watch cells being fixed, and to record the effects not only on the ground cytoplasm, but on every constituent part. It was probably the transparency of living cells that held back this kind of investigation in the early days of microtechnique. As one optical method after another has been invented to overcome this difficulty, so enthusiasts have hastened to apply the new microscopes to problems of fixation. Not everyone who has engaged in this work has provided himself with the pre-requisite knowledge of the subject of fixation. It has seemed sufficient to put living cells under one of the kinds of microscopes that evade the difficulty of transparency, and simply watch what happens when the fixative is added. It is not always realized that this is a test of preservation, not of fixation. It has already been remarked (p. 28) that the classical work of this kind was done by Strangeways and Canti,[491] by dark-ground microscopy.

The most convenient cells are those that will flatten themselves against a coverslip and adhere firmly to it. The amoeboid mesodermal cells of various kinds that move outwards from cultured tissue-fragments fulfil these requirements well.[491, 94, 134, 141] Their thinness makes them readily observable by dark-ground or phase-contrast microscopy, and their adherence makes it easy to replace the culture-fluid by a fixative. The replacement can be achieved by quite simple apparatus,[491] but a special perfusion-chamber has been designed [142] that would lend itself well to this kind of work.

The extreme thinness of cultured amoeboid cells, when stretched over the surface of a coverslip, presents disadvantages. Such cells are far from being representative of cells in general, and their firm attachment and unusual shape prevent them from shrinking, or at any rate from showing clearly that they shrink (apart from the retraction of pseudopodia). Loose cells have their merits, despite the technical difficulties they present. Among those that have been used are salivary corpuscles,[402] polymorphs,[407] cultured neurones,[242] and amphibian spermatogonia.[134]

The results obtained by research of this kind will be incorporated briefly in chapters 5 and 6. By the common consent of all engaged in it, osmium tetroxide gives the most faithful preservation of the living condition of the cell (fig. 8, A, B). Chromium trioxide may be cited as an example of a fixative that alters the structure considerably (fig. 8, C, D). The pseudopodia are blunted; cytoplasm and nuclear sap coarsely coagulated; lipid droplets fused (probably by the stresses of coagulation); mitochondria destroyed. In dividing cells, however, the chromosomes are rendered clearer than they were in life.

One would wish to obtain quantitative information about the chemical changes wrought by fixatives on tissues and cells. A piece of tissue may be analysed chemically, as a whole, before and after fixation,[329] or the fixative solution may be analysed for dissolved tissue-constituents after fixation,[493] but the substances

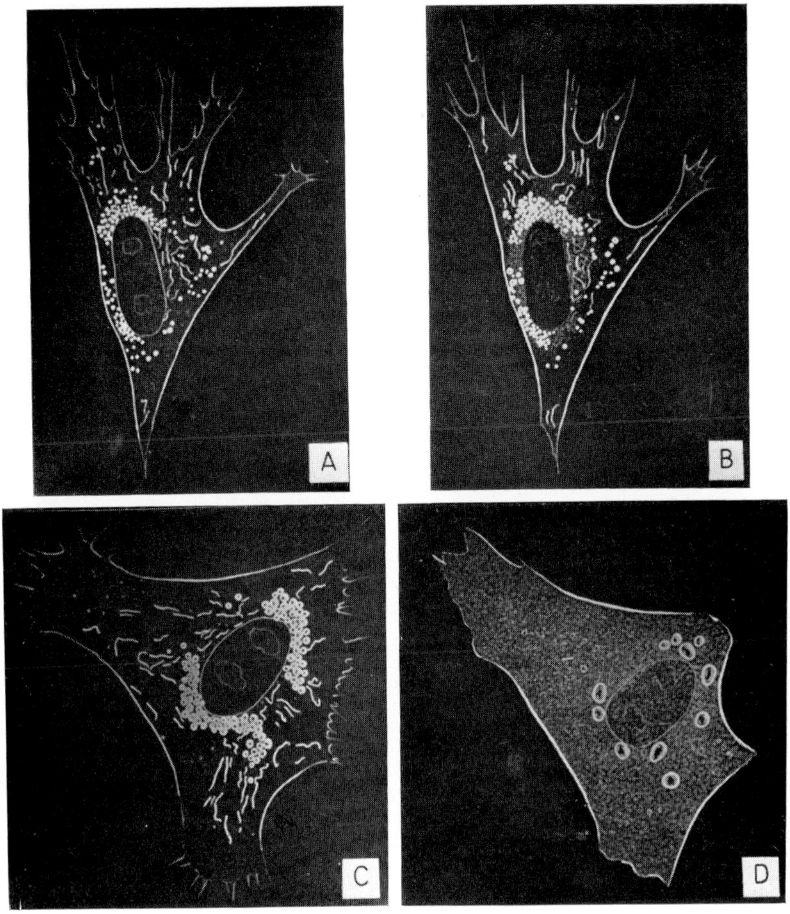

FIG. 8. The effect of fixatives on cultured cells from the chorioid or sclerotic coats of the eye of the chick embryo.

A, living cell. B, the same cell in 2% osmium tetroxide solution. C, living cell. D, the same cell in 0·5% chromium trioxide solution. Dark-ground microscopy. Drawings by Dr H. Fell.

(From Strangeways & Canti,[491] by kind permission of Dr H. Fell and the Company of Biologists, Ltd.)

revealed in these ways may be hard to trace to their sources. An estimate of the amount of nucleic acids lost by cells on fixation can be obtained by measurement of the optical density of parts of the cell before and after fixation.[141] This method depends on the opacity of nucleic acids to ultra-violet light of wave-length 265 mμ. A photomicrograph may be taken with this light before and after fixation, and the density of the images compared. Allowance must be made for change in volume of the part measured. Results with cultured fibroblasts from the chick embryo suggest that 4% neutral formaldehyde may reduce the amount of nucleic acids (DNA + RNA) in the nucleus by between 10% and 35%. It would be interesting to have comparable figures for all primary (unmixed) fixatives.

In order to investigate fixative as opposed to preservative action, it is necessary to examine cells that have been through the complete routine of microtechnique, and to compare these with living cells. Buchsbaum [94] in Chicago has worked in this way with cultured amphibian macrophages. He studied them in life by phase-contrast microscopy, and then fixed, washed, dyed, and mounted the same individual cells, making photographic records at every stage. The results showed that the changes caused by fixation were more fundamental, in this particular case, than any that resulted from subsequent treatment; but embedding was omitted, and this is often a destructive process.

All the studies of this kind so far mentioned are open to one major objection. The cells were separate from one another, and the fixative came directly up against them at its full concentration. In ordinary microscopical preparations the superficial cells are often damaged while those lying below the surface are well fixed (p. 28). Unless it so happens that information is particularly required about separate or superficial cells, the test should be designed in such a way that its result does not depend on the response of such cells, and it should involve the whole of the after-treatment that is to be used in practice. Since dyeing does not ordinarily introduce serious distortions, any dye or dyes that will reveal the structure of the object clearly may be used; but for a full test of any fixative it would be necessary to try all the usual embedding and mounting media.

Ideally each primary (unmixed) fixative should be tried with each embedding and mounting medium, and each fixing/embedding/mounting method should be assessed from the point of view of general micro-anatomy and histology, and also separately for its results with each cellular constituent (cell outline, ground cytoplasm, mitochondria, lipid globules, nuclear membrane, nuclear sap, nucleoli, chromosomes, etc.).

A comprehensive study of this kind has never been made, but a certain amount of work has been done. More than half a century ago Tellyesniczky [497–499] in Budapest studied the spermatogonia and spermatocytes of the salamander in paraffin sections of testes that had been fixed in various ways. Unfortunately he deliberately omitted to make careful comparisons with the living cell. 'So far as possible', he wrote, 'we avoid the question of *Lebenstreue*.' [497] The Austrian cytologist Pischinger [402] studied the effects of various fixatives on the nuclei of mammalian liver, with subsequent embedding in paraffin. His results, though interesting, were marred by his belief in the homogeneity of living nuclei.

For any test of this sort, it is of great importance to choose a suitable test-object. There are some organs that are reasonably well fixed even by indifferent fixatives. The intestine of the newt is an example. Others are susceptible to distortion in various degrees. One needs an organ that is difficult to fix well, so that the defects of fixatives may expose themselves clearly. It seems likely that protoplasm is easily fixed when it contains a high proportion of protein, and conversely; but this can only be proved when the interference microscope has given more information about the protein-content of cells. The distinguished cytologist, Bělař,[53] particularly recommended the testes of grasshoppers as test-objects for fixatives. The former are only available during a limited season, but crickets (*Acheta domesticus*) are convenient laboratory animals,[254] and their testes are ripe at all times of year. (See p. 329.)

As Bělař remarked, the testis of the laboratory mouse is *ein sehr heikles Object* for tests of fixatives. The kidney-cortex of the same animal is equally sensitive. A test of fixatives has recently been devised [38] in which these two organs are the objects of study. The test involves the use of only a single embedding medium and a single mountant, and the results are judged only from the point of view of the histologist. The test is made as thorough as these limitations permit. It could easily be made wider in scope.

The testis is cut into four parts and the kidney cortex also into small pieces. After appropriate washing the fixed tissues are passed through graded ethanols and toluene into paraffin wax.

This method of embedding was chosen partly because it is so much used, partly because it has a strong tendency to distort and thus provides a stringent test. Sections are cut at 8μ and dyed with Hansen's haematein (so-called *Trioxyhämatein*).[210] When a piece of tissue has been dehydrated, passed through toluene or other antemedium, and embedded in paraffin, it has probably undergone all the distortion that the processes of microtechnique can wreak upon it, and the choice of mounting medium is therefore of little significance for the purpose of the test. Canada balsam was chosen, mainly because it seems to remain the most popular mountant.

It is to be wished that we had some means of estimating the quality of fixation objectively. The degree of shrinkage or swelling can be measured (p. 75), but we have no other numerical data and for the present it is necessary to rely on subjective impressions. Anyone who proposes to judge fixatives should equip himself for the work by prolonged experience in the study of living cells.

People are often prejudiced in their beliefs about the value of different fixatives. To prevent this from influencing the results of tests, it is essential that the judge should never know what fixative he has been judging until after he has given his opinion. He should examine many preparations fixed in different ways and report fully on each before being told which is which. Preparations fixed in fluids that contain osmium tetroxide are usually darkened, and they should therefore be bleached before dyeing, so as not to be recognizable.

Written records should be made under a standardized set of headings before a definite opinion of the value of a fixative is formed. The following will serve as examples of headings in tests carried out with the testis of the mouse:—

- outlines of seminiferous tubules (whether smooth as in life, or wrinkled by shrinkage);
- spaces between tubules (whether artificially enlarged or distorted);
- cohesion of spermatogenetic cells (whether maintained or lost);
- cytoplasm of spermatogenetic cells (whether homogeneously fixed, coarsely coagulated, or disintegrated);
- chromosomes (whether fixed in life-like form in the meiotic phases);
- dyeing (whether intense or weak, differential or diffuse).

The report on the kidney-cortex of the mouse may be made under these headings:—

- spaces between convoluted tubules (as above, under seminiferous tubules);
- cytoplasm of convoluted tubules (as above, under spermatogenetic cells);
- free border of epithelium of first convoluted tubules (whether smooth or ragged);
- nuclei (whether smoothly rounded or distorted);
- red blood corpuscles (whether they retain their natural shape or are swollen into spheres or otherwise distorted);
- dyeing (as above).

The most delicate parts of the two organs are the cytoplasm of the primary spermatocytes and the free border of the first convoluted tubules. Really good fixation of these is rarely seen in paraffin sections.

It is convenient to compare fixatives by assigning them to different grades. It is an arbitrary matter to decide the number of such grades, but everyone who undertakes work of this kind will agree that two are too few (because more accurate distinctions can be made with confidence) and ten too many (because, if the assessment were repeated with the same slides, it would often happen that the same preparation was not assigned to the same grade). It is perhaps reasonable to make five (grade I the best, grade V the worst). To prevent waste of time from prolonged indecision, one may sometimes record the result as I–II, II–III, etc. Examples of grade I and grade V fixation are shown in fig. 9. Some of the results of this test with simple fixatives and mixtures will be mentioned in chapters 5, 6, and 7.

The fact that a fixative falls into a low grade in this test by no means condemns it. Grade V fixatives are usable with many organs that are less delicate than mammalian testis and kidney-cortex. They may also have particular virtues of their own. Altmann gives blocks of tissue that crack easily with paraffin embedding; there is considerable shrinkage and distortion of the cells; chromatin is dissolved away. Mitochondria, however, are excellently fixed, and Altmann remains a useful fluid despite its manifest defects. (The cracking does not occur when Altmann tissue is embedded in collodion.) As a general rule, nevertheless, it is obviously best to

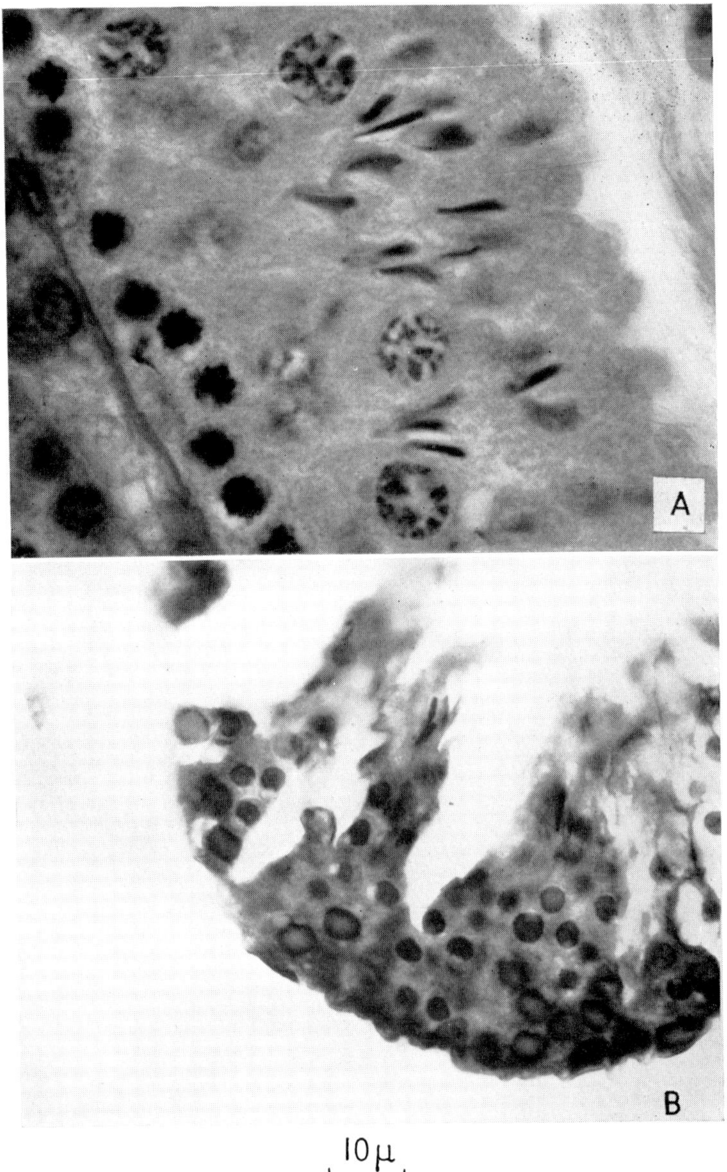

FIG. 9. Sections of the testis of the mouse, to show good and bad fixation. Both are 8μ paraffin sections dyed with Hansen's haematein.[38]

A, fixed with Allen's 'B.15'.[3] Grade I fixation. B, fixed with osmium tetroxide, 1%, buffered at pH 7·4. Grade V fixation. (With methacrylate embedding, the same fluid gives good fixation for electron microscopy.)

rely on fixatives in the higher grades, when the embedding and mounting media to be used are the same as in the test.

It is wrong to speak of good and bad fixatives without mentioning the after-treatment. A fixative that stabilizes structure against embedding in paraffin is likely to give good results with milder after-treatment, but one that works well with mild after-treatment may give poor results with paraffin. Flemming's weak fluid [172] seems a very poor fixative (grade IV–V) when paraffin sections are used. It should be remembered that the great cytologist left minute pieces of tissue or separate cells in his fluid for half an hour and then examined them in water without any other treatment whatever. There was no question of embedding, and he admitted that his preparations lost some of their delicacy if mounted in glycerine. In fact, he used his weak fluid rather as a preservative than as a fixative. Ordinary formaldehyde/saline (formaldehyde at 4% in 0·7% aqueous sodium chloride solution) gives poor results (grade III–IV) with paraffin sections, but quite good (grade II) with collodion. An extreme example of the same kind is provided by Palade's buffered osmium tetroxide solution.[380] This fixative is one of the best available to the electron-microscopist when tissues are embedded in methacrylate, but the results with paraffin embedding are very poor (fig. 9, B).

Non-quantitative tests, including the one we have been discussing, have considerable value. Thus it is important to know that one fixative destroys mitochondria, a second leaves them undamaged but unfixed, and a third stabilizes them against paraffin embedding. A multitude of similar examples could be quoted. Still, it is to be wished that we had more knowledge that could be expressed in numbers. Our main fund of quantitative information concerns shrinkage and swelling.

The length and breadth of a whole organism may be measured before and after fixation, and at various subsequent stages of treatment. This can be useful in various ways (for instance, by enabling us to judge the initial size and therefore the age of embryos that have been fixed); [388] but it provides little knowledge about the action of fixatives, since shrinkage or swelling may affect mainly the cells or mainly the body-cavity and other intercellular spaces, or a vertebral column may interfere with changes in length that would have occurred in its absence. Soft organisms with relatively small internal cavities do, however, provide useful information, for shrinkage is usually uneven and therefore produces obvious changes of form. The ctenophore *Pleurobrachia* is a delicate organism, well suited to observations of this kind. If fixed in a solution of formaldehyde in sea-water, it retains its shape well. If it now be transferred to 10% ethanol and thence slowly through 20%, 30%, 40%, and so on up to absolute ethanol, the main shrinkage occurs in 70%, with obvious deformation.[38] The lower ethanols, up to 60% or thereabouts, appear not to cause much shrinkage of properly fixed tissues, but there is always a stage in dehydration by ethanol, somewhere in the higher concentrations, against which no known fixative will protect the tissues.

Accurate measurements may be made of changes in volume undergone by whole organs,[490, 63] or by large cubes cut from whole organs.[548] The liver and spleen are especially suitable because they are fairly homogeneous in structure and contain no large empty spaces. The volume may be recorded when the organ is fresh and again at any number of subsequent stages up to and including infiltration with paraffin. The results of experiments of this sort with liver are shown in table 4. It will be noticed that the change in

TABLE 4

The effect of fixation and subsequent treatment on the volume of whole livers (? mammalian).

(Means of several observations in each case; rearranged from the data of Berg.[63])

	Mean volume expressed as % of fresh organ		
	after fixation	*after dehydration with absolute ethanol*	*after infiltration with melted paraffin wax*
mercuric chloride, sat. aq.	91	80	70
formaldehyde, 4% aq.	99	83	68
chromium trioxide, 1% aq.	78	68	64
ethanol, 96%	82	76	55
potassium dichromate, 3% aq.	100	64	49
picric acid, sat. aq.	74	64	42

volume cause by fixation itself gives little indication of the total shrinkage that will have occurred when the organ has been infiltrated with paraffin wax. Potassium dichromate, for instance, causes no change of volume but allows the tissue to be excessively shrunk by after-treatment. Picric acid shrinks strongly and permits much subsequent shrinkage. Mercuric chloride and formaldehyde gave less final reduction in volume than any other primary (unmixed) fixative that was tested.

Hooks may be inserted in each end of a piece of liver or other tissue, and a string from one of them may be attached to a lever; this will record on a revolving drum the progressive changes in linear dimensions that occur at every stage from fixation up to embedding in melted paraffin.[496] Experiments carried out in this way show that in routine microtechnique there are three stages at which shrinkage chiefly occurs after fixation. There is sudden shrinkage of mammalian organs on passing from 70% to 90% ethanol (reminiscent of the shrinkage of *Pleurobrachia* on passing from 60% to 70%), a gradual but pronounced shrinkage in xylene, toluene, or other antemedium, and a further prolonged shrinkage in hot paraffin. (Liquid paraffin at room temperature also shrinks.) These, then, are the particular stages in embedding against which fixatives do not give protection. There are obvious advantages in methods of embedding that evade some or all of the stages at which shrinkage chiefly occurs. It is to be noted that absolute ethanol does not have much effect on the volume of fixed tissues that have already been soaked in 90%.

It is particularly important that we should know how fixation and the subsequent processes of microtechnique affect the volumes of individual cells. Those of regular form naturally commend themselves for work of this kind, because they are easy to measure. The eggs of cows,[256] rabbits,[256] and echinoids [230] have been chosen for this reason; also the fully-grown primary spermatocytes of the snail (*Helix aspersa*).[428] The possible sources of error in this kind of investigation have been best appreciated by Ross,[428] who, in his recent work at Oxford, has made every effort to avoid them and has even taken into consideration the change of magnification of objectives when mounting media of different refractive indices are used. He has expressed his results in the form of medians and shown the scatter of his observations by the use of frequency polygons, while the others have relied on the means of relatively few measurements.

The diameters of the cells can be measured while they are alive in sea-water or other appropriate fluid, and at any number of subsequent stages in the routine processes of microtechnique. Hertwig's [230] investigation was the most complete in this respect.

Paraffin embedding has always been used. It is desirable that comparative studies should be made with other media.

Fixation does not necessarily shrink cells, and may indeed swell them strongly, but dehydration always reduces the volume below that of the living cell, and immersion in xylene carries the process still further. Some typical examples are given in tables 5 and 6.

TABLE 5

The volumes of unfertilized eggs of Arbacia pustulosa *at various stages of embedding in paraffin, expressed as percentages of the volumes while alive in sea-water. Each figure is calculated from the mean diameter of* 10 *to* 15 *eggs. (Data of Hertwig.*[230]*)*

Fixative	*Volume in*			
	fixative, 24 *hrs.*	*ethanol,* 70%	*ethanol, abs.*	*xylene*
mercuric chloride, sat. aq.	130	64	50	49
formaldehyde, 4% in sea-water	105	70	50	48
ethanol, abs.	48	—	—	41

TABLE 6

The volumes of the primary spermatocytes of the snail, Helix aspersa, *fixed in various simple fixatives, embedded in paraffin, dyed, and mounted in Canada balsam. They are expressed as percentages of the volume of the living cell. (Rearranged from the data of Ross.*[428]*)*

Formaldehyde, 4% aq. (neutralized)	34%
Mercuric chloride, sat. aq.	30%
Chromium trioxide 0·75% aq.	29%
Acetic acid, 5% aq.	28%
Potassium dichromate, 5% aq.	23%
Picric acid, sat. aq.	20%
Ethanol, abs.	19%

Shrinkage always occurs when cells fixed in aqueous media are transferred to 70% ethanol, and further shrinkage at each subsequent stage, at any rate up to xylene. We unfortunately have no data on the change of volume of whole cells on passing from xylene to melted paraffin.

A comparison of table 4 with tables 5 and 6 suggests that individual cells are more shrunken than whole organs by the processes of routine microtechnique. It is doubtful whether most histologists and cytologists realize the extreme degree of compression that a cell has undergone when it is examined in an ordinary microscopical preparation. (See fig. 10.)

The evidence suggests that nuclei generally shrink less than whole cells. Some of Hertwig's results with nuclei are shown in fig. 11. It will be noticed that infiltration with melted paraffin causes further shrinkage of nuclei beyond that caused by dehydration and soaking in xylene.

Shrinkage is harmful partly because it always involves distortion, partly because structural detail that approaches the limit of resolu-

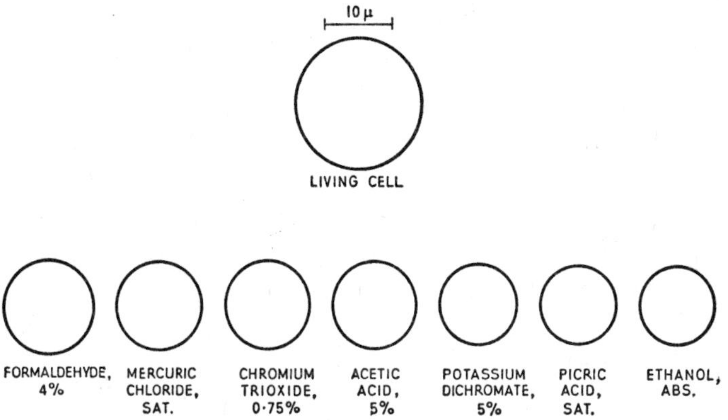

FIG. 10. Outlines of the fully-grown primary spermatocyte of the snail, *Helix aspersa*, to show the effect of fixation and subsequent treatment on the size of the cell. All the cells except the one at the top have been fixed, embedded in paraffin, sectioned, dyed, and mounted in Canada balsam.

(Diagrammatic but accurately to scale: from the data of Ross.[428])

tion of the microscope may pass beyond it. It is not surprising that paraffin wax is so much favoured by microtechnicians, for serial sections can be obtained very easily and their attachment to slides is quick and simple; but there is probably no other method of embedding that involves so much shrinkage. Our familiar fixatives may perhaps owe their survival to the fact that they give the best (or least bad) results with paraffin embedding. (See p. 148.)

There is room for much experiment here. We need new fixatives that will stabilize tissues better against existing after-treatments, and new after-treatments that will cause less shrinkage, or preferably none at all. The stearates of diethylene glycol present considerable advantages in this respect.[489, 120] The polyethylene glycols ('carbowax') [419, 166, 81] are particularly promising embedding media, because tissues can be passed directly into them from water and shrinkage seems to be slight; but there are still practical difficulties in their use, especially in the flattening of sections and their attachment to slides.

'For the good working of many reagents it is of the greatest importance that they should be dissolved in a medium that cannot itself produce disorganization by osmotic disturbances.

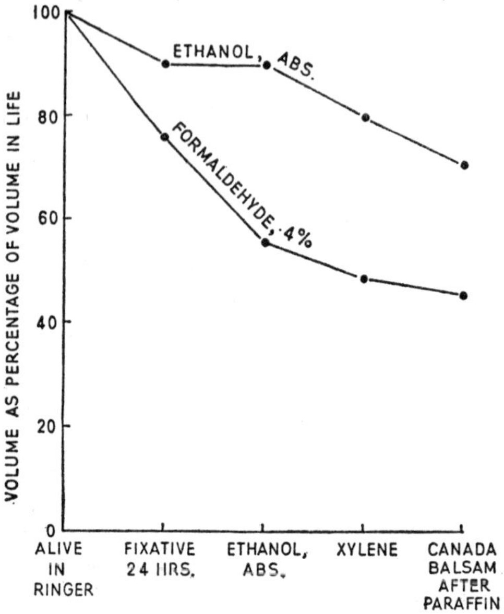

FIG. 11. Graph showing the effect of fixation and subsequent treatment on the volume of the nuclei of cartilage-cells (scapula of *Salamandra maculosa*). Each point represents the mean of 20 measurements.

(From the data of Hertwig.[230])

Therefore, for instance, for the preservation of marine algae alcoholic solutions and solutions in distilled water are to be absolutely rejected. Parallel experiments with picric acid, osmic acid, and iodine dissolved in alcohol, in distilled water, and in sea-water show without exception that only the solutions in sea-water give really good results, when one uses delicate objects.'

These statements were published in 1882 by Berthold,[67] who, as we have seen (pp. 40 and 66), was one of the first to approach the problems of fixation scientifically. We do not yet understand the causes underlying the results he obtained. The facts may be briefly stated thus. The best fixation is often obtained if the fixative substance is dissolved, not in distilled water, but in a solution of an 'indifferent' or non-fixative salt, such as sodium chloride or sulphate.

Müller [362, 363] used sodium sulphate nearly a century ago, in a mixture with potassium dichromate, and this salt is used in making up Zenker [554] and Helly [224] to this day, because these fixatives were based on Müller's. Heidenhain [222] long ago dissolved mercuric chloride in a solution of sodium chloride, and the same salt is commonly used with formaldehyde.

The advantage of adding an indifferent salt is shown particularly clearly by the cells of marine algae and marine invertebrates. Young [550] investigated this matter with the neurones of the cuttlefish, *Sepia officinalis*. He cut the stellate ganglion in two and put one half into a solution of a fixative in distilled water, and the other into a solution in sea-water. The two halves were then embedded together in paraffin and sectioned. When the fixative was picric acid, chromium trioxide, formaldehyde, osmium tetroxide, or potassium dichromate, the half fixed in the solution in sea-water was greatly superior to the other (fig. 5, B–E, opposite p. 67). It appeared that when the cells were placed in a solution of the fixative in distilled water, they swelled, burst, and then collapsed with serious shrinkage and distortion, though this sequence of events was not actually observed; in the presence of sea-water they more or less retained their shape. Fixation by mercuric chloride and acetic acid was, however, scarcely affected by the presence of the salts of sea-water.

Hertwig [230] made a quantitative study of the effect of sea-water on the fixation of the eggs of sea-urchins. Some of his results are shown in fig. 12. It will be seen that when formaldehyde was dissolved in distilled water, the eggs increased greatly in volume during fixation, but they did not burst. Increase in volume was slight when sea-water was used as solvent, and the eggs remained smaller than the others throughout dehydration and in xylene.

The cells of fresh-water and terrestrial organisms are less sensitive to the presence of indifferent salts than those of marine algae and marine invertebrates, but similar effects have been observed in them by Sjövall,[475] Carleton,[104] and others. For many references to the early literature of the subject, see Schaffer.[439] When the vertebrate kidney is fixed in formaldehyde dissolved in distilled water, the tubules appear shrunken in paraffin sections; but when this fixative is dissolved in 0·9% sodium chloride solution, the epithelium of the tubules sometimes swells so much that the lumen is almost obliterated in the final preparation.[104] Some impression of the effect of indifferent salts on the fixation of

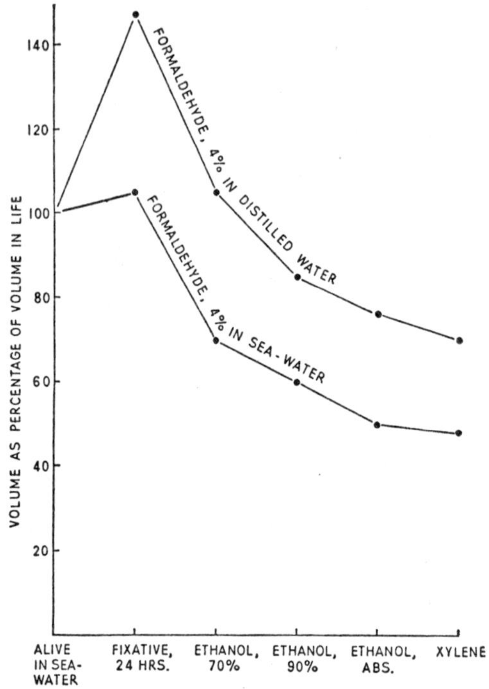

FIG. 12. Graph showing how the volume of the eggs of *Arbacia pustulosa* is affected by the addition of non-fixative salts to formaldehyde solution, at various stages in paraffin embedding. Each point represents the mean of 10–15 measurements. (From the data of Hertwig.[230])

mammalian cells can be obtained by means of the system of grading described on pp. 72 to 75. It will be remembered that the grading is done while the judge does not know what fixative was used in the preparation of the slide: his judgement is therefore impartial. The results with several fixatives are shown in table 7. In no case was a fixative placed in a higher grade when used without an indifferent salt than when used with it. The first three fixatives in the table were markedly improved by the addition of an

TABLE 7

Grading of simple fixatives with and without the addition of indifferent salts.[38] *The fixatives were used at the concentrations shown on p. 24.*

Fixative	*Grade without indifferent salt*	*Grade with indifferent salt*	*Indifferent salt used*	*Concentration of indifferent salt, %*
chloroplatinic acid	IV	II	sodium chloride	0·75
chromium trioxide	III	II	sodium chloride	0·7
formaldehyde	IV–V	III	calcium chloride (anh.)	1·0
picric acid	IV–V, V	IV	sodium chloride	0·7
mercuric chloride	IV	III–IV	sodium chloride	0·6
osmium tetroxide	IV–V, V	III–IV, III–IV, IV	sodium chloride	0·7
potassium dichromate	V	IV–V	sodium chloride	0·7
trichloracetic acid	III–IV	III–IV	sodium chloride	0·7
acetic acid	III	III	sodium chloride	0·7

indifferent salt, and most of the others slightly; with trichloracetic and acetic acids, however, there was no difference.

The facts, then, are clear enough. Indifferent salts do improve fixation by certain fixatives, especially the fixation of the cells of marine algae and marine invertebrates. The reason for this is not obvious. More than half a century ago the Swedish investigator Sjöbring [474] laid it down that formaldehyde should be used at such a concentration as to make the fluid isotonic with the tissues to be fixed. For mammalian tissues he used formaldehyde at 8 to 10%. He should actually have used it at about 4% to achieve his purpose,[235] but the question is whether his principle was correct. The evidence against it is conclusive.[439, 235, 26] Fixatives may be made up so as to be hypotonic, isotonic, or hypertonic to the body-fluids: the results show that there is no virtue in isotonicity.[235] Acetic acid at 5% *swells* tissues, yet it exerts an osmotic pressure of 20 atmospheres, about three times that exerted by mammalian blood. Chromium trioxide at $\frac{1}{2}$%, with the very small osmotic pressure of 1·4 atmosphere, shrinks cells strongly. Similar examples could be multiplied.[26] It is clear that the total osmotic pressure exerted by the constituents of a fixative is unrelated to the swelling or shrinkage of cells.

According to several authors,[67, 475, 104, 550] fixative substances should be dissolved in a solution of an indifferent salt giving an osmotic pressure equal to that of the intercellular fluids of the tissues that are to be fixed. It follows that the fixative solution as a whole should be hypertonic to these fluids.

It is claimed that when a piece of tissue is placed in a solution of a fixative substance dissolved in distilled water, the cells in the centre of the piece are affected as they would be if it had been placed in distilled water. The ions of the intercellular fluid diffuse rapidly outwards into the fixative solution, while the fixative substance penetrates more slowly inwards. The intercellular fluid therefore becomes hypotonic to the cells, and the latter accordingly swell and burst. If the fixative is dissolved in a suitable saline medium, the mobile ions of the latter counteract this process by diffusing into the tissues ahead of the fixative substance.[550] It is to be noticed that this theory postulates the presence of an intercellular fluid. Now we have already seen that echinoid eggs swell strongly when placed in a formaldehyde solution devoid of indifferent salts, but scarcely at all when this fixative is dissolved in sea-water. Thus the initial swelling, postulated as the prime cause of the damage, occurs also when there is no intercellular fluid. Beyond this, it is to bc held in mind that no one has proved that cells do in fact burst when placed in fixative solutions lacking indifferent salts. They appear *shrunken* in the final preparation. It is not clear why a burst cell must shrink.

It is argued that fixative substances can pass easily through cell-membranes, and that they are therefore not able to exert osmotic pressure on the cell-contents.[550] It has also been argued that the cell-surface is so much altered by fixation that it is no longer able to act as an osmotic membrane when a fixative has reached it.[500] Hertwig's experiments with echinoid eggs do not appear to support these opinions. Formaldehyde comes up against the cell-membrane directly the eggs are put in the solution, yet there is swelling unless the fixative was dissolved in sea-water. This suggests that the cell-surface continues to act as an osmotic membrane.

If the indifferent salt acts osmotically, it must perform its function while remaining outside the cells. We have no proof that in fact it remains outside. On the contrary, it is conceivable that it enters the cells with the fixative and then produces the observed results. Perhaps the salt acts on the proteins of the cell in such a way as to affect the imbibition of water. Some protein gels have a tendency to swell in the presence of non-fixative salts, in much the same way (though not so much) as they swell in the presence of acetic acid; this would counteract the shrinking effect of certain fixatives. It is relevant to remark that when acetic acid is included in a fixative, the presence or absence of an indifferent salt has little effect.[550] The fact must be kept in mind, however, that echinoid eggs *swell* in the *absence* of an indifferent salt.

If an indifferent salt acts within the cell, it does not necessarily do so by affecting imbibition. Its influence may be on the reaction of the fixative substance with the proteins of the cell. We have already seen how the action of mercuric chloride on proteins is affected by the presence of sodium chloride (p. 54). A still more striking instance is provided by ferric sulphate, which is rather a useful fixative.[38] By itself, a $\frac{1}{2}\%$ solution of the anhydrous salt does not coagulate egg-albumin; in the presence of ammonium sulphate it coagulates it instantly (see Appendix, p. 316). Ammonium sulphate alone is devoid of the ability to coagulate egg-albumin at any concentration, and it is therefore an indifferent salt.

If indifferent salts exert their effects after entering cells with the fixative substance, it does not follow that their osmotic pressure is a matter of indifference. Such a salt may diffuse rapidly into the piece of tissue and act osmotically, in the way suggested, until the fixative substance, diffusing more slowly, reaches the cells in the interior; it may then enter the cell with the fixative when the latter has damaged the cell-surface and stopped it from acting as a semi-permeable membrane.

It follows from what has been said that the use of indifferent salts is reasonable, though acetic acid may in some cases replace them; that they should not be used indiscriminately, without consideration of their possible effect on the coagulative powers of fixative substances; and that they should be present at such a concentration that they exert about the same osmotic pressure as the intercellular fluids of the tissues.

It has not been proved that the osmotic pressure of the indifferent salts should be exactly the same as that of the body-fluids. A slightly lower concentration seems to give better results. Thus sodium chloride works well at 0·7% with the tissues of mammals, and calcium chloride at 1%.[31]

When living cells are put in hypotonic solutions, those that are dividing have a tendency to swell strongly, and mitosis often comes to a halt at prometaphase or metaphase; the chromosomes are widely dispersed.[241] Cells may be fixed in this condition, which is especially favourable for the counting of chromosomes.[239, 240, 527,] An aqueous solution of sodium chloride at 0·3 to 0·5% is suitable for mammalian cells.[527]

There is no evidence that it is ever advantageous to dissolve fixative substances in complex physiological saline solutions.

Urea [3] and saponine [278] are occasionally added to fixative solutions. Their effects, if any, have not been critically analysed. There is no concrete evidence that the reduction of surface tension is helpful.

It is desirable that the whole subject of non-fixative substances in fixative solutions should be reinvestigated. Some of the most important problems remain unanswered. In the absence of indifferent salts, do cells in the interior of pieces of tissue actually swell, burst, and then shrink? Do the indifferent salts exert their effects outside or inside the cell, or first outside and then inside? We simply do not know.

In the early days of microtechnique it was important that tissues should be hardened so that they could be sectioned easily by hand. With the introduction of effective embedding media, hardening became less important. Nevertheless, the physical properties of fixed tissues remained and still remain important. It is necessary that the fixative should give them such a consistency that when they have been embedded, they can be cut into sections easily. They must not be friable, brittle, or very hard. Much depends on the process of embedding. Some tissues, such as the lens of the eye or thick pieces of voluntary muscle, are difficult to section in paraffin but easy in collodion. In such cases the process of embedding makes more difference than fixation.

It would be useful to have quantitative information about the physical properties of various tissues at all stages of microtechnique up to embedding. Unfortunately the available data are meagre. Wetzel [535] studied the elasticity of the belly-muscle of the cat. He cut out a long piece and put it in a fixative solution. He then attached it by one end in such a way that it was held horizontally. A weight attached to the free end caused it to sag. The coefficient or modulus of elasticity was calculated from a formula involving the dimensions of the piece of tissue, the weight applied at the free end, and the amount of sag. The higher the coefficient, the more rigid the tissue. Unfortunately he made only one or two observa-

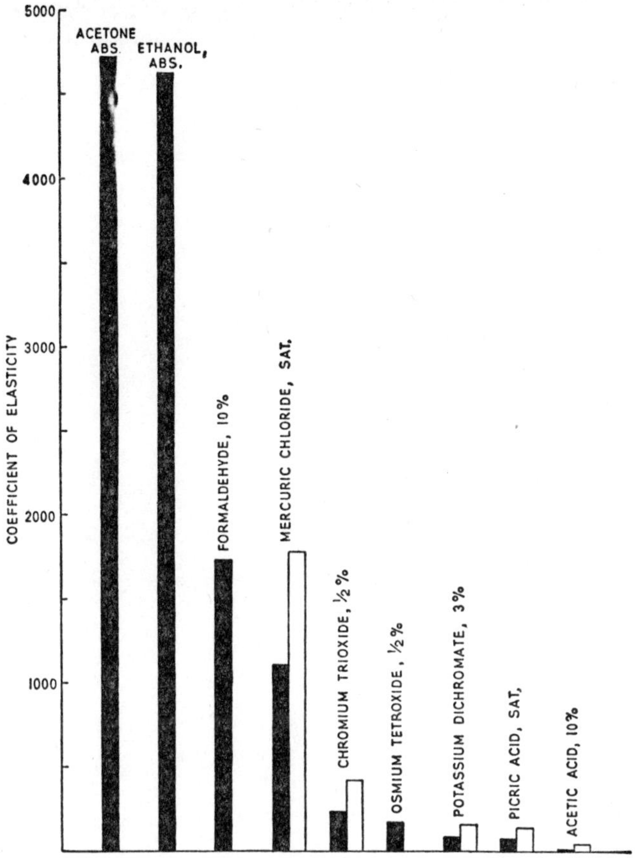

FIG. 13. Diagram showing the coefficient of elasticity of the belly-muscle of the cat, fixed in various ways. The black columns show the coefficient after fixation, the white columns after fixation and subsequent soaking for 4 days in 80% ethanol.

(From the data of Wetzel.[585])

tions with each fixative. Sometimes he transferred the tissue from the fixative to 80% ethanol and measured the elasticity once more. Some of his results are reproduced graphically in fig. 13. The excessive hardening effect of acetone and ethanol will be noticed. No other fixative approaches these two. Potassium dichromate, picric acid, and acetic acid leave the tissue very soft and incapable of being hardened to any considerable extent by subsequent soaking in 80% ethanol.

The effect of fixation on dyeing will be considered later in the book (p. 202), when the nature of dyes and the way in which they attach themselves to tissue-constituents have been discussed.

CHAPTER 5

Primary Fixatives Considered Separately 1. Coagulants

The term 'primary fixatives' is used here to mean single fixative substances as opposed to mixtures of two or more in a single solution. Primary fixatives may be used either absolutely, as ethanol, for instance, is sometimes used; or in solution in distilled water; or in solution in water with an indifferent salt.

The action of fixative mixtures can only be understood through a process of analysis. It is for this reason that the preceding chapters have been concerned almost entirely with primary fixatives. In this chapter and the next such fixatives will be discussed one by one. The various facts that have already been related about them will here be brought together in summary form, and further information will be added so as to afford as many-sided a view of each as possible. Certain important facts about fixation are not well suited to the kind of general treatment adopted in the earlier chapters, but lend themselves to separate exposition under the heads of particular fixatives.

It is thought that the principles of fixation can be explained better by a fairly full account of a few primary fixatives than by a sketchy one of many. The eight described in these two chapters have been chosen partly because they are components of familiar mixtures, partly because they are diverse in action and illustrate many important points about fixation. The selected primary fixatives are these:—

coagulant

non-coagulant

formaldehyde	Chapter 6, p. 111
osmium tetroxide . . .	,, p. 118
potassium dichromate . .	,, p. 126
acetic acid	,, p. 134

These fixatives will be considered one by one under a standardized set of headings. The headings are listed below, with short comments where necessary.

Standard concentration for fixation. Most of the simple fixatives can be used over a fairly wide range of concentrations, but it is a great convenience to select a rather arbitrary 'standard' concentration, somewhere within this range. It is better to know something about 5% acetic acid, for instance, from all points of view, than to have disjointed information about its pH at 1%, its rate of penetration at 2·5%, and its swelling effect at 10%. The standardization of concentrations also makes it possible to avoid unnecessary repetition. Wherever in this book a primary fixative is mentioned, it is to be understood, unless the contrary is distinctly stated, that the information refers to its use at the standard concentration, or at a concentration so close to this that the difference is insignificant (see pp. 24 and 32).

Description. Information about the appearance of each substance and its melting- and/or boiling-point, solubility, etc., is collected under this heading. Industrial uses are also mentioned.

Ionization. The figures given in the literature for the pH of aqueous solutions at the concentrations suitable for fixation are not always in good accord. This must be attributed partly to the impurity of the chemicals commonly used in biological laboratories. Two of the fixatives are used at saturation, and temperature may have a considerable effect on the concentration of such solutions and hence on pH; it may also affect the pH of other fluids.

The data on change of pH during fixation are those of Freeman and his collaborators [176] and Casselman.[112] Freeman generally allowed 1 g of spinal cord or liver of cat to 25 ml of fixative solution.

Oxidation-potential. The oxidation-potentials of fixatives have been investigated by Casselman,[112] whose figures are quoted in this chapter.

Manufacture. For fuller information about processes of manufacture than is given in these chapters, see especially Thorpe and Whiteley [504] and Kirk and Othmer.[267]

Introduction as fixative. Care has been taken to provide accurate information on this subject. Historical matters are often handled carelessly in scientific works.

Reactions with proteins. Readers who want further information on this subject, beyond the summaries given here, should turn back to chapters 2 and 3. In the present chapter and the next, brief notes are given under this heading about the effects of fixatives on solutions of nucleic acids, because this follows naturally on the discussion of their effects on nucleoproteins.

Reactions with lipids and *Reactions with carbohydrates.* The information under these headings is not given in any other part of the book.

Rate of penetration. The remarks made on this subject in these chapters summarize what has already been said about it in chapters 2 and 4; some additional facts are added. The K-values given are based on the following data:—

25 hours' penetration into gelatine/albumin gel; [38]
12 hours' penetration into mammalian liver; [500]
25 hours' penetration into liver of rabbit.[38]

It follows from what has been said in earlier chapters that the K-values would not have differed much if other periods had been chosen.

It is important for the reader to remember that fixatives penetrate much faster into gelatine/albumin gel than into liver. Thus a K-value of 0·8 is very slow for penetration into the gel, but moderate for penetration into liver.

Shrinkage or swelling. The figures for the shrinkage or swelling of gelatine/albumin gels represent the volumes after 18 hours' fixation. The results obtained by different authors with different organs and cells are not entirely concordant. The attempt is made to present a general picture of the effect of each fixative in causing change of volume and in leaving the tissues subject to such change on subsequent treatment. For further information on this subject, see pp. 36 and 75.

Hardening. There is unfortunately no quantitative information beyond the rather meagre data of Wetzel [535] (see p. 86).

Immediate effects on particular constituents of the cell. Under this heading an attempt is made to integrate the results obtained with various cells by several authors, especially Strangeways and Canti [491] and Policard, Bessis, and Bricka.[404] The effects described are those produced by the fixative itself, without subsequent dehydration or other treatment.

Methods of washing out. No comment is necessary here.

Effects on dyeing. The effects of different fixatives on the affinity of tissue-constituents for dyes have only been briefly touched on in the preceding part of the book, because it is thought better to discuss this subject in the part dealing with dyes (p. 202). The information is nevertheless summarized here, so that no one who uses chapters 5 and 6 for reference will be inconvenienced by the absence of any mention of this important aspect of fixation.

Effects on the histological picture seen in paraffin sections. The grading of fixatives, mentioned under this heading, has been fully described on pp. 72 to 75. It is important to remember that this system of grading is based on judgements made from the point of view of routine histology. The fact that a fixative falls in one of the lower grades by no means necessarily condemns it. It may have virtues that are only brought out by mixture with other fixatives, or it may be useful for some particular purpose in cytology or histochemistry, or give good results when some other method of embedding is used.

The attempt is made under this heading to present a general picture of what is seen in the finished preparation, by integrating the results of the grading-test with the findings of Tellyesniczky,[497] Zirkle,[556, 557] Pischinger,[402] and Casselman.[111] These authors did not all use the various fixatives at exactly the same concentrations; but despite this, and despite the diversity of the cells studied, it is possible to draw certain general conclusions.

Compatibility with other fixatives. No comment is necessary here.

ETHANOL (ethyl alcohol)

Standard concentration for fixation. Absolute (100%).

Formula and formula-weight. C_2H_5OH. 46·0.

Description. Ethanol is a light fluid (specific gravity 0·791 at 20° C), miscible with water in all proportions. It boils at 78° C and solidifies at about −112° C. If free from water it gives no cloudiness on mixture with benzene and does not produce gas (acetylene) on contact with calcium carbide. Ethanol is a powerful dehydrating agent.

Ionization. Not ionized.

Oxidation-potential. Ethanol at 95% shows an oxidation-potential of 0·45 volt.[113] Of the eight fixatives considered in this chapter and the next, only formaldehyde shows a lower potential.

Manufacture. It is usually prepared from malt or molasses by the action of yeast, with subsequent distillation. Fractional distillation does not give a higher concentration than 95%. Distillation of this strong ethanol at 81° C in the presence of anhydrous calcium sulphate or some other suitable dehydrating agent will give ethanol at 99·95%.

Ethanol is azeotropic with benzene: that is to say, it forms with benzene a binary mixture with sharp boiling-point (68° C). It also forms a ternary azeotropic mixture with benzene and water, with even lower boiling-point (65° C). If benzene is added to ethanol containing some water and the mixture is distilled, the ternary mixture comes off first, then the binary: nearly pure ethanol is left behind, slightly contaminated with benzene. The latter can be removed by filtering through active charcoal. It is called 'azeotropic' ethanol, though actually it consists of that fraction of the ethanol that did not form an azeotropic mixture.

Introduction as fixative. Wine was used in embalming by the ancient Egyptians, but not as the principal preservative. Ethanol appears to have been first used as a preservative for specifically anatomical purposes by Robert Boyle at Oxford in 1663.[201] The use of spirits of wine for the preservation of animals was well established before the middle of the eighteenth century. It was the examination of a spirit-specimen of *Alcyonium*, collected by Bernard de Jussieu on the coast of Normandy, that convinced Réaumur that this and other zoophytes were in fact animals,[413] and Réaumur himself [414] subsequently recommended the use of the same fluid for the preservation of birds. Henry Baker [24] preserved hydra in spirits of wine, but subsequently allowed his specimens to dry. The special virtue of ethanol when mixed with acetic acid was discovered by Clarke [127] in 1851.

Reactions with proteins. Ethanol is a non-additive, coagulant fixative. Coagulates are digestible by pepsin and typsin. In denaturing proteins it shifts their iso-electric points less than other fixatives do. Its capacity to denature is much affected by temperature: there is no measurable denaturation below −15° C. It slowly gelatinizes histones.[402] It does not denature zein or gliadin, nor does it stabilize gelatine gel against solution by warm water. It converts haemoglobin into kathaemoglobin;[459] that is to say, it denatures the globin without splitting off the haem.

Ethanol is a non-coagulant of nucleoprotein. Nucleic acids are precipitated from solution,[169, 402] but not rendered insoluble in water.

Reactions with lipids. Ethanol does not fix lipids in the sense of rendering them insoluble in lipid-solvents.

Tripalmitin and tristearin are almost insoluble in cold ethanol, triolein somewhat soluble; similarly palmitic and (especially) stearic acid have only slight solubility in the cold fluid, while oleic acid is readily soluble. Cholesterol is slightly soluble, the common cholesteryl esters nearly insoluble. Lecithin is soluble; the ethanolamine and serine esters of phosphatidic acid, together constituting 'cephalin', are insoluble, except in the presence of lecithin. Sphingomyelin and the cerebrosides are insoluble.[303, 95] Plasmalogen is soluble in ethanol, but when associated with elastic fibres it is not extracted by this solvent.[165]

Thus among the common lipids, only triolein, oleic acid, lecithin, and plasmalogen show more than slight solubility in cold ethanol. Most lipids are soluble in hot ethanol, but this is not used for fixation.

Dilute ethanol is an 'unmasking' agent, capable of splitting lipids off from certain lipoprotein complexes.[124]

Reactions with carbohydrates. Ethanol does not fix carbohydrates. Glycogen is insoluble in it and is therefore precipitated unchanged.

Rate of penetration. Ethanol penetrates slowly into clotted blood-plasma,[337] but it goes rather quickly into tissues.[500, 510, 462] Tellyesniczky's[500] and Seki's[462] results with mammalian liver agree closely. The K-value obtainable from their data is almost exactly 1·0.

Shrinkage or swelling. Ethanol shrinks gelatine/albumin gel more strongly than any other fixative except acetone.[38] It also shrinks whole livers strongly, though not so much as chromium trioxide and picric acid do.[63] It shrinks *Arbacia* eggs to less than half their original volume. The primary spermatocytes of *Helix aspersa* are shrunk to 19% of their original volume when paraffin sections of ovotestes fixed in ethanol have been mounted in

Canada balsam. No other fixative studied by Ross [428] gave so much shrinkage of cells in final preparations. There is evidence that nuclei are not so badly shrunk as whole cells.[230]

Hardening. Ethanol hardens tissues extremely, more so than any other fixative except acetone. Wetzel's [535] figure representing rigidity (elasticity) is about 4600 (about 20 times as great as the figure for chromium trioxide).

Immediate effects on particular constituents of the cell. A coarse coagulum is produced throughout the cytoplasm; mitochondria are destroyed; lipid globules tend to fuse and may dissolve. A coarse coagulum appears also in the nucleus; the nucleolus is shrunken.

Methods of washing out. Ethanol is miscible in all proportions with toluene and similar antemedia for paraffin embedding, and also with ether (and with water). It follows that no special washing out is necessary.

Effect on dyeing. Ethanol has less effect on the dyeing of proteins than most fixatives have. Albumin remains readily colourable by both basic and acid dyes: protamine, strongly basic before fixation, remains acidophil afterwards.[459] Fixed cytoplasm reacts to dyes like fixed albumin; chromatin is rendered strongly colourable by basic dyes, but its distribution in the cell is not stabilized, and ethanol is therefore a very poor fixative for chromosomes at all stages.

Effects on the histological picture seen in paraffin sections. By itself, ethanol is a poor fixative (grade IV). Cellular aggregates tend to shrink apart from one another, leaving rather large spaces in between; cells also tend to separate from one another; cytoplasm contracts strongly. Red blood-corpuscles are not badly fixed, partly because haemoglobin is preserved (as kathaemoglobin [459]), partly because the chief distortion to which they are subject is swelling, and ethanol prevents this. Nuclei are fixed without much distortion of shape; a nuclear network is not produced. Chromosomes are not distinctly seen.

The cytoplasm often contracts in such a way that it is piled up against the cell-membrane on the side opposite to that from which the fixative penetrated. Any glycogen present in the cell will be found in the displaced mass of cytoplasm. The contents of the nucleus are often piled up against the nuclear membrane in the same way. This curious orientation of the protoplasm has not been satisfactorily explained. It is presumably connected with the considerable reduction in volume that takes place when ethanol is mixed with water.

Compatibility with other fixatives. Ethanol is compatible with picric acid, mercuric chloride, formaldehyde, and acetic acid, but it reacts with the latter substance to produce ethyl acetate if sulphuric acid be added to the mixture. Since ethanol tends to be oxidized through aldehyde to acetic acid, it is best not to mix it with chromium trioxide, potassium dichromate, or osmium tetroxide. It does not react quickly with potassium dichromate, however, unless the mixture be acidified. The reaction with osmium tetroxide to produce a black precipitate is also slow at room-temperature.

Unclassified remarks. So far as is known, ethanol is the only non-aqueous fluid, other than methanol and acetone, that gives reasonably good results when used undiluted as a fixative,[462] though some might make a claim for dioxane.

PICRIC ACID (trinitrophenol)

Standard concentration for fixation. Saturated aqueous solution.

Formula and formula-weight.

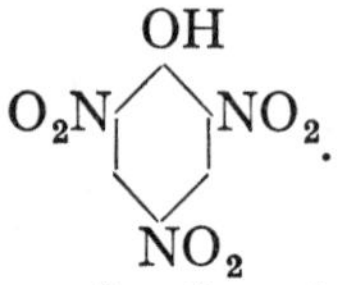

. 229·1.

Description. Picric acid consists of yellow leaflets or prisms, melting at 122° C. It is slightly soluble in water (1·4% w/w), more so in ethanol (4·9%), still more in benzene (10%). Its solubility in ethanol and ether often results in the yellow coloration of collodion blocks containing tissues fixed in picric acid.

Picric acid decomposes explosively when heated suddenly or detonated, and is therefore usually damped for storage. Until displaced by the related trinitrotoluene, it was the chief high explosive used for military purposes. It was manufactured at Lydd in Kent and received the name of lyddite. At the present day it is still used as an explosive in the form of its potassium, sodium, and ammonium salts.

This is the only substance that is used in microtechnique both as a fixative and as a dye (p. 185).

The name is derived from the Greek work meaning bitter.

Ionization. This is a much stronger acid than phenol. The pH of the saturated solution has been given as 1·33[285] and 1·6.[459]

Oxidation-potential. The saturated solution shows an oxidation-potential of 0·82 volt,[113] a little more than that shown by a 5% solution of mercuric chloride.

Manufacture. Picric acid may be prepared by the direct action of nitric acid on phenol, but this is wasteful because part of the phenol is lost by oxidation. In industry the ortho- and para-sulphonic acids are first formed, by treating melted phenol with concentrated sulphuric acid. The resultant phenolsulphonic acids are dissolved in water and treated with nitric acid; NO_2 groups now occupy the ortho- and para-positions.

Introduction as fixative. Picric acid was mentioned as a hardening agent by Ranvier [412] in 1875, but it is not known who first introduced it into microtechnique. It was much used by Flemming [170, 172] from 1879 onwards in his pioneer work on chromosomes.

Reactions with proteins. Picric acid is a coagulant fixative. It is a particularly strong coagulant of the histone of the nucleus.[402] It does not stabilize gelatine gels. It forms crystalline compounds with amino-acids and is probably to be regarded as an additive fixative of proteins, though its precise points of attachment have not been disclosed. It is an alkaloidal reagent, and the claim is made that it fixes basic proteins while allowing acidic ones to escape by solution.[459] Coagulated protein is digestible by pepsin and trypsin.

With nucleoprotein solution picric acid produces fine flocculi, which are eventually precipitated. DNA is either left in solution, or precipitated [402] in a water-soluble form.[169] The fact that picric acid may precipitate the protein of nucleoprotein while leaving nucleic acid in solution has been used in the preparation of DNA.[298]

Reactions with lipids. None has been recorded. Lipids are not dissolved by aqueous solutions of picric acid.

Reactions with carbohydrates. Picric acid is not a fixative of carbohydrates, but is strongly recommended by Lison [310] as a constituent of fixatives for glycogen. The latter substance is apparently bound to protein (sometimes quite firmly as 'desmo-glycogen'), and picric acid not only does not set it free, but acts on the protein in such a way as to cause the glycogen to resist solution.

Rate of penetration. Picric acid penetrates into gelatine/albumin gel more slowly than any other fixative that has been tried ($K = 0{\cdot}8$). It also goes rather slowly into liver ($K = 0{\cdot}5$) [38] (see fig. 5, A, opposite p. 67). Tellyesniczky [500] and Underhill [510] both tested it at about half-saturation and found slow penetration into liver ($K = 0{\cdot}4$).[500]

Shrinkage or swelling. Picric acid has little effect on the volume of gelatine/albumin gel, but it shrinks whole livers to 74% of the original volume; after infiltration with paraffin the volume is reduced to 42%. This represents greater shrinkage at the paraffin stage than that produced by the other fixatives studied by Berg.[63] It is evident that the shrinking effect of coagulation outweighs any tendency to swell as a result of acidity. The total shrinkage of the primary spermatocyte of *Helix aspersa* caused by fixation in picric acid and subsequent treatment up to the mounting of paraffin sections in Canada balsam, is excessive: the cell only occupies 20% of its original volume.[428] Ethanol is the only fixative that produces greater final shrinkage than this.

Hardening. It is strange that this fixative, which rivals ethanol in the shrinkage it produces, comes almost at the opposite end of the scale as a hardener. Ethanol hardens tissues excessively: picric acid leaves them very soft, and incapable of being much hardened by 80% alcohol. Wetzel's figure for rigidity is 69. That for chromium trioxide is about 3·4 times as great. The value of picric acid as a fixative lies partly in the soft consistency it gives to tissues.

Immediate effects on particular constituents of the cell. Pseudopodia tend to be constricted into separate globules; the ground cytoplasm is coagulated with varying degrees of coarseness; mitochondria are not destroyed, but sometimes become moniliform; lipid droplets may fuse. The nuclear sap is coagulated, the nucleolus somewhat shrunken. Chromosomes are rather well preserved.

Methods of washing out. It is sometimes said that protein coagulated by picric acid is soluble in water, and that care should therefore be taken to transfer tissues directly to ethanol. This does not appear to be true, but nevertheless there is generally no advantage in washing out in water, since the excess of fixative is more readily removed by ethanol or indeed by benzene or other antemedium used in paraffin embedding. Any yellow colour remaining in sections may be removed by the action of an aqueous solution of lithium carbonate,[250] or replaced by the action of some other acid dye.

Effect on dyeing. Picric acid renders egg-white acidophil. The effect on cytoplasm is similar: very little affinity for basic dyes is retained. When nucleoprotein is coagulated by picric acid, the

DNA remains in solution (compare Levene [298]). Something similar appears to occur in the fixation of tissues, for chromatin fixed by picric acid has little affinity for basic dyes, while some forms of it are rendered quite strongly acidophil.[38] It must be supposed that in these cases a basic protein is separated from DNA and coagulated, while the DNA dissolves.

Effects on the histological picture seen in paraffin sections. Even with the help of an indifferent salt, picric acid by itself is a grade IV fixative. There is little tendency for cellular aggregates to shrink away from one another, but cytoplasm generally assumes a very coarsely reticular form, or contracts into a mass surrounding the nucleus, or even disintegrates. Red blood corpuscles swell and become spherical. The shape of nuclei is rather well preserved; the nuclear membrane is clearly shown; a nuclear network is produced.

Compatibility with other fixatives. Picric acid is very tolerant of mixture with other fixatives. It may be used with any of the others described in this chapter and the next.

MERCURIC CHLORIDE

Standard concentration for fixation. Saturated aqueous solution.

Formula and formula-weight. $HgCl_2$. 271·5.

Description. The crystals are long, white needles. The molecule has a linear shape, the chlorine atoms being at opposite poles of the mercury atom. The melting-point is 275° C, but the boiling-point is not far above (301°) and the substance sublimes easily. At room-temperatures mercuric chloride dissolves in water at from 6·6 to 7·1% w/w. It is also readily soluble in ethanol and benzene.

This substance is used as an antiseptic and disinfectant, in the dressing of furs, in the intensification of photographic negatives, and in the preparation of various mercuric compounds.

Since mercuric chloride is usually prepared by sublimation and in strong solution is corrosive to the tissues of the mouth, it used to be called corrosive sublimate. When taken into the body in small quantities it may cause acute nephritis, the tubules and sometimes the glomeruli being affected; the formation of urine is reduced and may cease, with consequent death.

Ionization. The electrical conductivity of solutions of mercuric chloride is low, because of limited ionization. Solutions of this substance are extremely complex. Partial hydrolysis is said to give $(HgCl)_2O$ or HgClOH, with hydronium and chloride ions. $[HgCl_4]^-$, $[HgCl]^+$, and the mercuric $[Hg]^{++}$ ion are present, in addition to undissociated $HgCl_2$.[317] $[HgCl_4]^-$ is particularly readily formed in the presence of other chlorides, and mercuric chloride is more soluble in solutions of sodium chloride than in distilled water (about 10% in 0·75% sodium chloride).

The hydronium ion produced by hydrolysis results in moderate acidity. The figures given for the pH of saturated aqueous solutions vary from 2·8 to 3·6.[176, 285, 397, 459] A careful reading of the pH of a 5% solution gives 3·25.[113]

Solutions of mercuric chloride become more acid during fixation.[176] All other acidic fixatives maintain their pH or become less acid during fixation.

Oxidation-potential. The mercuric ion is capable of reduction to the mercurous $[Hg_2]^{++}$ ion, and mercuric chloride can therefore act as an oxidizer. A 5% aqueous solution shows an oxidation-potential of 0·75 volt. This is a little less than the figure for 2·5% potassium dichromate and a little more than that for 2% osmium tetroxide. Thus mercuric chloride is a moderately strong oxidizer.

Manufacture. Metallic mercury is dissolved in concentrated sulphuric acid; the solution is heated and sodium chloride is added, with a little manganese dioxide. Mercuric chloride is formed. The temperature is increased and it sublimes. It is caught on the interior surfaces of glass funnels.

Alternatively mercuric chloride may be made by the direct chlorination of mercury. The metal 'burns' in an atmosphere of chlorine. The product sublimes and is deposited on the walls of the chamber in which the combination occurs.

Introduction as fixative. Mercuric chloride appears to have been first used as a preservative for anatomical specimens by the French surgeon and anatomist, Chaussier (1746–1828).[11] Goadby included it in his preservative mixture, which was recommended by Quekett;[409] but the substance was so dilute (about 0·023%) that it cannot have acted as a fixative, and Goadby's fluid must have relied upon its other constituents for its preservative qualities. In 1847 the distinguished French morphologist Blanchard[79] added a solution of mercuric chloride to sea-water to fix marine Turbellaria. Corti[133] used it as a fixative in his histological study of the inner ear, and so did Remak[416] in his work on the multinucleate cells of the liver.

Lang, the celebrated Swiss zoologist and microtechnician,

alighted by chance on Blanchard's paper and decided to try mercuric chloride as a fixative, first [280] with marine Turbellaria in 1878, afterwards [281] with a wide variety of marine invertebrates. He used it both as a simple saturated solution and also in mixtures. The popularization of mercuric chloride as a fixative was due to Lang. Unfortunately he gave no exact reference to Blanchard's paper, and Whitman,[537] in his well-known textbook of microtechnique, gave a wrong one. These facts have clouded Blanchard's priority of 31 years. Ranvier [412] mentioned mercuric chloride in 1875 as a substance used in histological technique, but did not say for what purpose.

Reactions with proteins. This is a powerful coagulant of soluble proteins. The coagulate is not so readily digestible by pepsin or trypsin as that produced by ethanol or picric acid. Histone is gelatinized.[402] Gelatine gels are not stabilized.

Mercuric chloride reacts with the sulphydryl groups of the cysteine component of proteins, forming links between protein chains; but in the circumstances of ordinary fixation much more mercury is taken up in other ways. On the acid side of the iso-electric point of the protein it is taken up chiefly by amino-groups, as $[HgCl_4]^{-}$. The combination is very loose; it is promoted by the presence of added chloride (for instance, sodium chloride). On the less acid side of the iso-electric point the molecule is taken up as a whole, as $HgCl_2$. It associates with amino-groups, by which it is loosely held; the combination is opposed by chloride and by iodide. These reactions with amino-groups are coagulative. It is perhaps relevant that mercuric chloride is said to be less antiseptic in the presence of sodium chloride.[274] In alkaline solution mercuric ions are taken up by carboxyl groups in the protein.

Mercuric chloride is not a very powerful coagulant of nucleoprotein: fine flocculi are produced. It is also a weak precipitant of nucleic acids.[169, 402]

Reactions with lipids. Mercuric chloride is not known to react with triglycerides. There is evidence,[78] however, that these colour less strongly with Sudan III and IV (p. 299) and with Nile red (p. 301) after treatment with this fixative than after treatment with formaldehyde. Mercuric chloride forms compounds with phospholipids,[95] but the solubilities of such compounds appear not to have been investigated. The solubility of certain conjugated lipids in lipid-solvents is supposed to be reduced by treatment with salts of cadmium (compare Ciaccio [124]), and since this metal is related to mercury, we may perhaps expect that mercuric chloride will be found to be a fixative for these lipids.

The salts of the three related metals, mercury, cadmium, and zinc, are said to separate certain lipids from lipoprotein complexes and thus 'unmask' them.[124]

The most striking property of mercuric chloride in relation to lipids or lipid-like substances is its capacity to hydrolyse plasmalogen in such a way as to separate plasmal (chiefly palmitic and

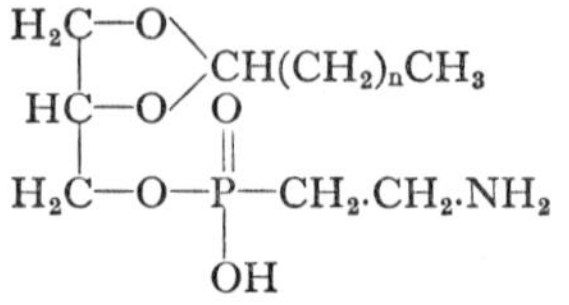

Plasmalogen

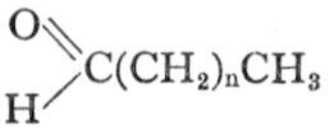

Plasmal

stearic aldehydes) from glyceryl-phosphoryl-ethanolamine. The plasmal can then be made visible by the use of Schiff's aldehyde-reagent (p. 308).[165, 163, 501, 502] It is not fixed by mercuric chloride, but retains its solubility in lipid-solvents (unless it chances to be adsorbed on elastic fibres).[165]

Reactions with carbohydrates. None is described. Mercuric chloride is regarded as a particularly good fixative for certain mucopolysaccharides called by histologists 'mucin'.[105]

Rate of penetration. Mercuric chloride penetrates at moderate speed into gelatine/albumin gel (K = 2·2). There is good agreement between different investigators about the rate of penetration into mammalian liver. The observations give K-values of 0·78 [500] and 0·84.[38] This also is moderate speed. For a photograph showing the penetration of mercuric chloride into liver, see fig. 5, A (opposite p. 67).

Shrinkage or swelling. Mercuric chloride only produces a small reduction in the linear measurements of gelatine/albumin gels and whole livers: the volume is reduced by less than 10% in both cases. There is, however, further contraction to 70% of the original volume on embedding liver in paraffin. Cells are much more variable in volume after the action of this fixative than gels or whole organs are. *Arbacia* eggs swell considerably in a saturated aqueous solution of mercuric chloride, but by the time they are in xylene they occupy only 49% of their original volume. Spermatocytes of *Helix* occupy 30% of their original volume when paraffin sections have been mounted in Canada balsam. It must be remembered that there is more shrinkage than this after the action of most fixatives (see fig. 10, p. 79).

Hardening. Mercuric chloride hardens moderately. Wetzel's figure is about 1100, nearly 5 times that for chromium trioxide. There is considerable further hardening on subsequent soaking in 80% ethanol.

Immediate effects on particular constituents of the cell. The shape of the cell, including any pseudopodia, is rather well preserved. The cytoplasm is sometimes much more finely coagulated than one would expect of a fixative of this sort. Indeed, mercuric chloride is particularly recommended by Policard and his colleagues [404] on this account. Cytoplasmic inclusions of various sorts, including mitochondria and the neutrophil granules of polymorphs, are preserved, though lipid globules may fuse. The nuclear membrane is clearly seen; nuclear sap is finely coagulated, the nucleolus very distinct. The mitotic spindle appears fibrous. Mercuric chloride distorts the cell less than any other coagulant fixative.

Methods of washing out. Since mercuric chloride is readily soluble in ethanol, there is no purpose in washing in water if the tissue is going to be dehydrated.

Mercuric chloride has a troublesome tendency to produce extrinsic artifacts. These appear in the final preparation as small, black, amorphous particles, down to about 1μ in diameter, and also needle-shaped, birefringent crystals, up to 25μ long, usually with a lump of the black material at each end. Both artifacts have a tendency to be deposited eventually on the surfaces of the slide and coverslip, in contact with the tissue.

Mayer [334] concluded that the black material was metallic mercury, but he could not determine the chemical composition of the crystals. It is unlikely that they consist of calomel, since they are also seen after the use of Millon's reagent,[348] which contains no chloride. Mayer was unable to produce such crystals from mercuric chloride or Millon's reagent except by their action on the tissues of organisms. The crystals, unlike the black material, are not present as such until the tissue is brought into a mounting medium, such as Canada balsam; cedarwood oil is particularly apt to produce them.

The black artifact is removed by the action of iodine in alcoholic solution. This was discovered in 1886.[525, 331] Presumably the mercury is oxidized to mercuric iodide, which is soluble in ethanol. The precursor of the crystals is also removed by this treatment.

The removal of mercury deposits may be delayed until sections have been prepared, but it is best to treat the tissue in bulk with iodine solution, as this makes it less brittle and therefore easier to cut. When sections have been prepared but not yet dyed, a glance through the microscope will show whether a second treatment with iodine is necessary.

Since protein coagulates may, in certain circumstances, be soluble in iodide solutions (p. 54), it is best not to dissolve the iodine with potassium iodide.

The colour of iodine is removable by soaking in 70% ethanol, or almost instantly by the action of sodium thiosulphate. This method

$$\underset{\text{thiosulphate}}{2[S_2O_3]^{=}} + I_2 = \underset{\text{tetra-thionate}}{[S_4O_6]^{=}} + \underset{\text{iodide}}{2I^-}$$

of removing iodine was introduced into microtechnique by Heidenhain (1909).

Effect of dyeing. Within the range of pH at which it is ordinarily used in fixation, mercuric chloride leaves the tissues more receptive of dyes in general than any other fixative. Cytoplasm will accept both basic and acid dyes, particularly the former. Chromatin is rendered very strongly colourable by basic dyes and dye-lakes.

Effects on the histological picture seen in paraffin sections. Mercuric chloride by itself gives grade IV fixation. Cell-aggregates are not seriously distorted. Cytoplasm is rather homogeneously fixed, but it is badly shrunken and there is a tendency for cells to separate from one another. Red blood-corpuscles are not badly fixed; haemoglobin is retained within them (perhaps as methaemoglobin [459]). The nuclear membrane is clearly shown; a nuclear network is not produced; chromosomes are not well fixed.

Compatibility with other fixatives. Mercuric chloride may be mixed with any other fixative. In certain mixtures, acetic acid appears to reduce its solubility.[554]

CHROMIUM TRIOXIDE

Standard concentration for fixation. 0·5% w/v aqueous solution.

Formula and formula-weight. CrO_3. 100·0.

Description. Chromium trioxide consists of brownish-red crystals, readily deliquescent by absorption of water. Their melting-point is about 198° C. They are extremely soluble in water, a saturated solution having a concentration of 62·4% w/W.

Ionization. For a careful study of this subject, see Casselman.[111] When chromium trioxide is placed in water, it forms chromic acid; this may be regarded as H_2CrO_4, but no such substance can be isolated. The acid ionizes to form hydronium ions and three different chromic anions. These are the orange-red dichromate $[Cr_2O_7]^{=}$ and hydrogen chromate $[HCrO_4]^{-}$ ions, with traces of the yellow chromate $[CrO_4]^{=}$ ion: a certain amount of chromic acid remains undissociated. The relative proportions of the four substances are shown in fig. 14 (left-hand column).

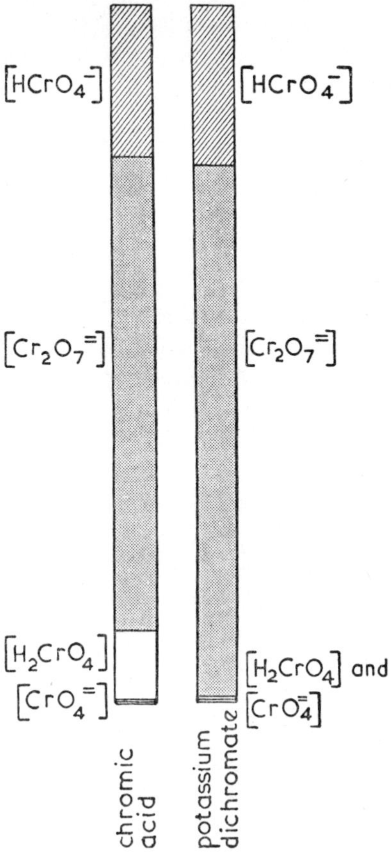

FIG. 14. Graphical representation of the ions present in a 2·5% aqueous solution of potassium dichromate and in a solution of chromium trioxide containing the same weight of chromium.

(From Casselman,[112] by kind permission of himself and of the Company of Biologists, Ltd.)

The hydronium ions released by the ionization of chromic acid make the solution strongly acid. A 1% solution of analytical-grade chromium trioxide has a pH of 1·20.[112] (Lassek [285] gives 1·12.) If chromium trioxide of 'laboratory reagent grade' is used, the pH of solutions is significantly higher.[111]

There is a very small rise in pH (about 0·1 unit) during fixation, almost entirely in the first 3 or 4 hours.[111]

Oxidation-potential. The dichromate ion can readily be reduced to give the chromic ion Cr^{+++}. The oxidation-potential of a 1% solution of chromium trioxide is 1·08 volt.[113] Data do not exist for a 0·5% solution, but extrapolation from available figures [112] makes it unlikely that the potential would differ significantly from this. Chromium trioxide is a much stronger oxidizer than any other fixative.

The oxidation-potential does not change measurably even during prolonged fixation.[111]

Manufacture. Chromium trioxide is prepared by mixing excess of concentrated sulphuric acid with a saturated aqueous solution of potassium or sodium dichromate.

Introduction as fixative. Chromium trioxide was introduced into microtechnique by Hannover in 1840. He had been staying in Copenhagen with a Professor Jacobson, who had experimented with the use of this substance in medicine. On the day of Hannover's departure, Jacobson had shown him a divided mammalian eye, preserved in a solution of chromium trioxide. Hannover was struck by the state of preservation of the organ, and on his return home he tried hardening various tissues in 5% or even stronger solutions and then cutting sections for microscopical study. He found that the consistency was good for cutting. He mentions a large number of tissue-constituents that were well preserved, including blood-corpuscles, cartilage-cells, medullated nerve-fibres, and ciliated epithelium from the mouth of the frog. His paper [209] takes the form of a letter to Jacobson.

Chromium trioxide was used by Corti [133] in 1851 for the fixation of the inner ear, and by Müller [362] in 1860, in a mixture with potassium dichromate and sodium sulphate, to fix an abnormal human eye. The discovery that it is useful in the fixation of chromosomes was made by the Polish cytologist Mayzel [335] in 1878. Strasburger [492] in 1879 found a 1% solution the best fluid for maintaining plant chromosomes in their living form, and he made much use of it in his subsequent work. Flemming exploited Mayzel's discovery from 1879 onwards.[170, 172] Much of his pioneer work on chromosomes was done with chromium trioxide at 0·1 to 0·2%.

Reactions with proteins. This is a powerful coagulant of albumin, but it does not stabilize gelatine gels. It is exceptional among fixatives in rendering proteins wholly resistant to digestion by pepsin and trypsin. Chromium cannot be removed from tissues even by prolonged washing, and it is probable that this is an additive fixative; but very little is known about the chemical changes that take place when solutions of chromium trioxide react with proteins at room-temperature. Tyrosine, trytophane, and histidine present side-groups that are capable of oxidation, and it is noteworthy that after prolonged fixation by chromium trioxide, histochemical tests for these amino-acids give feeble results.[111] This suggests that oxidation has in fact occurred, but it does not explain the firm binding of chromium.

Chromium trioxide is a coagulant of nucleoprotein. DNA is precipitated from its solutions in an insoluble form; [169] it is hydrolysed, with conversion of the pentose sugar to aldehyde, and Schiff's reagent (p. 308) therefore gives a positive reaction without the necessity for hydrolysis by hydrochloric acid on the slide.[111]

Reactions with lipids. It was stated long ago by Smith and Mair [481] that the fat of adipose tissue (triglyceride) can be rendered insoluble in lipid-solvents by prolonged treatment with chromium trioxide. This has recently been confirmed.[111] The reactions with lipids have not been fully investigated. Unsaturated fatty acids are first oxidized at the double bonds, with production of hydroxy-acids.[323] Oleic acid is eventually split, with production of pelargonic and azelaic acids; these are somewhat soluble in cold water. Presumably the reactions with lipids in general are similar to those of potassium dichromate (p. 128), but quicker and apt to go too far. It may be for this reason that potassium dichromate is nearly always used when tissues are 'postchromed' (p. 129) to render lipids insoluble in lipid-solvents by partial oxidation.

Reactions with carbohydrates. When chromium trioxide acts on polysaccharides, it works as an oxidizer, converting them to aldehydes. The exact site of this reaction in the polysaccharide molecule has not been determined.[310] The aldehydes formed in this way can be exhibited by means of Schiff's aldehyde-reagent (p. 308). This is the basis of Bauer's [49] histochemical test for glycogen and other polysaccharides. It is usual to fix tissues in a mixture of ethanol and acetic acid or some other fixative that will not dissolve water-soluble polysaccharides, and then to treat sections with a solution of chromium trioxide. If a simple aqueous solution of chromium trioxide be used as fixative, a certain amount of glycogen can subsequently be revealed in sections by the direct application of Schiff's reagent.[111]

If the proteins of a cell are fixed in such a way that the escape of glycogen is hindered, and a solution of chromium trioxide is subsequently allowed to act, the aldehyde produced will not dissolve out if the section be placed in distilled water. Nevertheless, it is doubtful whether chromium trioxide can be regarded unequivocally as a fixative for glycogen.

Polysaccharides that are not naturally chromotropic are rendered so by the action of chromium trioxide. This subject will be considered in the chapter devoted to metachromasy (see especially p. 247).

Penetration. Chromium trioxide penetrates slowly into gelatine/albumin gel (K = 1·0), and slowly also into liver (K = 0·25). Tellyesniczky's data give a higher K-value for liver, but he used a 1% solution. It will be seen from fig. 5, A (opposite p. 67) that chromium trioxide penetrates liver more slowly than picric acid.

Shrinkage or swelling. Gelatine/albumin gels only shrink slightly (to 91% of their original volume) in chromium trioxide solution, and tissues in general are moderately shrunk. The volume of the whole liver is reduced to 78% of the original and there is further reduction to 64% in paraffin. The spermatocytes of *Helix*, in paraffin sections mounted in Canada balsam, have 29% of their original volume. In comparison with the other fixatives tested by Ross, this is moderate shrinkage.

Hardening is moderate. Wetzel's figures show that acetone and ethanol leave tissues about 20 times as hard as chromium trioxide does, while the latter leaves them about 25 times as hard as they are left by 10% acetic acid.

Immediate effects on particular constituents of the cell. As we have see (p. 81), fixation by chromium trioxide is much improved by the addition of an indifferent salt. It is unfortunate that in the experiments of Strangeways and Canti and of Policard and his colleagues, no such addition was made. Their results can be briefly summarized thus. Pseudopodia, if present, are blunted; the ground cytoplasm is coagulated, sometimes rather coarsely; mitochondria are rendered invisible (and perhaps destroyed); lipid droplets tend to fuse together. The nuclear membrane is rendered clearly visible, the nuclear sap coarsely coagulated, the nucleolus somewhat shrunken. Chromosomes are seen more clearly than in life. The appearance of a cell before and after the addition of chromium trioxide solution is shown in fig. 8, C, D (opposite p. 70).

Methods of washing out. It is important to get rid of the excess of the fixative, lest there should be reduction at some later stage to green chromic oxide, Cr_2O_3, which is insoluble in ordinary solvents and remarkably resistant to acids and other reagents. Since ethanol might cause this reduction, tissues are usually washed in running water before dehydration. Virchow [526] showed that no insoluble precipitate was formed on direct transference of tissues to 95% ethanol, if light was excluded. Overton [377] recommended the use of sulphurous acid after fixation. He thought that chromic sulphate was formed: this, being stable and soluble, would be harmless, and indeed would act as a mordant for certain dyes. Direct transference from solutions of chromium trioxide to 50% ethanol containing 2% of sulphuric acid is probably safe.[38]

Effect on dyeing. Cytoplasm is rendered strongly acidophil by chromium trioxide, while chromatin is fairly easily colourable by most basic dyes. Differential dyeing of cytoplasm and chromatin is therefore easy, though there is some tendency for acid dyes to colour chromatin.

Effects on the general histological picture seen in paraffin sections. The excellence of chromium trioxide for the fixation of the neurones of *Sepia*, when dissolved at 1% in sea-water, was noted by Young.[550] In the presence of an indifferent salt at an appropriate concentration, this substance gives a better general picture than any other primary fixative, with the possible exception of chloroplatinic acid.[38] These two fixatives, with sodium chloride at 0·75%, fall into grade II. This grade is not reached by any other primary fixative than these (though it could be reached easily enough if paraffin embedding were not employed). In the absence of sodium chloride, chromium trioxide falls into grade III.

The general picture given by chromium trioxide is as follows. Cellular aggregates are fixed without serious distortion; cytoplasm is sometimes contracted round the nucleus, or coarsely coagulated so as to leave wide meshes; mitochondria are not fixed; red blood-corpuscles are swollen. The form of the nucleus is well preserved; the nuclear membrane is clearly seen; the nuclear sap is rather coarsely coagulated; the nucleolus is well fixed and has a strong affinity for both iron haematein and acid fuchsine; chromosomes are very well shown. The mitotic spindle appears fibrous.

The concentration at which this fixative is used makes remarkably little difference. Even at 0·01% (with sodium chloride at 0·85%), chromium trioxide gives much the same picture as when used at the ordinary concentrations.[111]

Compatibility with other fixatives. It is best not to mix chromium trioxide with substances that will reduce it, such as formaldehyde or ethanol. The reaction with formaldehyde is very quick. The reaction-products in the different mixtures containing chromium trioxide and formaldehyde have not been identified. In Sanfelice's fluid,[438] which contains these substances and acetic acid, the colour of the reaction-products is brown; in Allen's 'B.15',[3] which contains picric acid and urea in addition, it is green. The latter colour, combined with solubility, is suggestive of reduction to the ter-positive ion in one of its hydrated forms.

Chromium trioxide is compatible with all the primary fixatives mentioned in this chapter and the next, other than formaldehyde and ethanol, but when it is mixed with potassium dichromate, the latter substance does not exhibit its own character (see p. 128).

Unclassified remarks. Bright light is destructive of cells fixed in solutions of chromium trioxide.[64] If the light of a dark-ground condenser is focused on a cell while it still lies in the fixative, the following changes may be observed. The outline of the cell becomes hazy; the particles of coagulated cytoplasm begin to show Brownian movement and soon disintegrate, leaving only the lipid globules and nucleus; the latter then dissolves, the nucleolus resisting solution longer than the rest of it; finally nothing is left except the lipid globules.[491] The result is the same when heat-rays from the microscope-lamp are filtered off.

The action of chromium trioxide as a fixative will be referred to again in the section on potassium dichromate (p. 128).

CHAPTER 6

Primary Fixatives Considered Separately 2. Non-coagulants

FORMALDEHYDE (methanal)

Standard concentration for fixation. 4% w/v aqueous solution.

Formula and formula-weight. $\mathrm{H_2C{=}O}$. 30·0.

Description. Formaldehyde is a colourless gas. If liquefied it boils at −19° C. It is very soluble in water and is commonly sold in aqueous solution as 'formalin'. This is usually a 37% w/w solution or thereabouts. The specific gravity of formalin is about 1·08, and 100 ml therefore contain almost exactly 40 g of formaldehyde. Thus formalin is approximately a 40% w/v solution. Commercial formalin contains a little formic acid (generally less than 0·05%) and a considerable amount of methanol (6 to 15%).

Monomeric formaldehyde probably exists in water as $HOCH_2OH$. It has a strong tendency to polymerize as dimer, trimer, etc., having the general formula $HO(CH_2O)_nH$. In the 40% solution, only a small part (11%) of the formaldehyde is monomeric, but when this is diluted to 4%, the monomer predominates. Paraformaldehyde tends to be deposited from concentrated solutions of formaldehyde as a white powder. This is a highly polymeric form (n = 100 or more). The methanol in commercial formalin hinders polymerization.

As Bethe [70] pointed out 60 years ago, a considerable amount of confusion is caused by the loose usage of the trade-name 'formalin'. When authors say '10% formalin', they presumably mean formalin diluted with 9 times its volume of water or salt-solution to give 4% w/v formaldehyde, but there is always the possibility that they mean 10% formaldehyde. It is best to restrict the name formalin to the commercial product at 40% and to express the concentration of diluted fluids in terms of their formaldehyde content.

The only disadvantage in doing so is that the concentration of the formalin from which solutions are made up is not always exactly 40% w/v, and the concentration of formaldehyde in diluted solutions is therefore not exactly known. Any error from this source, however, is likely to be small.

The trade-term 'formol' is objectionable, for the ending in *-ol* is unsuitable. 'Formal' is not allowable, because it is the accepted name of another substance.

Ionization. Formaldehyde ionizes to a minute extent to form hydronium ions,[299] but acidity from this cause is negligible. The acidity of formaldehyde solutions is due to oxidation to formic acid by atmospheric oxygen. The pH of formalin is said to vary from 3·1 to 4·1.[176] The pH of the 4% w/v solution, made by diluting formalin with 9 times its volume of distilled water, is given by various authors at figures varying from 3·4 to 4·6.[112, 397, 176, 459] The addition of 5 ml of pyridine to 100 ml of 10% formaldehyde brings the pH to 7·0.[98] The 4% solution 'neutralized' by excess of calcium carbonate has a pH of 6·4;[425] if basic magnesium carbonate is used the pH is 7·6.[397] In the presence of 1% of calcium chloride and excess of calcium carbonate the pH is 5·7.[38]

Oxidation-potential. Formaldehyde is capable of being reduced to methanol according to the equation $HCHO + 2H^+ + 2e^- = H_3COH$. It can thus act as an oxidizer, though a weak one. It was claimed by Kingsbury[266] long ago, in an important contribution to the theory of fixation, that formaldehyde acts as an oxidizing fixative. The oxidation-potential of the 4% solution is 0·23 volt. This is the lowest of the oxidation-potentials of fixatives measured by Casselman.[112]

Manufacture. Formaldehyde is made by passing a mixture of air and vaporized methanol over a heated catalyst. Silver gauze at 635° C is suitable. Water formed by the reaction passes over and dissolves the gas; also some unchanged methanol. The amounts of these two substances are adjusted to give a solution at about 40% w/v, with as much methanol as is considered appropriate by the manufacturer for protection against excessive polymerization.

Introduction as fixative. Formaldehyde was introduced into microtechnique in 1893, later than any other important fixative. The discovery of its fixative properties was made accidentally by Blum,[82] who had previously introduced it as an antiseptic. One day he slit up a mouse that was infected with anthrax and left it overnight in a solution of formaldehyde (presumably as a disinfectant). In the morning he was surprised to find that it felt as hard as though it had been preserved in ethanol. He then undertook a systematic study of the effects on tissues, trying it at 4% on various organs, including liver, kidney, the mucous membrane of the stomach, and brain. He found that it hardened tissues faster than ethanol and preserved their external form better, so far as could be seen with the naked eye. He embedded various organs in collodion, sectioned them, and found the cells well fixed and capable of being dyed with haematein and synthetic dyes. It is perhaps fortunate that he used collodion, because formaldehyde unmixed with other fixatives gives rather poor results when tissues are embedded in paraffin (p. 118).

Reactions with proteins. Formaldehyde does not coagulate albumin; it renders this protein not coagulable by ethanol.[26] Unlike the coagulative fixatives described in the last chapter, it stabilizes gelatine gels very perfectly, so that they do not dissolve in water at 37° C but retain their form and transparency. There is a slow gelatinizing action on histone.[402] Haemoglobin is retained in blood-corpuscles, perhaps with conversion to methaemoglobin.[459] Fixation by formaldehyde does not affect the action of pepsin or trypsin on the proteins of blood plasma. There is contradictory evidence about the ability of trypsin to digest collagen fixed by formaldehyde.[339, 202]

The chemistry of the reactions with proteins has been thoroughly investigated. Formaldehyde reacts with the $-NH_2$ of the side-groups of certain amino-acids, probably forming methylene bridges that link protein chains together, lysine to lysine or lysine to glutamine. The reaction is slow, especially below pH 3: the greatest binding of formaldehyde occurs at about pH 7·5 to 8.

Nucleoproteins are not coagulated; extremely minute flocculi may appear eventually. DNA is not precipitated from its solutions.[402] The amount of nucleic acids (DNA + RNA) in cells is reduced by 10 to 35% by formaldehyde fixation.

Certain enzymes are not wholly inactivated by formaldehyde. The enzymes in the liver of the rat have been studied in this connexion.[463] β-glucuronidase retained its activity best; then sulphatase, acid phosphatase, and esterase; alkaline phosphatase least well.

Reactions with lipids. In general, formaldehyde is a good preservative of lipids, including cholesterol and its esters, unless fixation is very prolonged. Frozen sections are generally used, and fixation of lipids (as opposed to their preservation) is therefore unnecessary.

It has been known for many years that there is a gradual loss of lipids if tissues are left in solutions of formaldehyde for a long time.[208, 349, 532, 90] Cholesterol, triglycerides, and cerebrosides appear to remain unaffected, but certain phospholipids begin to disappear. Lecithins are well retained, but phosphatidyl serine, phosphatidyl ethanolamine, and sphingomyelins become reduced in amount. There appear to be two factors in the loss of phospholipids. On one hand there is thought to be a splitting off of glycerophosphoric acid, the fatty acids remaining in the tissues; on the other there may be a direct removal of lipid in colloidal solution.[208, 90] In this connexion it is to be remarked that phosphatidyl serine dissolves in 4% aqueous formaldehyde to form a clear colloidal solution.[90]

Phospholipids have a strong tendency to take up water and extend their surface by growing outwards in worm-like 'myelin forms'. This can easily be seen by smearing some lecithin across the bottom of a cavity-slide and watching through the microscope on the addition of water (fig. 15, A). The outgrowth of myelin forms from the lipid constituents of cells would be damaging morphologically and would also favour gradual solution. It was shown by Leathes [290] that calcium ions have a remarkable effect in preventing these outgrowths (fig. 15, B). The idea of adding calcium chloride to formaldehyde solution therefore suggested itself.[33] The salt has a double function: it checks the distortion and solution of certain lipids, and it also improves fixation in the same way as sodium chloride or other indifferent salts.

After treatment with formaldehyde, phospholipids are less soluble in lipid solvents and therefore less extractable by these from tissues.[352, 265, 28] Experiments on this subject may be carried out as follows.[28] Phospholipids are incorporated in elder-pith, and the latter then placed in formaldehyde solution; frozen sections are cut and treated with various lipid solvents, and then with one of the colouring-agents for lipids (p. 299). In a series of experiments of this sort, with formaldehyde at 4% in 1% (anhydrous) calcium chloride solution as fixative, lecithins, 'cephalins', and sphingomyelins were all rendered insoluble even in boiling ethanol followed by boiling ether, and also in paraffin wax at 60° C preceded and followed by xylene. This shows that formaldehyde is a fixative for phospholipids, though not necessarily for mixtures of these substances with other lipids. The chemistry of this process is not known, but reaction between formaldehyde and the $-NH_2$ group of phosphatidyl ethanolamine has been suggested.[265]

If, after fixation with formaldehyde, less of a particular lipid is extractable by lipid solvents than was present in the fresh organ,

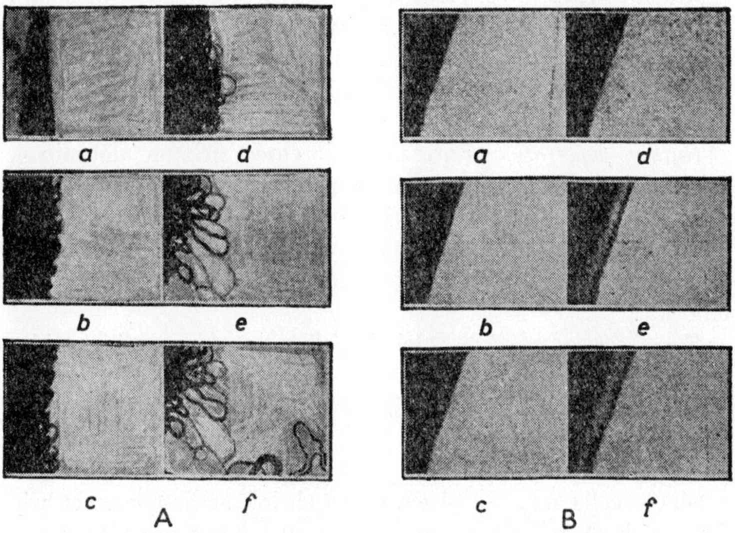

FIG. 15. Photomicrographs of lecithin smeared on glass.

A, in distilled water, showing the outgrowth of myelin forms. B, in a concentrated solution of calcium chloride, showing absence of myelin forms.

In each case the photomicrographs were taken *a*, before the addition of fluid; *b–f* after the addition, at intervals of $2\frac{1}{2}$, 9, 17, 33, and 57 minutes respectively.

(From Leathes,[290] by kind permission of the Editors and Publishers of *The Lancet*)

two very different explanations are possible. On one hand the lipid may have partly dissolved out into the formaldehyde solution; on the other it may have been rendered insoluble in lipid solvents and therefore not accessible to ordinary methods of analysis.

It is claimed by Wolman and Greco[543] that formaldehyde reacts with unsaturated lipids at the double bond $(-\underset{H}{C}{=}\underset{H}{C}-)$, producing carbonyl groups that will react with Schiff's aldehyde-reagent (p. 308). The capacity of the lipid to take up iodine is at the same time reduced, and this suggests that the double bond has been attacked by formaldehyde. If reagents that block carbonyl groups are used, there is no longer a positive result with Schiff's test.

Glycerides and fatty acids are sometimes crystallized and thus rendered anisotropic by the action of formaldehyde. Cholesteryl esters are transformed from liquid spherocrystals, showing the cross of polarization, into simply anisotropic, solid crystals.[310]

The colouring agents for lipids (p. 299) are said to act particularly strongly when formaldehyde is used as fixative.[78]

Reactions with carbohydrates. Formaldehyde does not fix soluble carbohydrates, but it has a remarkable capacity to fix proteins in such a way that the escape of glycogen by solution in water is hindered.[310] It may be remembered that picric acid has a similar property (p. 97).

Rate of penetration. Formaldehyde enters gelatine/albumin gel faster than any other fixative except the strong mineral acids, at a K-value of 3·6. The penetration of non-coagulant fixatives into tissues is hard to measure. Tellyesniczky's data give a K-value of 0·78. This is only a moderate speed, equal to that of mercuric chloride.

Shrinkage or swelling. Gelatine/albumin gels swell considerably (to 123% of their original volume) in formaldehyde solution. The volume of whole liver remains almost unchanged (99%), but there is subsequent shrinkage to 68% by the time the organ is in paraffin.

Single cells may be observed to shrink at first contact with formaldehyde solutions, especially if the latter are rather concentrated; this is usually followed by expansion at some time during the first hour to a considerably greater volume than the original, and later by a second shrinkage that leaves the cell somewhat larger than it was in life.[134] The expansion after the original shrinkage is less marked if the formaldehyde is used in saline solution, and the cell may eventually return to its original size. The pulsation of nuclei is less than that of the cytoplasm, and not exactly synchronized with it.

Most observers have not noticed this curious pulsation, but have simply recorded the final size when fixation is complete. *Arbacia* eggs swell strongly (to 147%) in formaldehyde dissolved in distilled water. They swell slightly (to 105%) when it is dissolved in sea-water, but the volume is down to 48% in xylene. *Helix* spermatocytes have a volume of 34% of the original in paraffin sections mounted in Canada balsam. It is a sobering thought that this represents less shrinkage than that obtained with any other primary fixative.

Hardening. Formaldehyde hardens strongly. Wetzel, using a 10% solution, obtained a rigidity-figure of about 1700. This is $7\frac{1}{2}$ times the figure for chromium trioxide and is exceeded only by ethanol and acetone.

Immediate effects on particular constituents of the cell. The cell outline is often well preserved, but there is a strange tendency for blebs of cytoplasm to separate from the cell. The ground cytoplasm is not so homogeneously fixed as experiments with gelatine and gelatine/albumin gels would lead one to expect: there is a tendency for a certain amount of granulation to occur. Mitochondria are preserved, often rather well, though sometimes they become moniliform. When once they have been fixed by formaldehyde, they are no longer subject to destruction by acetic acid.[556] Lipid globules usually remain as in life, but a wide variety of these has not been studied.

The shape and structure of the nucleus are on the whole well preserved, though there is some tendency for the nuclear sap to become granular. The heterochromatic segments of the chromosomes remain more or less as in life. The nucleolus is less clearly seen than before fixation.

There can scarcely be any doubt that formaldehyde (with an indifferent salt) preserves the structure of the living cell better than any other primary fixative except osmium tetroxide.

Methods of washing out. As a general rule, no special washing out is necessary. Tissues may be transferred to water or to ethanol of any grade.

Dark brown, birefringent crystals are sometimes seen in tissues that are rich in blood, especially spleen. This so-called 'formalin-pigment' appears to arise by the reaction of formaldehyde with the haematin of haemoglobin that has escaped from red blood-corpuscles either before or after death. It is not formed if short fixation in formaldehyde solution is followed by prolonged soaking in 5% mercuric chloride solution.[295] Once formed, it can be dissolved by a 1% solution of potassium hydroxide in 80% ethanol, or by picric acid dissolved in ethanol.[229] The chemistry of these processes has not been worked out.

Effect on dyeing. Formaldehyde renders proteins and cytoplasm more acidic (basiphil) than any other fixative, exceeding mercuric chloride in this respect: cytoplasm retains little affinity for acid dyes unless fixation is short. Chromatin is strongly coloured by basic dyes.

Effects on the histological picture seen in paraffin sections. Formaldehyde is a poor fixative for tissues that are to be embedded in paraffin (grade IV--V if dissolved in distilled water, III–IV if sodium chloride is added at 0·7%).

Cellular aggregates tend to be widely separated from one another; cytoplasm shrinks towards nuclei, and cells may lose contact with one another; mitochondria are sometimes retained, especially if the fixative be used in strong solution (10% or more).[42] Red blood-corpuscles tend to swell into spheres, but are rather well preserved if an indifferent salt is mixed with the fixative. The interphase nucleus is fixed in a remarkably life-like form, with no coarse network in it; [371] the nuclear membrane, heterochromatic segments of the chromosomes, and nucleolus are well shown. The mitotic and meiotic chromosomes, however, are very poorly fixed.

Formaldehyde gives good results if frozen sections are used, or tissues embedded in collodion.

Compatibility with other fixatives. Formaldehyde reduces chromium trioxide quickly, potassium dichromate much more slowly. The reaction with osmium tetroxide is very slow at room-temperature.[23] Formaldehyde is compatible with ethanol, picric acid, and acetic acid, and is generally regarded as compatible with mercuric chloride.

OSMIUM TETROXIDE

Standard concentration for fixation. 1% w/v aqueous solution.

Formula and formula-weight. OsO_4. 254·2.

Description. Osmium tetroxide occurs as pale yellow crystals. These melt at 41° C. The liquid boils at 131° C, but much vapour comes off before this temperature is reached. The vapour arises also from the crystals and from aqueous solutions; it is damaging to the epithelium of eyes, nose, and mouth. This vapour has a not unpleasant smell. The name of the metal is derived from the fact that the tetroxide smells. The volatility of osmium tetroxide makes it imperative to keep and use solutions in tightly-stoppered vessels.[381]

The solubility in water at 25° C is 7·24% w/W.[10] (The figure given by Thorpe and Whiteley,[504] who are usually so reliable, is grossly in error.) Osmium tetroxide is extremely soluble in carbon tetrachloride at 25° C (about 375% w/W [10]) and soluble also in liquid paraffin and certain lipids.

Most compounds of osmium are dark or black. On reduction, osmium tetroxide is converted to the grey or black anhydrous dioxide, OsO_2, or to the brown, gelatinous, hydrated dioxide, $OsO_2.2H_2O$ or $Os(OH)_4$.[387] Reduction takes place readily when solutions of osmium tetroxide are exposed to light, though it is claimed that this is prevented by the complete exclusion of dust.[65] Reduction by light is prevented by strong oxidizers (p. 125), and is also hindered, curiously enough, by sodium chloride.[38]

Ionization. Osmium tetroxide scarcely ionizes. It was recognized long ago that the electrical conductivity of its solutions was extremely low. Hofmann and his colleagues [237] found that the specific conductivity of a 1% solution in distilled water was $1{\cdot}09 \times 10^{-5}$; that of the water used was $0{\cdot}5 \times 10^{-5}$. 'Osmium tetroxide', they remarked, 'is therefore chemically neutral and no acid.'

On solution, osmium tetroxide takes up a molecule of water and becomes H_2OsO_5.[549] This substance cannot be isolated. A minute amount of ionization occurs, according to the formula $H_2OsO_5 \rightleftharpoons H^+ + HOsO_5^-$. The ionization constant is $8{\cdot}0 \times 10^{-13}$. [549] It follows from this that 100 ml of a 1% solution of osmium tetroxide contain 0·000000035 g of hydrogen ion. To call such a substance an acid seems rather far-fetched, though a sodium salt, $NaHOsO_5$, can in fact be obtained. Hydrogen peroxide has an ionization constant of $1{\cdot}78 \times 10^{-12}$, [449] more than twice that of H_2OsO_5, but it is not called hydroperoxidic acid. The pH of a solution of osmium tetroxide is almost exactly that of the distilled water used to make it. The values published in the biological literature give us information about the distilling apparatus or glassware used in various laboratories, but not about osmium tetroxide.

The proper name for H_2OsO_5 is not easy to determine. It cannot be osmic acid, for on the analogy of manganic acid (H_2MnO_4), osmic acid is H_2OsO_4; and this is a real entity, as we have seen (p. 62), though it is not formed when osmium tetroxide is dissolved in water. Permanganic acid is $HMnO_4$, and H_2MnO_5 would be per-permanganic acid. On this analogy H_2OsO_5 would be per-perosmic acid,[549] if it were unequivocally an acid. It seems best to call it hydrogen per-perosmate.

Oxidation-potential. The 2% solution has an oxidation-potential of 0·64 volt, somewhat less than that of a 5% solution of mercuric chloride.

Manufacture. Metallic osmium occurs naturally as an alloy with iridium, usually in association with platinum. Deposits are found in Alaska, the Ural Mountains, and in South Africa. There is an ounce of osmium in 1,200 tons of Witwatersrand gold ore. The metal is extremely heavy (density about 22·5). It is also very hard and its alloys are used in making pivots for scientific instruments and tips for nibs of fountain pens.

Osmium tetroxide is made by heating spongy metallic osmium in a current of air or oxygen. On account of the rarity of the metal, this fixative is extremely expensive. A gram costs £3. 10. 0; thus 1 ml of a 2% solution costs 1/4¾, and a single drop of it about ¾*d.*

Introduction as fixative. We are indebted to Franz Schulze of Rostock, the inventor of that invaluable histochemical reagent, chlor-zinc-iodide,[448] for the introduction of osmium tetroxide into microtechnique. Unfortunately it is impossible to discover what led him to try it. He noticed that different tissue-constituents differed in their capacity to reduce it to the dark lower oxide. He sent a weak solution (0·1 or 0·2%) to his friend and former pupil, Max Schultze, with the request that he should try it in histological investigations. Schultze did so in 1864.[444] He plunged the male of the beetle *Lampyris splendidula*, alive and shining, into Schulze's fluid, and made a microscopical study of the phosphorescent organ. To his surprise, the tracheal end-cells were blackened and thus showed up strongly against the parenchymal cells.[444, 445] In collaboration with Rudneff [447] he next tried it on a variety of tissues of plants and animals. He noted especially the reduction of osmium tetroxide by fat, myelin, and tannic acid. It is interesting that his emphasis was at first on the darkening of particular objects, not upon delicacy of fixation. Retaining his interest in phosphorescence, however, he tried it on the marine protozoon *Noctiluca*, and was now struck by the life-like preservation.[446]

Reaction with proteins. Osmium tetroxide gives no coagulum with albumin solutions. It renders albumin not coagulable by ethanol or by heat.[65] It sets undiluted egg-white and strong solutions of serum albumin, serum globulin, and fibrinogen into gels. It stabilizes gelatine gels against solution by water at 37° C.

This is an additive fixative. It probably reacts at the double bonds of the side-groups of tryptophane and histidine, linking protein chains together through these. The failure of acid dyes to act after fixation by osmium tetroxide suggests the blocking of amino-groups, but there is no positive evidence of this.

There is no coagulation of nucleoprotein, nor is DNA precipitated from solution.[169]

Reactions with lipids. As we have seen (p. 120), Schultze and Rudneff [449] had already in 1865 recognized the capacity of fats and other lipids to reduce osmium tetroxide. It was shown by Altmann [5] that whereas oleic acid and olein are blackened by this substance, palmitic and stearic acids and their triglycerides are not. This led to the understanding that osmium tetroxide reacts with the double bonds in lipids. Since lipids generally occur in nature as mixtures, and some of the constituents of these mixtures are usually to some degree unsaturated, most lipids as they occur in organisms will sooner or later be darkened by the action of osmium tetroxide. As a general rule, mixed triglycerides in the form of storage-fat blacken with osmium tetroxide more easily than conjugated lipids.

When osmium tetroxide reacts with lipids, three possibilities present themselves. The tetroxide may be reduced to the dark lower oxide; or it may combine with the lipid to form a dark compound; or both may occur. If a compound is formed, alteration in solubility is to be expected. It is well known that the sites of storage fat are often black in paraffin sections mounted in Canada balsam. Lipids are unlikely to survive such embedding and mounting, unless profoundly altered. The possibility presents itself, however, that the lipid has gone and only insoluble osmium dioxide remains to mark its former sites.

Comments on this subject are scattered through the literature. Recently the Chinese cytologist Chou,[121] working at Oxford, has made a systematic study of the subject. The subcutaneous fat of the mouse was used in his experiments. This was fixed in 1% osmium tetroxide or in Flemming's strong fluid,[173] which contains osmium tetroxide at 0·4%. The fixed tissue was dehydrated with ethanol, left for 30 min. in an antemedium, and embedded in paraffin. Sections were soaked for 5 min. in the same fluid that had been used as antemedium, brought down to water, and mounted in an aqueous medium. The antemedia used were xylene, toluene, benzene, and chloroform.

The fat-sites were in all cases black. If, however, the sections were left in xylene or toluene for 40 instead of 5 min., the fat-sites were colourless: the globules appeared empty. They remained black, however, if benzene or chloroform was used for the same period.

When sections that had been 5 min. in any of the antemedia were brought down to water and bleached, it could easily be shown that the lipid still remained in them, for it could be coloured black by Sudan black. If, however, sections were left for 40 min. in the antemedia, the ones treated with xylene or toluene were proved to contain no lipid, while those that had been in benzene or chloroform still contained it.

Sections that had been 5 min. in the antemedia were brought down to water, bleached with hydrogen peroxide, taken up to the antemedia again, and left there for 30 min. On subsequent treatment in the usual way with Sudan black, they were shown to contain no lipid. This applied in all cases, whether the antemedium was xylene, toluene, benzene, or chloroform.

It is to be noted that so long as the lipid-sites were still blackened by osmium, they contained lipid. The evidence suggests strongly that the black substance is a compound of lipid with osmium. This compound is resistant to solution by benzene and chloroform, but dissolves slowly in xylene or toluene; when bleached by hydrogen peroxide, it is soluble in any of the lipid solvents. Xylene is capable of acting as an oxidizing agent,[322] and the same may perhaps apply to toluene. It is to be supposed that osmium tetroxide reacts with the double bonds of lipids in much the same way as it does with those of the substances discussed on pp. 61–2, but no direct evidence on this subject is available.

The possibility that osmium tetroxide sometimes simply oxidizes lipids, without forming an additive compound, must be kept in mind. Hofmann [236] believed that this was the way in which it ordinarily reacted with lipids. He made a careful study of oxidation by osmium tetroxide, and reached the conclusion that it can act as an adjuvant to other oxidizers. Thus, certain substances that cannot be oxidized by potassium chlorate alone, can be oxidized if osmium tetroxide is present as well. The latter oxidizes the substrate and is itself reduced: the chlorate re-oxidizes it, and the process begins again. Wolman [541] suggests that osmium tetroxide may act as an oxidative catalyst in this way in microtechnique, when mixed with other oxidizers.

In the presence of a strong oxidizer, such as chromium trioxide, the mixed triglycerides of adipose tissue are generally blackened by osmium tetroxide, while conjugated lipids as a rule are not. This is useful as a rough pointer in preliminary histochemical work. Since the fatty acid component of conjugated lipids is often highly unsaturated, it is not clear why they remain colourless. Perhaps they tend to be simply oxidized instead of forming additive compounds.

Since osmium tetroxide is soluble in certain lipids, it can be taken up without change by fully saturated ones, and then reduced by subsequent soaking of the tissue in ethanol (compare Starke [485]). Thus the fact that a lipid is black in a paraffin section does not prove that it is unsaturated.

Reactions with carbohydrates. It appears that osmium tetroxide does not react with most hexoses or pentoses or their polymers at room-temperature, though sucrose is very slowly oxidized to oxalic acid.[23] There is some darkening if glycogen is treated with osmium tetroxide for long periods at 50° C.[23]

Rate of penetration. Osmium tetroxide penetrates slowly into gelatine/albumin gel. The K-value for 25 hours is 0·85, which is nearly as low as the figure for picric acid. The K-value gradually falls off with time. During the first 16 hours it is 1·0, but during the period 16 to 144 hours it is only 0·31. It must be supposed that the osmium deposited in the gel (whether in combination with protein or in the form of the dioxide) presents an obstacle to diffusion. A measurable fall-off in K-value is not known to occur with any other fixative.

Tellyesniczky's data [500] for penetration into liver (12 hours) give K-values of 0·29 for the 0·5% solution and 0·58 for the 2%. This indicates that the 1% solution would be one of the more slowly penetrating fixatives.

Shrinkage or swelling. Gelatine/albumin gel shrinks very slightly (by less than 10% of its volume) in osmium tetroxide solution. The change of volume of whole livers has not been measured, presumably because the experiment would be too expensive. It is unfortunate that there are no satisfactory numerical data for the shrinkage or swelling of cells. Kaiserling and Germer [256] found that mammalian eggs increased somewhat in diameter when transferred from saline to osmium tetroxide dissolved in distilled water. The saline was hypotonic and the eggs had already swollen somewhat in it.

Hardening. Osmium tetroxide leaves tissues rather soft. Wetzel's figure is 171; the figure for chromium trioxide is 1·4 times greater. Tissues fixed in osmium tetroxide are crumbly in paraffin and do not section well.

Immediate effects on particular constituents of the cell. By common consent of all who have studied the subject, osmium tetroxide preserves the structure of the living cell better than any other primary or mixed fixative. Fig. 8, A, B (opposite p. 70) gives a good impression of its action. The nuclear sap and the ground cytoplasm in the vicinity of the nucleus become less perfectly homogeneous than they were in life; the nucleus may retract slightly from the cytoplasm; nucleoli become difficult to see; lipid globules gradually darken. Until the latter change has taken place, one might almost suppose that the cell was still alive, except that any Brownian movement will have ceased. Mitochondria are perfectly preserved.

Methods of washing out. Osmium tetroxide is washed out in running water, because if any were left in the tissues, it might be gradually reduced by ethanol at subsequent stages, with consequent darkening.

Effects on dyeing. Osmium tetroxide leaves cytoplasm readily colourable by basic dyes (after bleaching), but scarcely at all by acid ones. The nuclear sap also tends to be made basiphil, and this interferes with the differential dyeing of the component parts of the nucleus.

Effects on the histological picture seen in paraffin sections. It is sad to turn from the magnificent view of a cell still lying in osmium tetroxide solution, to look at a paraffin section of a piece of tissue fixed in the same fluid. The fixation is poor (grade III–IV or IV), even with the addition of 0·7% of sodium chloride to the fixative.

Cellular aggregates are severely shrunken, so that they are separated by wide artificial spaces; cracks often run at random across the section; ground cytoplasm, though fairly homogeneous, is strongly contracted, and often condensed round nuclei. The shape of nuclei is well retained. Pischinger [402] considered that the nucleus as a whole was well fixed. He thought that there was no nuclear membrane in the living cell but only a physical interface, and that while other fixatives thickened the interfacial region to form an artificial nuclear membrane, osmium tetroxide provided an approximation to the living condition. The nuclear sap is rather homogeneously fixed, but the objects contained in it (especially the meiotic chromosomes) are very poorly shown.

Mammalian testis fixed in osmium tetroxide solution buffered at pH 7·4 is shown in fig. 9, B (opposite p. 74).

Compatibility with other fixatives. Osmium tetroxide is compatible with all the fixatives mentioned in this chapter and the preceding one, except formaldehyde and ethanol. The reaction with formaldehyde is slow: no darkening occurred within 24 hours at 20° C in the circumstances of Bahr's experiments.[23] Mercuric chloride, chromium trioxide, and potassium dichromate prevent the reduction of osmium tetroxide by daylight.

Unclassified remarks. It was first pointed out by Altmann [4] in 1889 that lipid globules are sometimes only blackened by osmium tetroxide on the outside, so that they appear as rings in optical section. He called these *Ringkörner*. He found that they were not seen initially, but only when tissues fixed in osmium tetroxide had been brought into ethanol. He attributed their formation to partial solution of the lipid droplet by ethanol.[5]

FIG. 16. Three *Ringkörner* and a cap or hood (*Kapuze*) formed by partial solution of lipid globules: osmium preparations.
(From Starke.[485])

This subject was carefully investigated by Starke.[485] He found that when lipid droplets that had been treated with osmium tetroxide were set free in ethanol, they never became *Ringkörner*, but shrank into irregular shapes, blackened all through. When similar droplets were treated with osmium tetroxide while still contained in the tissues, the result was different; for when the tissues were subsequently placed in ethanol, *Ringkörner* were formed. Starke concluded that lipid droplets consisted of a part that was rendered insoluble in ethanol by osmium tetroxide, and a part that was not. When the latter was dissolved out by ethanol, the droplet shrank if it could; but if it were surrounded by fixed cytoplasm it could not shrink, and a spherical hole was left, to the walls of which the fixed and blackened lipid material attached itself (fig. 16).

The black rings and crescents commonly seen in osmium preparations are in many cases to be attributed to the cause suggested by Starke. His paper, published more than 60 years ago, has unfortunately been overlooked by many authors.

Pieces of tissue that have already been fixed in another fixative (generally a mixture containing osmium tetroxide) may be soaked for several days in a simple aqueous solution of osmium tetroxide, to darken certain cytoplasmic inclusions. This process of 'postosmication' was introduced by the Polish cytologist Weigl.[531] It is useful for directing attention to a particular part of a cell, but it should be used with caution. *Ringkörner* are often seen in postosmicated preparations. There is a tendency for a black material (presumably osmium dioxide) to be deposited on the surfaces of granules or other cytoplasmic inclusions, and especially to fill up the spaces between crowded granules. The appearances given can be very misleading morphologically, and should not be trusted unless they can be confirmed by study of the living cell.

POTASSIUM DICHROMATE

Standard concentration for fixation. 1·5% w/v aqueous solution.

Formula and formula-weight. $K_2Cr_2O_7$. 294·2.

Description. Potassium dichromate crystallizes readily in large, orange-red prisms or tables. These melt with decomposition at 396° C. They are soluble at about 10% w/w in water at room-temperature (18% at 30° C), but insoluble in absolute ethanol.

Potassium dichromate is more expensive than the sodium salt, but the fact that it is anhydrous and not deliquescent gives it an advantage for certain industrial purposes. It is used in making matches and fireworks and in the chrome tanning of leather; dissolved with sulphuric acid it acts as a bleaching agent for tallow and palm oil.

It is wrong to call this substance potassium bichromate, because the name would only be applicable to potassium hydrogen chromate, which does not exist.

Ionization. The ionization of potassium dichromate has been carefully considered by Casselman.[111] The ions are the same as those produced by chromium trioxide, but in different proportions. The ions in solutions of the two substances containing the same weight of chromium are compared in fig. 14 (p. 105). In both solutions by far the greater part of the chromium is in the form of dichromate $[Cr_2O_7]^=$ and hydrogen chromate $[HCrO_4]^-$, the former predominating in both cases, especially in the solution of potassium dichromate. The chromate ion $[CrO_4]^=$ is present in minute quantities in both. Undissociated chromic acid, H_2CrO_4, is present in considerable amount in the solution of chromium trioxide, but there is scarcely any of it in the solution of potassium dichromate.

There is a striking difference between the hydronium-ion concentration of the two solutions. A 2·5% solution of potassium dichromate has a pH of 4·05; in a solution of chromium trioxide containing the same amount of chromium as the dichromate solution the pH is 0·85.[111]

If a solution of potassium dichromate be acidified to the same pH as a solution of chromium trioxide containing the same weight of chromium, the ions present in the two solutions will be the same, except that the former will contain potassium ions and the anions of the added acid. If hydrochloric acid be used, one has a fluid almost identical with a solution of chromium trioxide to which some potassium chloride has been added. Since potassium and chloride ions are inactive in fixation, it follows that an acidified potassium dichromate solution will act like a solution of chromium trioxide.

Casselman does not give the pH of a 1·5% solution of potassium dichromate, but a 1% solution (pH 4·10) differs only slightly from the 2·5% solution. Lassek [285] gives pH 4·0 for Müller's fluid, which is 2 or 2·5% potassium dichromate with 1% sodium sulphate (see below).

Oxidation-potential. The oxidation-potential of a 3% solution of potassium dichromate is 0·79 volt.[113] A 2·5% solution has almost exactly the same oxidation-potential.[112]

Manufacture. Chromium occurs naturally as chromite, $FeO.Cr_2O_3$. It is treated with sodium carbonate to produce sodium chromate, and this with sulphuric acid to give sodium dichromate. Potassium chloride is added to a strong, warm solution of the latter, and large crystals of potassium dichromate separate out on slow cooling, or small ones if the tank is shaken.

Introduction as fixative. Potassium dichromate was introduced into microtechnique in 1860 by H. Müller,[362] who used it in studies of the human eye. His first fluid consisted of this salt and sodium sulphate (presumably Glauber's salt, crystallized with 10 molecules of water), both at about 1·5%, with *etwas* chromium trioxide. Later in the same year [363] he mentioned another fluid, from which the chromium trioxide was omitted; the concentration of the other components was not stated, but was presumably the same as before. The fluid called *Müller'sche Augenflüssigkeit* consists of the same two salts, at 2–2½% and 1% respectively.[358] It forms the basis of Zenker's [554] and Helly's [224] fluids.

The fact that potassium dichromate is unsuitable for use in studies of mitosis was first made known by Mayzel.[335]

Reactions with proteins. This is a non-coagulant of albumin solution, but it very gradually renders undiluted egg-white more viscous and eventually transforms it into a weak, semi-transparent gel.

Since, as we have seen (p. 126), the chrome anions are almost the same whether potassium dichromate or chromium trioxide be dissolved, it must be supposed that the striking differences between the effects of the two substances on proteins must be due to the large difference in pH. If potassium dichromate be acidified, it reacts with proteins like chromium trioxide: that is to say, it becomes a strongly coagulant fixative. The change-over from one behaviour to the other occurs in the pH-range 3·4 to 3·8.[111] As Casselman points out,[111] this is near to the iso-electric points of many proteins (though somewhat below that of most). It must be supposed that in the region of the iso-electric point, the proteins change radically in their reactions to the chrome anions. The chemical changes concerned in the slow gel-forming process that occurs above the critical pH-range have not been investigated.

If gelatine that has been impregnated with potassium dichromate is exposed to light, it becomes insoluble in warm water. This fact is used in the 'Autotype' process of photographic printing. It is interesting to notice that chromium trioxide, on the contrary, makes protoplasm soluble on exposure to bright light (p. 110). In the ordinary circumstances of fixation, gelatine/albumin gel is not stabilized against warm water by the action of potassium dichromate.

Nucleoprotein solution is not coagulated by potassium dichromate, and DNA is readily dissolved by this salt; the histone of the nucleus, however, is strongly gelatinized by potassium dichromate.[402] There is a marked contrast here with the effect of acetic acid, which precipitates DNA but dissolves histone (see p. 135).

Reactions with lipids. Potassium dichromate is able to attach chromium to certain lipids, and to render them insoluble in lipid solvents. The chromium can subsequently be made to react with haematein to give a black lake. This is the essence of Weigert's [530] method for myelin. Tissues were fixed in Müller's fluid, embedded in collodion, and treated with a solution of haematein (often called Weigert's 'haematoxylin', but haematoxylin is not a dye (see p. 173)). Myelin was coloured black. Other tissue-constituents were darkened by unmordanted haematein; this was removed by bleaching with alkaline potassium ferricyanide, which did not affect the lake.

Better results were obtained by short fixation in some other fluid and subsequent 'postchroming' in potassium dichromate solution. It is usual to postchrome tissues for quite a long time, often days or weeks, sometimes at 37° or even 60° C. This process, as a method of fixing particular constituents of cells, was introduced by the celebrated German cytologist Benda,[55] who used it in his pioneer research on mitochondria. He named it *Postchromirung*. It had previously been used only to harden tissues for easier sectioning by hand. Benda sometimes used chromium trioxide in the same way, but potassium dichromate is nearly always used nowadays.

Smith [481, 483] modified Weigert's method by introducing a preliminary fixation in formaldehyde, followed by postchroming and the cutting of frozen sections. This technique was adapted by Dietrich [143] and made into a histochemical test for phospholipids. The acid haematein test [31, 32] is its modern version.

The chemistry of the action of potassium dichromate on lipids has been studied especially by Kaufmann and Lehmann [260–262] and by Lison.[310] It would appear that a wide variety of unsaturated lipids can be rendered insoluble in lipid solvents by the prolonged action of potassium dichromate. There is no action on saturated ones. Different periods of postchroming are suitable for different unsaturated lipids.

The evidence suggests that three processes can be involved in the action of potassium dichromate on lipids, and that they need

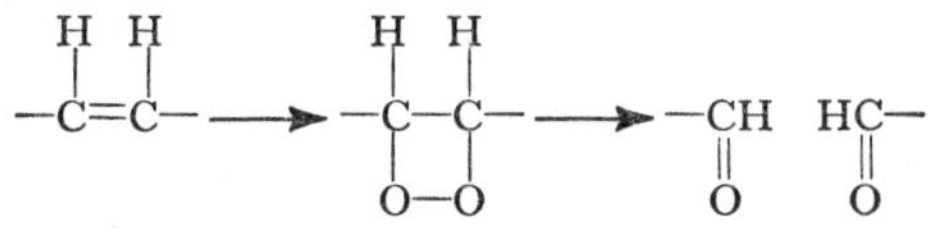

Double bond *Peroxide* *Aldehydes*

not all occur together. These three are simple oxidation, polymerization with loss of solubility in lipid solvents, and binding of chromium (additive fixation).

Simple oxidation at the double bond occurs particularly when there is only one such double bond in a fatty acid radicle. The fatty acid chain is split at the double bond, with formation of two aldehydes.

Polymerization is particularly apt to occur when a double bond lies near the opposite end of a fatty acid chain to the carboxyl group. Oxidation proceeds as far as the peroxide stage, and molecules then associate to form a polymeric, insoluble substance.

When a fatty acid radicle is highly unsaturated, the double bond nearest to the carboxyl group behaves in a special way. Oxidation proceeds as before to the peroxide stage; there is then a passage

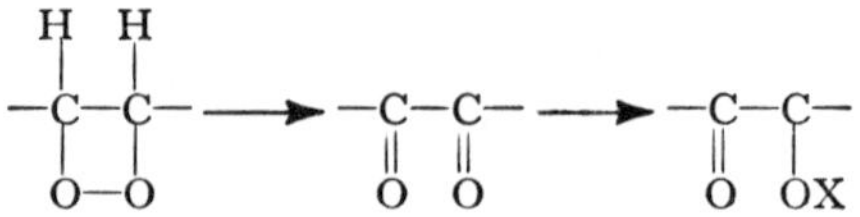

Peroxide *Diketone* *Chromium compound*

through dihydroxyketone to diketone, and chromium is then taken up as oxide. This oxide, of unstated composition, is shown as X in the formula given here.

In Lison's view,[310] phospholipids take up chromium because their fatty acid components tend to be particularly highly unsaturated.

One of the most valuable properties of potassium dichromate is its ability to fix mitochondria by rendering their lipid components insoluble in lipid solvents.

It is important to bear in mind that triolein, according to Smith and Thorpe,[483] can take up chromium if postchroming is prolonged, and become insoluble in alcohol, xylene, and ether; even storage-fat can be made to give a black reaction with haematein.[481] Kaufmann and Lehmann [262] claimed that all unsaturated lipids could be rendered insoluble in lipid solvents by potassium dichromate. Chou [121] has recently obtained some confirmatory results. He has shown that if the subcutaneous adipose tissue of the mouse be fixed in Ciaccio's fluid [123] and then left in a saturated solution of potassium dichromate at 37° C for 49 days, the fat is rendered insoluble in certain lipid solvents. The tissue can be dehydrated and brought through xylene into paraffin: the fat globules can be deeply coloured with Sudan black in sections of such material. Nevertheless, potassium dichromate does not compare with osmium tetroxide in ability to render most lipids insoluble, and storage fat is not ordinarily preserved in paraffin sections of tissues fixed with potassium dichromate.

It seems probable that lipids are more widely dispersed in the ground cytoplasm than is usually supposed, and potassium dichromate may fix partly by acting on these.

Reactions with carbohydrates. Potassium dichromate is not a fixative for glycogen. Chromium is not known to be taken up from solutions of potassium dichromate by any carbohydrate or related substance, with the possible exception of lignin.[31] It is to be presumed, however, that acidified potassium dichromate will react towards carbohydrates in the same way as chromium trioxide (p. 107).

Penetration. Tellyesniczky's data give a high K-value (1·33) for the penetration of the 3% solution into liver, but it is doubtful whether this means much. The term *rate of penetration*, as used in this book, means the *rate of penetration with fixative effect.* Now potassium dichromate does not coagulate proteins, nor does it gelatinize most of them in the ordinary period of fixation. Figures for rate of penetration do not seem to be applicable to this substance, though no doubt it runs quickly through protein gels [337] and tissues.

Shrinkage or swelling. Gelatine/albumin gels swell strongly (to 160% of their original volume) in potassium dichromate solution. Whole livers remain unchanged in volume in a 3% solution, but are shrunken by subsequent dehydration and retain only 49% of their original volume when brought into paraffin wax. Primary spermatocytes of the snail are reduced to 23% of their original volume when paraffin sections of the ovotestis fixed in a 5% solution have been mounted in Canada balsam. This represents greater final shrinkage than that which follows fixation by most primary fixatives.

Hardening. Tissues are left very soft. Wetzel's figure for rigidity after fixation in a 3% solution is 171. Chromium trioxide leaves tissues 2·7 times as rigid as this. After subsequent soaking in 80% ethanol tissues are still very soft.

In the old days, when potassium dichromate was used as a hardening agent before sectioning by hand without embedding, tissues were left in the solution for long periods.

Immediate effects on particular constituents of the cell. The shape of the cell is rather well preserved, though there may be some retraction of small pseudopodia. The ground cytoplasm becomes somewhat granular. Mitochondria are preserved, but transformed from threads into ovoids and short rods: their form would probably be better maintained in the presence of an indifferent salt. Lipid globules tend to run together. The nucleus retains its form but may be somewhat retracted away from the cytoplasm; its membrane is very clearly seen; the nuclear sap is finely granular; the nucleolus is shrunken.

Washing out. An insoluble precipitate (presumably of chromic oxide, Cr_2O_3) tends to be formed in tissues if they are transferred directly from potassium dichromate to aqueous ethanol. It has already been mentioned that the salt is insoluble in absolute ethanol. Potassium dichromate is therefore usually washed out in running water. The experiments of Virchow [526] suggest that it may be safe to transfer tissues directly from potassium dichromate solution to 95% alcohol if light be excluded. Overton [377] advised washing tissues in sulphurous acid after fixation in potassium dichromate (see p. 109).

Effect on dyeing. Seki [459] claims that potassium dichromate renders proteins and cytoplasm acidophil, but in fact cytoplasm can be coloured quite strongly by certain basic dyes after the action of this fixative. Chromatin is left strongly colourable by basic dyes,[38] but it is not fixed in its original position within the cell. Since the nuclear membrane is well fixed, the chromatin cannot escape, but distributes itself almost at random within the nucleus. This is the last fixative one would choose for studies of chromosomes (unless acidified).

Effects on the histological picture seen in paraffin sections. It was on paraffin sections of root-tips of maize (*Zea mays*) that the American cytologist Zirkle first clearly demonstrated the effect of pH on fixation by potassium dichromate. He showed that when the hydronium-ion concentration was on the more acid side of a certain range, the fixation-image was that of chromium trioxide; on the less acid side the image was completely different. Simple solutions of potassium dichromate fall on the less acid side. Zirkle put the change-over range at pH 4·2 to 5·2, but Casselman, in a recent careful study with mammalian tissues, put it at pH 3·4 to 3·8.

By itself, potassium dichromate is a very poor fixative for paraffin sections (grade V). Cellular aggregates shrink apart from one another, leaving wide artificial spaces; cytoplasm is rather homogeneously fixed, but tends to shrink round the nuclei and sometimes even disintegrates partially, so that cells become separated from one another; mitochondria are retained, though often somewhat rounded up; red blood-corpuscles are swollen and irregular. The shape of nuclei is fairly well retained; the nuclear sap is homogeneously fixed without net-like coagulum, but may retract from the membrane; the nucleolus is shrunken and often surrounded by a halo, and may be subdivided; the heterochromatic segments of the chromosomes are not seen in the interphase nucleus, and the definitive chromosomes of mitosis and meiosis are unfixed.

In brief summary and at some risk of over-simplification one may express the effect of pH on the action of potassium dichromate thus. On the less acid side of the critical range (pH 3·4 to 3·8), the cytoplasm, nuclear sap, and mitotic spindle are homogeneously fixed; mitochondria are retained; the nucleolus is partly dissolved; chromosomes are scarcely visible. On the more acid side potassium dichromate acts like chromium trioxide: that is to say, cytoplasm and nuclear sap are coarsely coagulated and the mitotic spindle is fibrous; mitochondria are non-existent; the nucleolus and chromosomes are well fixed. These may be called respectively the less acid and more acid fixation-images of the chrome anions, as seen in paraffin sections.

Compatibility with other fixatives. Potassium dichromate is compatible with picric acid, mercuric chloride, and osmium tetroxide. If mixed with more than a very small amount of chromium trioxide, it shows the more acid fixation-image.[556, 111] It also shows this image if mixed with more than a very small amount of acetic acid or any other acid used in fixation. It reacts rather slowly with formaldehyde, and mixture with this substance is allowable if it is done immediately before use; when the colour changes, the fluid should be renewed. Potassium dichromate solution should not be mixed with ethanol, lest chromic oxide be deposited in the tissue.

Unclassified remarks. It has been known since the end of the last century that different dichromates give different fixation-images,[97] but the explanation awaited the work of Zirkle.[556] The subject has now been re-investigated by Casselman.[111] Those dichromates that show, at fixative concentration, a pH on the more acid side of the critical range, fix like chromium trioxide: barium, calcium, mercuric, and silver dichromates are examples. A saturated solution of mercuric dichromate is particularly strongly acid (pH 1·05). Those that show a pH on the less acid side fix like potassium dichromate. The ammonium, lithium, and sodium salts do this. The last-named is the least acid, a solution of the same molecular concentration as 2·5% potassium dichromate (pH 4·05) showing a pH of 5·10.[111] Ammonium dichromate presents the advantage that it does not swell mitochondria, as the potassium salt does.[556]

ACETIC ACID

Standard concentration for fixation. 5% v/v aqueous solution.

Formula and formula-weight. $H_3C.C\begin{smallmatrix}/\!\!/\mathrm{O}\\ \diagdown\mathrm{OH}\end{smallmatrix}$. 60·0.

Description. Acetic acid is a colourless liquid with a pungent smell. It boils at 118° C; the crystals formed on cooling melt at 16·6° C. The ease with which it may be frozen has given rise to the name of 'glacial' acetic acid, which is applicable only when the acid is free of water. The acid is miscible with water and ethanol in all proportions.

Beyond its use as vinegar and for pickling, acetic acid is important in the production of cellulose acetate plastics.

Ionization. Acetic is a moderately weak acid, with ionization constant $1{\cdot}80 \times 10^{-5}$. The pH of the 5% solution is 2·32 [285] (Seki [459] gives 2·4).

Oxidation-potential. Acetic acid can act as an oxidizer on reduction to acetaldehyde, or oxidized by strong oxidizers to carbon dioxide and water. The oxidation-potential of the 5% solution is 0·77 volt.[113]

Manufacture. The best culinary vinegar is made by the oxidation of the ethanol in wine or other alcoholic liquors through the action of bacteria (generally *Acetobacter* spp.). Acetic acid is also made by the destructive distillation of the sawdust of beech and other hard-woods under a pressure of several atmospheres. The distillate is wood-vinegar or pyroligneous acid, which contains acetic acid at about 5%, with creosote and other contaminants. The weak acetic acid made by either of these processes may be treated with lime or soda to make calcium or sodium acetate. The salt is distilled with sulphuric acid, and the distillate fractionally distilled to give the glacial acid.

Acetic acid is also made by the oxidation of acetaldehyde with atmospheric oxygen in the presence of cobalt acetate as catalyst; the acetaldehyde is prepared from acetylene.

Introduction as fixative. Vinegar appears to have been used for pickling vegetables from remote times, and it is remarkable that the regular use of acetic acid in microtechnique is scarcely more than a hundred years old. In the eighteenth century Henry Baker [24] had tried vinegar as a preservative for hydra, but without much success. Quekett, in his book published more than a century later,[409] does not even mention acetic acid. The softness of tissues fixed with this substance probably counted against it, for the microscopists of the time were more interested in hardening agents than in fixatives. Corti,[123] who experimented freely with fixative fluids, tried it in his study of the inner ear in 1851, and in the same year Clarke [127] used it in a mixture with ethanol for the treatment of tissues that had already been soaked in the latter fluid. It was subsequently used by Remak [416] in 1854 and Auerbach [22] 20 years later. Flemming [171] mentioned that it made the nuclear membrane very refractive and tended to distort it: he preferred chromium trioxide and picric acid. Acetic acid appears to have been valued in the seventies and eighties chiefly for showing nuclei clearly and making connective tissue transparent; pyroligneous acid was sometimes preferred, because it hardens somewhat.[179]

Reactions with proteins. Acetic acid (at the standard concentration) does not coagulate albumin, does not set egg-white into a gel, and has no fixative effect on gelatine/albumin gel or on haemoglobin. Histone can be extracted from tissues by acetic acid. Its most evident effects are to swell protein gels and fibres and to produce a precipitate with nucleoprotein.

The undissociated acid is thought to break the linkages between amide groups of contiguous protein chains, by associating with these: this would permit swelling. The dissociated acid splits the salt-links (amino to carboxyl) that also hold protein chains together, and this again permits swelling. Water is drawn into the protein by attraction to the hydrophil groups exposed by these reactions. The hydronium ion has a preservative effect, because it checks autolysis and stops the growth of putrefactive bacteria.

Acetic acid gives a thick precipitate with nucleoprotein solution. This is attributed to the action of the acetate ion in splitting off DNA from the protein. DNA is precipitated from solution by acetic acid.

Reactions with lipids. Certain lipids are miscible with glacial acetic acid, or soluble in it: sphingomyelin, the ricinolein of castor oil, and cholesterol are examples (though the latter is only slightly soluble). These facts, however, are not of much significance for microtechnique, since lipids are not ordinarily soluble in acetic acid at the usual fixative concentration of 5% or thereabouts. Phospholipids can form colloidal solutions in water, but their solubility in acetic acid of fixative strength does not appear to have been determined. Acetic acid is not known to fix any lipid.

Reactions with carbohydrates. Acetic acid neither fixes nor destroys carbohydrates.

Rate of penetration. Acetic acid penetrates at moderate speed into gelatine/nucleoprotein gel ($K = 2{\cdot}75$). It may be remembered that its rate of penetration into gelatine/albumin gel cannot be measured, because it does not fix this gel. For the same reason the rate of its penetration into tissues cannot be measured in such a way as to give a K-value comparable with the others quoted in this book (see under potassium dichromate, p. 131). No doubt it runs quickly through the tissues, as Tellyesniczky's data suggest ($K = 1{\cdot}2$), but it penetrates without fixing proteins, precipitating nucleic acids as it goes.

Shrinkage or swelling. Acetic acid swells protein gels far more than any other fixative, for reasons that have been discussed (p. 64). A simple aqueous gelatine gel (15% w/W), placed in acetic acid solution, expands to about 13 times its original volume in a week. Under the standard conditions of measurement, gelatine/albumin gel expands to 455% of its original volume (see fig. 1, p. 36). Tissues and cells also swell in acetic acid, but if they are not stabilized in the swollen state by the action of some other fixative, they shrink strongly on dehydration and subsequent treatment. Thus the spermatocytes of the snail retain only 28% of their original volume when paraffin sections have been mounted in Canada balsam. They retain a considerably larger volume if formaldehyde be used as fixative, though this causes very much less initial swelling of protein gels.

Hardening. Acetic acid leaves tissues much softer than any other fixative. Wetzel's figure for rigidity is only about 9. The figure for chromium trioxide is 25 times as great. After subsequent soaking in 80% alcohol, tissues remain extremely soft.

Immediate effects on particular constituents of the cell. The cell-outline becomes rather indistinct; any thin pseudopodia tend to be transformed into rows of globules; ground cytoplasm loses its original homogeneity; mitochondria are transformed into faint rows of granules [301, 491] and generally disappear; lipid globules are sometimes well retained, but the neutrophil granules of polymorphs disappear. The nucleus sometimes retracts from the cytoplasm; the nuclear contents are transformed into a lumpy network; the nucleolus sometimes becomes irregular in shape.

Methods of washing out. Since acetic acid is perfectly miscible with ethanol and has no tendency to produce insoluble extrinsic artifacts, no special washing out is necessary.

Effect on dyeing. Cytoplasm is rendered rather strongly acidophil, though it will also take basic dyes. The chromatin of interphase nuclei colours rather feebly with basic dyes, and scarcely at all with acid ones (probably because it is represented only by DNA, the protein constituent having dissolved away). Metaphase and anaphase chromosomes colour strongly with basic dyes. The nucleolus is not readily coloured by dye-lakes.

Effects on the histological picture seen in paraffin sections. Zirkle [557] showed that the acetic anion only produced its characteristic fixation image if used on the acid side of pH 4.0 or thereabouts. At less acid pH than this, fixation does not occur and tissues macerate: little beyond the nucleoli can be identified in paraffin sections of the macerated material.

The typical 'acid' fixation-image may be summarized thus. Cell-aggregates tend to be widely separated from one another by artificial spaces. Cytoplasm is poorly represented: it is strongly contracted round the nuclei, or coarsely reticular. Mitochondria are not seen: this is particularly characteristic of acetic fixation. The shape of nuclei is fairly well retained, but the nuclear sap seems not to be fixed and there is only a coarse reticulum within the interphase nucleus, with a swollen, often vacuolate nucleolus. Definitive chromosomes are rather well fixed. The mitotic spindle appears fibrous.

Zirkle [557] considered that the fixation-image was similar to that given by chromium trioxide, and there are indeed similarities. Nevertheless, chromium trioxide (with sodium chloride) gives better general fixation. The cellular aggregates stand in more life-like relation to one another; chromosomes at all stages are better fixed; the nucleolus retains its original form.

It is not obvious why mitochondria do not appear in paraffin sections fixed with acetic acid alone. Their lipid content has not been proved to be soluble in the 5% solution, and indeed, as we have seen, they are not necessarily destroyed by the action of the fixative itself, though as a rule they are either destroyed in the fixative or else left in a condition that results in their destruction at a subsequent stage. In a few cases they can be seen in paraffin sections of material fixed in mixtures containing a considerable amount of acetic acid.[25] Despite the general belief to the contrary, it must be the acetate ion or the undissociated acid that acts unfavourably upon them, not the hydronium ion; for as Casselman and Jordan[114] showed, mitochondria can be quite well seen in paraffin sections of tissues fixed in 0·1N hydrochloric acid. This subject has recently been considered in some detail.[37a]

Compatibility with other fixatives. Acetic acid is compatible with all other fixatives, but when it is mixed with potassium dichromate, the fixation-image of chromium trioxide is given.

Unclassified remarks. The fixation-image given by acetates in paraffin sections is dependent mainly on their pH. This was shown by Zirkle,[557] who experimented with various salts containing the same amount of the acetate ion as 2% acetic acid. Those acetates that gave a pH less than 4·0 (bismuth subacetate, for instance) tended to produce the characteristic 'acid' fixation-image of acetic acid. Sodium acetate, on the contrary, and others that also gave a less acid pH than 4·0, generally macerated tissues. In some cases the cation affected the image.

The other short-chain fatty acids (formic, propionic, butyric, valeric) all give much the same fixation-image in paraffin sections as acetic; so do glycollic, glyceric, lactic, and gluconic.[558] Trichloracetic acid, however, acts in an entirely different way. It is a coagulant fixative; it leaves the nucleolus readily colourable by iron haematein, and mitochondria can be well fixed by mixtures containing it at 2%.[556, 558]

CHAPTER 7

Fixative Mixtures

The term 'fixative mixtures' is here used to mean mixtures of two or more substances, each of which acts as a fixative when used alone. A primary or unmixed fixative is regarded as remaining primary when nothing but an indifferent salt or other non-fixative substance is added to it.

The primary fixatives present the advantage that the interpretation of their effects—difficult enough though it may be—is much easier than that of mixtures. Still, the majority of successful fixatives used in routine work are mixtures. Ethanol is a poor fixative (grade IV), acetic acid an indifferent one (grade III): but mixed together in appropriate proportions in Clarke's fluid [127, 108] they produce a fixative that is not only very good (grade I) in routine histology, but also valuable in chromosome studies.

Most of the mixtures used today have come into being in a haphazard way. A study of the papers in which the formulae were first published will show this. One expects to find a careful consideration of the causes that led the author to choose certain primary fixatives and to mix them in particular proportions, but usually nothing of the kind is offered: the mixture is presented to the reader as a *fait accompli*, quite frequently in the form of a footnote. Occasionally the author tells us about the various mixtures he tried empirically, but the description of his experiments shows that he gave no consideration to the fact that the ingredients must necessarily interact.

It is clear that a process of natural selection has been at work. Many new mixtures have been thrown up more or less at random by processes analogous to mutation and recombination, and they have been tried out in practice by a number of independent workers. Only the ones that give reasonably good results continue to be used subsequently: many fall by the wayside in the struggle for existence, or drag out a futile old age in the pages of the recipe-books.

Some authors delight in making trivial changes in well-known formulae. Champy's fluid,[118] for instance, has this composition:—

potassium dichromate, 3% aq.	7 ml
chromium trioxide, 1% aq.	7 ml
osmium tetroxide, 2% aq..	4 ml

This is a useful fixative for certain cytoplasmic inclusions. Nassonov [364] used the proportions 4 : 4 : 2 (he calls it Champy's fluid, without comment). To this he adds a solution of pyrogallol, measured in drops: the total amount of pyrogallol added is about 0·1 mg to 10 ml of fluid. Pyrogallol could not exist for an instant in the presence of vastly greater amounts of two very strong oxidizers (chromium trioxide and acidified potassium dichromate) and one moderately strong one (osmium tetroxide): it must at once be changed to carbon dioxide and other oxidation-products. Yet many cytologists continue to believe that there is some special virtue in Nassonov's fluid. Actually there is none. This can be proved by getting a friend to fix one set of objects in Champy's fluid and another in Nassonov's, with secrecy as to which is which. It will not be found possible to distinguish the final preparations.

Fixatives are generally named after the persons who invented them. It has already been mentioned (p. 24) that it is often convenient to call them simply by the names of the inventors, without necessarily saying So-and-so's fluid. This works well when the inventor (Zenker, for instance) only introduces a single fixative. When someone introduces two or more, descriptive words are necessary. Thus one may refer to Flemming's weak [172] and strong [173] mixtures. It is desirable in such cases that the inventor should himself suggest suitable names. Heidenhain [221] named one of his fluids *Susa*, combining into a single word the first two letters of each of the words *Sublimat* and *Säure*. (Some authors have supposed Susa to be a person.) It is thoughtless of an inventor to call a fluid by the number that it happens to receive in his laboratory note-book ('B.15', for instance, or '2BD'), for this has no mnemonic value for others.

In some cases the reduction or omission of one constituent radically changes the nature of a fixative, and a change of name is then desirable. Flemming's strong fluid [173] contains 1 *Maastheil oder weniger* of acetic acid to 19 of other constituents. Benda [56] reduced the amount to 3 drops of acetic to 19 ml of other constituents: Lewitsky [302] omitted the acetic acid altogether. Flemming's fluid, as usually used, is a valuable fixative for the study of chromosomes and for the detailed histology of very small pieces; those of Benda and Lewitsky are quite different in character, being intended for work on cytoplasmic inclusions, and should be called by the names of the men who introduced them.

Fixative mixtures are not always ascribed to their actual inventors. Thus Clarke [127] introduced in 1851 a mixture of one volume of acetic acid with three of spirits of wine; in this he soaked tissues that had already been immersed in spirits of wine alone. This mixture was widely used as a direct fixative in the following years, by Beale [50] and others. Frey [178] quoted it in his well-known text-book in 1863, giving *die Clarke'sche Vorschrift* as 3 *Theile Alkohol mit* 1 *Theil Essigsäure*. He continued to quote it repeatedly in his various editions.[179] It is therefore rather strange that the mixture should nowadays be almost invariably attributed to Carnoy,[108] who gave this formula in 1886. It will here be called by Clarke's name, while Carnoy's will stand for the fluid of his own invention [109] (absolute ethanol, glacial acetic acid, and chloroform in the proportion of 6 : 1 : 3 by volume). It may be remarked that Clarke is the better fixative for routine paraffin sections: it falls in grade I, Carnoy in grade II.

Several authors have mixed a saturated solution of mercuric chloride with glacial acetic acid, but it does not seem possible to find out who first used the familiar mixture of the two substances in the proportion of 95 : 5 by volume. This fluid (here called mercuric/acetic) is useful in zoology, particularly in the preparation of whole mounts.

Another anonymous mixture is Zenker without acetic. The fluid is radically different from Zenker, because the pH lies on the opposite side of the critical range (p. 132), and proteins therefore react to the chrome anions in it in an entirely different way. This is the only fixative that is useful in cytoplasmic cytology and at the same time good enough for routine histology to reach grade I. Bensley [59] has recommended a fluid closely similar to Zenker without acetic, but not identical with it.

For the purpose of generalization it is necessary to choose a limited number of representative fluids. Twenty-five aqueous fixative mixtures have been selected for this purpose, and two not containing water. They are a typical selection of mixtures that are widely used in micro-anatomy, embryology, histology, and cytology. Different authors would have made different lists, but a

TABLE 8

Numerical and other information about 25 aqueous fixative mixtures. Concentrations are expressed as w/W% (acetic acid as v/W%). For full explanation see text.

	Year of publication	*Grade (general histological fixation)*	pH	*Authority for pH*	CONCENTRATION, %										*Chief use*
					Ethanol	*Picric acid*	*Mercuric chloride*	*Chromium trioxide*	*Formaldehyde*	*Osmium tetroxide*	*Potassium dichromate*	*Acetic acid*	*Other fixative ingredients*	*Non-fixative ingredients*	
Allen, B.15	1916	I	1·2	284		1·2		1·7	11			5·6		urea 2·2	histology, chromosomes
Altmann	1894	IV–V	4·0 4·2 3·9	112 397 285						1·0	2·5				cytoplasmic inclusions
Aoyama	1929	IV	4·8	38					6·4				cadmium chloride 1·1		" "
Benda	1903	III	1·1	176				0·79		0·42		0·19			" "
Bensley, AOB (weaker osmium)	1911	V	3·0	284						0·40	2·0	0·12			" "
Bouin	1897	III–IV	1·5 1·2 1·6	285 176 397		1·2			11			5·6			histology
Brasil	1905	I–II	2·2	38	180	1·4			36			23			"
Champy	1911	II–III	1·8 1·5 1·7	112 397 285				0·39		0·44	1·2				cytoplasmic inclusions
Flemming (strong; full acetic)	1884	III	1·4 1·2	112 397				0·79		0·42		5·3			histology, chromosomes
Gilson	1898	III	1·0	285	6·5		2·2					0·43	nitric acid 1·3		histology

Heidenhain, 'Susa'	1916	III	0·9 0·9	176 397			4·9		8·8			4·4	trichloracetic acid 2·2	sodium chloride 0·54	histology
Helly	1903	I–II	3·7 3·5 4·7	112 176 397			4·9		1·9		2·4			sodium sulphate 0·97	cytoplasmic inclusions, histology
Hermann (mammalian formula)	1889	II	1·5 1·6	284 176						0·42		5·3	chloroplatinic acid 0·79		chromosomes
Karpechenko-Navaschin	1924	I–II	1·5 2·9	176 284				0·43	1·4			2·9			"
Kleinenberg	1879	III	1·0	284		0·35							sulphuric acid 0·9; creosote to saturation		embryology, micro-anatomy
Lewitsky	1911	III–IV	1·1 1·2	284 397				0·79		0·42					cytoplasmic inclusions
Mann	1894	IV	4·9	397			5·0			0·50				sodium chloride 0·37	" "
Maximov	1909	I–II	2·5	284			4·3		3·5	0·17	2·3			sodium sulphate 0·86	" "
Mercuric/acetic		II	2·3	38			6·8					5·3			whole mounts, micro-anatomy
Orth (strong dichromate)	1896	II–III	3·8 4·0	284 176					3·8		2·4			sodium sulphate 0·94	cytoplasmic inclusions
Perényi	1882	III	0·3 0·4	176 284	43			0·21					nitric acid 4·0		embryology
Regaud	1910	II–III	4·7 4·0 4·8	112 176 397					8·7		2·6				cytoplasmic inclusions
Sanfelice	1918	I	1·6	38				0·77	15			4·8			chromosomes
Zenker	1894	I	2·5 2·2 2·3	112 176 397			5·0				2·5	5		sodium sulphate 1·0	histology
Zenker without acetic		I	3·6	112			5·0				2·5			sodium sulphate 1·0	cytoplasmic inclusions

considerable proportion of the 27 would almost inevitably have been chosen. The aqueous mixtures are listed in table 8. The two non-aqueous ones are Clarke and Carnoy.

The dates of introduction of these 27 fixatives have some interest from the historical point of view. The dates of invention of two are unknown. The formulae for the remaining 25 were first published in the several decades as follows:—

1850–59	1
1860–69	0
1870–79	1
1880–89	4
1890–99	6
1900–09	4
1910–19	7
1920–29	2

If the list of fixative mixtures commonly used today were considerably extended, it would probably be found that the great majority of them were invented between 1880 and 1919. The most ancient fixative mixture commonly used today is Clarke (1851).[127]

Table 8 gives various kinds of numerical information about the 25 aqueous fixatives. It was found impossible to make a direct comparison between all the fixatives in the table except by relating the amount of each ingredient to that of the only substance that occurs in all: namely, water. The concentrations are therefore expressed as percentages, w/W, except that acetic acid is expressed as v/W, since this substance is nearly always measured by volume in biological laboratories. (See Appendix, p. 313.)

Bouin's fluid [84] may be used as an example to explain how the concentrations were calculated. The composition is:—

formalin	25 ml
picric acid, sat. aq.	75 ml
glacial acetic acid	5 ml

The approximate composition of the formalin is 10 g of formaldehyde and 15 g of water. The water in the picric acid solution is very nearly 75 g. The total amount of water may therefore be taken as 15 + 75 = 90 g. Picric acid is soluble at about 1·4% in water: thus there is about 1·05 g in 75 ml of the saturated solution. The composition of the fluid is therefore:—

formaldehyde	10 g
picric acid	1·05 g
acetic acid (glacial)	5 ml
water	90 g

If each of these figures be increased by one-ninth, we have the composition expressed as percentages of the weight of water:—

formaldehyde	11 g
picric acid	1·2 g
acetic acid (glacial)	5·6 ml.

The concentrations of the solutes are therefore:—formaldehyde, 11% w/W; picric acid, 1·2%, w/W; acetic acid, 5·6% v/W. These are the figures given in table 8.

Care has been taken to make the table as accurate as possible, but perfect accuracy cannot be achieved. Some inventors of fixative mixtures give amounts in drops. It is impossible to be sure how much HNO_3 there was in Perényi's *Saltpetersäure* 10%. Saturated solutions are of varying concentrations. Authors do not tell us how much (if any) water of crystallization they introduce

TABLE 9

Analysis of 25 aqueous fixative mixtures: summary of part of the information contained in table 8.

	No. of mixtures in which each occurs	*Concentration, %*		
		Minimum	*Mean*	*Maximum*
ethanol	3	6·5	76·5	180
picric acid	4	0·35	1·0	1·4
mercuric chloride	8	2·2	4·8	6·8
chromium trioxide	8	0·21	0·73	1·7
formaldehyde	11	1·4	9·8	36
osmium tetroxide	9	0·17	0·47	1·0
potassium dichromate, scarcely or not acidified	7	2·0	2·4	2·6
potassium dichromate, acidified	2	1·2	1·8	2·5
acetic acid	13	0·12	5·2	23

with sodium sulphate. The attempt has been made to get the concentrations correct, where possible, to two significant figures. There would be no point in trying to go beyond this.

The table will enable the reader to make a direct comparison of any particular fixative mixture in the list with any other. Certain general conclusions can also be drawn. Some of these are shown in table 9. It will be noticed what a preponderating place is taken in our most familiar mixtures by the eight primary fixatives described in detail in chapters 5 and 6. The only other fixative substances that occur in any of these 25 mixtures are nitric acid, sulphuric acid, chloroplatinic acid, trichloracetic acid, cadmium chloride, and creosote. Of these, nitric acid occurs in two of the 25 mixtures, each of the rest in one only. It will be remembered that the two representative non-aqueous mixtures (Clarke and Carnoy) both contain ethanol and acetic acid; these substances therefore appear respectively in 5 and 15 of the 27 chosen mixtures.

The two occurrences of acidified potassium dichromate should properly be added to the 8 of chromium trioxide, since, as we have seen (p. 127), the ions are essentially the same.

The pH range of the 25 aqueous mixtures extends only from 0·3 to 4·9. It may be remembered that one of the primary fixatives (formaldehyde) is sometimes used in alkaline solution (p. 112).

It is not the purpose of table 8 to help in the making up of the various mixtures. It is best to keep stock solutions of most of the primary fixatives at suitable concentrations and to make up mixtures from these. In cytological work one requires very small amounts of fixatives. If a piece of tissue is only about a millimetre in greatest diameter, 2 ml of fixative are amply sufficient: 4 ml are enough for nearly all cytological purposes. It is convenient to have stock solutions from which several useful fixatives can be made up. Three valuable mixtures are Hermann, Flemming's strong fluid with full acetic, and a fixative called Lewitsky-saline (LS). The latter is Lewitsky's fluid with the addition of sodium chloride: it can be strongly recommended for studies of cytoplasmic inclusions.[36] These three fixatives can be made up from stock solutions in the way shown in table 10. If these small amounts of fixatives are used, the high cost of osmium tetroxide and chloroplatinic acid no longer forbids their frequent use.

Quite a number of commonly used mixtures are best made up a' the moment of use, since they contain ingredients that interact. Some part of the mixture may generally, however, be made, and the fixative will then be available at short notice. Thus for Sanfelice's fluid the chromium trioxide must be kept separate from the formaldehyde till the last moment, but there is no reason why the formaldehyde should not be mixed with the acetic acid.

Although certain very unstable fixatives work well, yet it is questionable whether they could not be replaced by more rational mixtures. Some of the changes that occur in unstable fixatives have been studied by Freeman and his colleagues [176] and by Casselman.[112] The fluids of Helly, Karpechenko, Orth, and Regaud all

TABLE 10

Method for making up small quantities of certain cytological fixatives. It is often convenient to halve these quantities.

	Flemming[173] (*strong, full acetic*) *ml*	*Hermann*[228] (*mammalian formula*) *ml*	*Lewitsky-saline*[36] *ml*
Distilled water . . .	1·6	1·6	2·0
Chromium trioxide, 5% aq.	0·6	—	0·6
Chloroplatinic acid, 5% aq..	—	0·6	—
Sodium chloride, 5% aq. .	—	—	0·6
Osmium tetroxide, 2% aq. .	0·8	0·8	0·8
Acetic acid, 20% aq. . .	1·0	1·0	—

rise gradually in pH from the time when they are made up. Thus Helly rises from pH 3·70 when fresh to pH 4·20 on the next day. The oxidation-potential falls at the same time; that of Sanfelice falls even more. Other fixative mixtures, such as Altmann, Bouin, Champy, Susa, Hermann, Perényi, and Zenker, maintain a constant or nearly constant pH, and at least some of them (Altmann, Champy, Zenker) a constant oxidation-potential; Flemming with full acetic becomes more acid on standing.

The products of the reactions in unstable mixtures appear not to have been worked out. The reduction of the chrome anions is likely to produce cationic chromium; this would itself act as a fixative, though quite differently from the anions. The use of cationic chromium in fixation has been suggested by Zirkle.[556] If unstable mixtures are really necessary, it is important that their reaction-products and the effects of these on the tissues should be known.

Of the fixatives listed in table 8, nearly all that are used chiefly in micro-anatomy, embryology, histology, and in studies of chromosomes contain acetic acid. This applies also to Clarke and Carnoy. The only exceptions are Kleinenberg (which contains sulphuric acid) and Helly. For chromosome studies acetic acid is almost invariably included, though exceptions can be quoted.[148, 373] Fixatives for chromosomes usually also contain chromium trioxide (or acidified potassium dichromate).

Many of the mixtures used for the purposes mentioned in the last paragraph contain, in addition to acetic acid, one or more coagulant fixatives and either formaldehyde or osmium tetroxide. This applies, for instance, to Allen's B.15, Bouin, Brasil, Flemming, Susa, Hermann, Karpechenko-Navaschin, and Sanfelice. This trio of ingredients (acetic acid + coagulant + non-coagulant fixative of ground cytoplasm) is so frequent in widely-used mixtures that the reasons for its success must be sought. Acetic acid does not in itself prevent great shrinkage in final preparations (p. 136), but it does prevent it during fixation. In mixtures it presumably antagonizes the shrinking effect of other fixatives while the latter are stabilizing the proteins in the unshrunken state; it also prevents excessive hardening. Ross [428] showed that mixtures have much less tendency than primary fixatives have to produce a badly-shrunk final preparation. The primary spermatocytes of the snail, fixed in 4% formaldehyde, retain in balsam only 34% of their original volume, and this appears to be the primary fixative that results in the least final shrinkage of any (with the doubtful exception of osmium tetroxide). With chromium trioxide at 0·75% the figure is 29% of the original volume; with 5% acetic acid, 28%. Yet when these three primary fixatives are mixed to produce Sanfelice, the final volume in the mounted paraffin section is 66% of the original. Sanfelice gives less final shrinkage than any other mixture studied by Ross. All the mixtures he studied (including Bouin, Helly, strong Flemming, mercuric/acetic, and Zenker) gave less final shrinkage than any of the primary fixatives (except perhaps osmium tetroxide).

This only explains in part why fixatives containing the trio are so successful. Why should not acetic acid and one other ingredient suffice? It seems that tissues are not readily infiltrated with paraffin unless they have been made spongy to some extent by the inclusion of a coagulant fixative. If, however, a coagulant is used alone, the cytoplasm and nuclear sap tend to be transformed into rather coarse meshworks. If formaldehyde or osmium tetroxide be included, a compromise is reached: the protoplasm is more smoothly fixed, but paraffin can still enter.

The requirements of paraffin embedding have in the past to a large extent controlled our choice of fixatives. The introduction of new embedding media is likely to result in less reliance being placed on coagulant fixatives.

It follows from what was said in chapter 6 that fixative mixtures intended for the study of cytoplasmic inclusions will contain formaldehyde, osmium tetroxide, or unacidified potassium dichromate, or more than one of these. Acetic acid is omitted or reduced to a minute amount. These generalizations apply to Altmann, Aoyama, Benda, Bensley's AOB, Champy, Helly, Lewitsky, Mann, Orth, Regaud, and Zenker without acetic. Of these, only Benda and AOB contain acetic acid, at great dilution (less than 0·2%). It must be emphasized once more that the omission or great reduction of acetic acid is not necessarily connected with pH. Thus Lewitsky differs from Flemming's strong fluid only in the absence of acetic acid from the former: yet they have the same pH.[397]

Acetic or any other acid, unless very dilute, will transform potassium dichromate into a coagulant fixative and in so doing abolish its ability to fix mitochondria.

Coagulants are often omitted from mixtures used for the study of cytoplasmic inclusions. This applies, for instance, to Altmann, AOB, Orth, and Regaud. Easy infiltration with paraffin is sacrificed for the sake of homogeneous fixation of the ground cytoplasm. It is particularly difficult to obtain good paraffin sections of tissues fixed in Altmann.

Each component of a fixative mixture should so far as possible compensate for a defect in another. Thus ethanol shrinks tissues strongly and does not fix chromatin: acetic acid compensates for both these defects. Acetic acid does not fix cytoplasm or nuclear sap: ethanol fixes both (though the former indifferently). When ethanol and acetic acid are mixed to make Clarke's fluid, an excellent fixative for routine histology and chromosome studies is produced.

The properties of Bouin's fluid can be analysed in a similar way. Formaldehyde fixes cytoplasm and nuclear sap, but hardens tissues unduly, prevents paraffin from penetrating easily, makes cytoplasm basiphil (so that acid dyes do not work well), and fixes definitive chromosomes poorly. Picric acid compensates for most of these defects. It leaves tissues soft, coagulates cytoplasm in such a way that it readily admits paraffin, makes it strongly acidophil, and fixes chromosomes rather well. It has, however, two serious defects: it shrinks tissues badly and makes chromatin acidophil. Acetic acid compensates for both these defects.

Unfortunately the good qualities of primary fixatives cannot always be combined in mixtures. Potassium dichromate tends to stabilize the cytoplasm and nuclear sap in a homogeneous state, but dissolves and disperses chromatin: acetic acid fixes chromatin but does not fix cytoplasm or nuclear sap. The attempt was naturally made to let each compensate for the defects of the other. Yet Tellyesniczky's fluid [497] did not and could not achieve its objects, because the acidification of the chrome anions causes them to act as though chromium trioxide had been dissolved instead of potassium dichromate, and this is a strongly coagulative fixative that does not stabilize cytoplasm or nuclear sap as a homogeneous gel.

In deciding which primary fixatives to mix, it is important to take into account not only their obvious mutual compatibilities and incompatibilities, but also their more subtle influences on one another's properties. Thus anyone who includes mercuric chloride in a mixture should remember that its coagulative power and the solubility of its coagulates are affected by acidification (p. 52).

The effect of so-called indifferent salts must also be considered. Sodium chloride, as we have seen (p. 54), can dissolve protein coagulates produced by mercuric chloride, in certain circumstances. Ammonium sulphate can transform ferric sulphate, at certain concentrations, from a non-coagulant to a coagulant fixative (p. 85). Indifferent salts can be very useful ingredients of mixtures, but it is unlikely that they improve any fixative that contains acetic acid at 5% or thereabouts.

All the fixative ingredients of an ideal mixture would penetrate at the same speed. This could be achieved by adjustment of their concentrations. It does not appear that any inventor of a fixative mixture has taken this into consideration. A section through a large piece of tissue often gives unmistakeable evidence that one ingredient reached the centre at fixative concentration in advance of the others. When a small piece of tissue is used, the unevenness due to this cause is minimized. The argument presented in detail on p. 69 applies here. When a piece of tissue has been placed in the fixative mixture, each ingredient must be supposed instantly to reach the opposite side of it from that at which it started, at infinitesimal concentration; from then onwards it will help to build up the concentration there faster than it would have been built up if the fixative had come from one side only. It follows that although perfectly even fixation by all the ingredients at the same time could only be achieved if they all had the same K-value, yet the smaller the piece of tissue, the more even the fixation will be.

PART TWO

DYEING

CHAPTER 8

Introduction to the Chemical Composition of Dyes

The constituent parts of cells and of intercellular material are usually transparent and colourless, and therefore not distinguishable from one another unless there are appreciable differences of refractive index. It was Leeuwenhoek (1719) [292] who first used a colouring agent in an attempt to overcome this difficulty. He was studying mammalian muscle, and tried the effect of an alcoholic solution of saffron.[300] His experiment was not very successful, and others were slow to follow his example. Iodine was used from time to time for the same purpose, but during a very long period cells and tissues were commonly examined in their natural, transparent state. During the decade 1848–1858 the use of colouring agents in microtechnique was repeatedly rediscovered,[29] and from that time to the present day enormous use has been made of artificially coloured preparations.

The colouring agents used in microtechnique are of diverse kinds, but the majority of them are dyes. It will be best to consider carefully what dyes are, before discussing the other kinds of colouring agents that can be used (p. 296).

If a porous body, such as a sponge, be placed in a vessel containing a solution of a coloured substance, some of the coloured substance will be imbibed by the porous body, and if the latter be lifted out of the solution, will be removed with it. There are now two possibilities about the solution that remains in the vessel. Either the coloured substance is at the same concentration as it was before the porous body was put in it, or else the concentration has been reduced (that is to say, the solution has been to some extent 'exhausted'). If exhaustion has occurred, the porous body has clearly shown a special affinity for the coloured substance in preference to the solvent. It is characteristic of the process of dyeing that this special affinity exists, and indeed in the commercial dyeing of textiles (though not in microtechnique), exhaustion is often nearly complete.

If, then, a substance acts as a dye, two problems present themselves: why is it coloured, and why is there an affinity between it and the object that it dyes? The present chapter is concerned with the first of these problems; the second will be considered in chapter 10.

All dyes, in the strict sense of the word, are organic compounds. It is a remarkable fact that very few aliphatic substances are coloured. Aromatic substances, on the contrary, have a tendency to absorb electro-magnetic waves. Whenever we can write two or more equally good structural formulae for a particular substance, it is supposed that there is a possibility of rapid change in the configuration of the molecule between the various possible states, and this change or *resonance* involves the absorption of electro-magnetic waves.

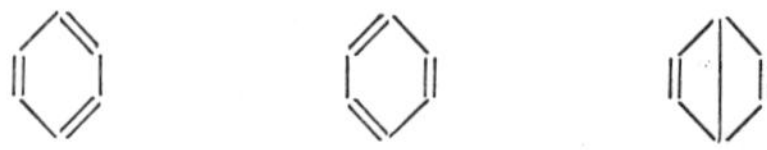

Alternative structural formulae for benzene

If our eyes were sensitive in the ultra-violet, benzene would appear coloured, for it has an absorption-band at wave-length 256 mμ; and other aromatic compounds have absorption-bands at various wave-lengths in the ultra-violet. There is a tendency for resonance to be transmitted through a path of alternate double and single valency-bonds. Indeed, aromatic compounds tend to absorb electro-magnetic waves at particular wave-lengths just because there is this alternation of bonds. Annatto (bixin), a plant-product used in the colouring of butter and cheese, is of particular interest in this connexion, for it is one of the rare aliphatic compounds that are strongly coloured, and its long molecule presents that alternation of double and single bonds that is so characteristic of aromatic substances.

As a general rule, the alternation of bonds does not in itself shake up the molecule sufficiently to extend the absorption-bands into the visible region and thus produce colour. Any configuration of the molecule that pushes the absorption bands downwards into longer wave-lengths is said to be *bathochromic*, and certain particular configurations are so effective in this respect that their presence is always associated with colour. One of these is the quinonoid arrangement. Quinone itself (parabenzoquinone) is coloured (yellow), but it is chiefly when a quinonoid ring is introduced into more complicated compounds that brilliant colours are produced. Since the paraquinonoid ring confers the property of colour, it is called a *chromophore* or colour-bearer. It occurs in many dyes that are important in microtechnique. Other dyes owe their colour to other chromophores, which will be mentioned in chapter 9.

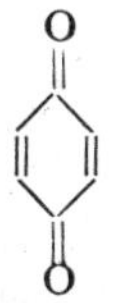

Parabenzoquinone

Quinone itself has no tendency to be exhausted from its solvent by attaching itself to a textile or to the substance of a microscopical preparation. In general, substances that have this tendency ionize in aqueous solution, and quinone does not. A chromophore by itself, then, does not confer upon a molecule the capacity of acting

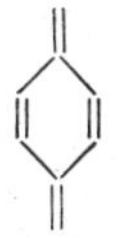

The paraquinonoid ring

as a dye. An ionizing group is required as well. Those ionizing groups that transform substances that are merely coloured into dyes are called *auxochromes*. As their name suggests, they have a tendency to increase the intensity of the colour, often very markedly, and it is sometimes convenient to think of the *auxochromophoric systems* in dyes, in order not to distinguish too sharply between the effects of chromophore and auxochrome.

One of the commonest auxochromes in dyes is the group $-NH_2$,

Aniline

and it is because aniline contains this group that it is of such particlar importance in dye-chemistry.

To make a dye, it is only necessary to combine a molecule that can spring into a chromophoric configuration with another that contains an auxochrome. We may start with two colourless substances and produce a dye in a single process. A very easy one to make is the dye known commercially as *pararosaniline*. (The *para-* in this name has no connexion with the chemical usage of the

Aniline

Aniline

Paratoluidine

prefix. It means that the dye is a modified form of another dye with the commercial name of rosaniline.) Practical instructions are given on p. 321. It is only necessary to bring together some aniline and paratoluidine and to heat them in the presence of chloride and a suitable oxidizing agent. The quinonoid chromophore then appears, and the $-NH_2$ groups are already present to act as auxochromes.

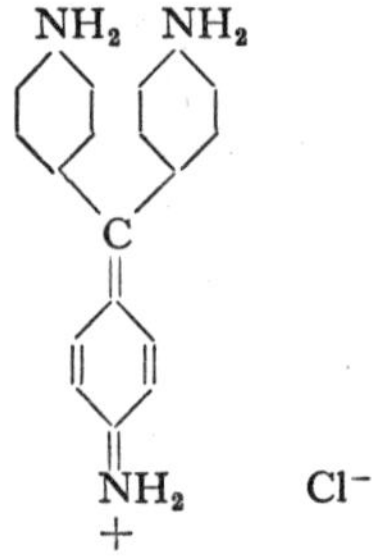

Pararosaniline

Pararosaniline is a solid, magenta substance, soluble in water and alcohol.

In this book attention will be directed to quinonoid chromophores by the drawing in of double bonds in quinonoid rings, but aryl rings will be represented by a simple hexagons.

It will be noticed that one of the three rings—the quinonoid one —has been shown with a positive charge on the nitrogen atom, a charge balanced by the negative charge on the chloride ion. It must not be supposed, however, that the position of this charge is constant. On the contrary, it changes continually. It may be associated with any of the three nitrogen atoms, or with the central carbon atom, and indeed there are other possible positions. It is the resonance between the particular positions that are available in a molecule that determines the colour. This fact will be strik-

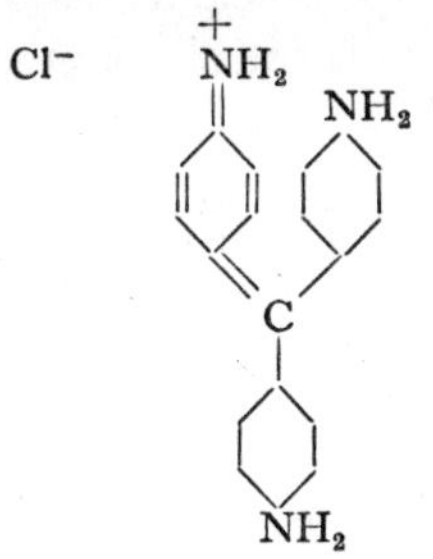

Pararosaniline: another resonance position

ingly illustrated below, by reference to a related dye, crystal violet (p. 168).

It is accurate to regard the positive charge as belonging to the ion as a whole, and therefore to write the formula in the way shown here.

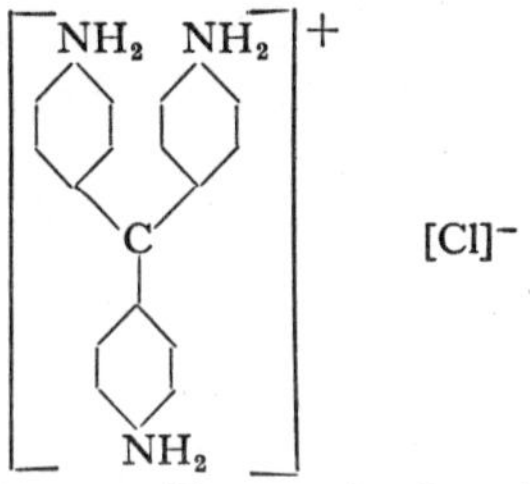

Pararosaniline: another formula

This kind of symbol will nevertheless not be used in this book, because it is convenient to draw special attention to the chromophore. The reader should remember that the formulae that will be used represent, as it were, instantaneous photographs of the dye-ion. He may decide to work out for himself the changes in the position of the electric charge involved in the process of resonance.

The colour of the dye is very slightly changed if, instead of taking two molecules of aniline and one of paratoluidine, we take one of aniline, one of orthotoluidine, and one of paratoluidine. The dye formed from these constituents is called *rosaniline*. It will be noticed that it differs from pararosaniline only in the possession of the methyl group introduced with orthotoluidine.

The methyl group affects the colour, for rosaniline is a very slightly bluer magenta than pararosaniline. Side-groups such as this are called *modifiers*.

The familiar dye called *basic fuchsine* is a mixture of pararosaniline with rosaniline; some specimens contain another related dye as well. Since pararosaniline is the simplest of the triarylmethane dyes, it could with advantage have been used as a type in the

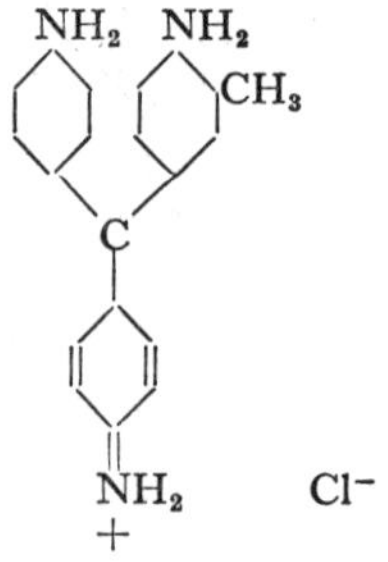

Rosaniline

experiments described later in this book (p. 324); but whereas one must make pararosaniline for oneself, basic fuchsine is readily available in commerce, and for this reason it has been chosen instead. Almost exactly the same results would be obtained if pararosaniline had been used.

Magenta is another name for basic fuchsine. Mere inspection of the solution suggests that it transmits much red light and some blue, but absorbs light of intermediate wave-length. For accurate comparison of the colours of dyes it is necessary to obtain figures that will show the transmission or absorption at various wave-lengths. This is done by use of a spectrophotometer. Light of known wave-length is shone through a solution of the dye contained in a flat-sided vessel of uniform thickness, and the intensity of the light that comes through is measured by means of a photo-electric cell. A measurement can also be made with exactly the same apparatus and the same solvent (water, for instance), in the absence of the dye. The amount of light that comes through when the dye is present with the solvent can then be expressed as a percentage of the amount that comes through when it is not. This percentage is known as the *transmission* at the particular wave-length used. It is usual to find the transmission at 10 $m\mu$ intervals throughout the visible spectrum. A curve like that shown in fig. 17 is then obtained. It is generally best to dissolve the dye at a concentration that will give a transmission of about 20% at the wave-length of maximal absorption. It will be observed from the graph that basic fuchsine transmits most of the red and orange light, and also a considerable amount of violet and blue, but absorbs strongly in the green, especially at about 540 mμ. Different specimens of this mixed dye naturally differ somewhat in the wave-length of greatest absorption.

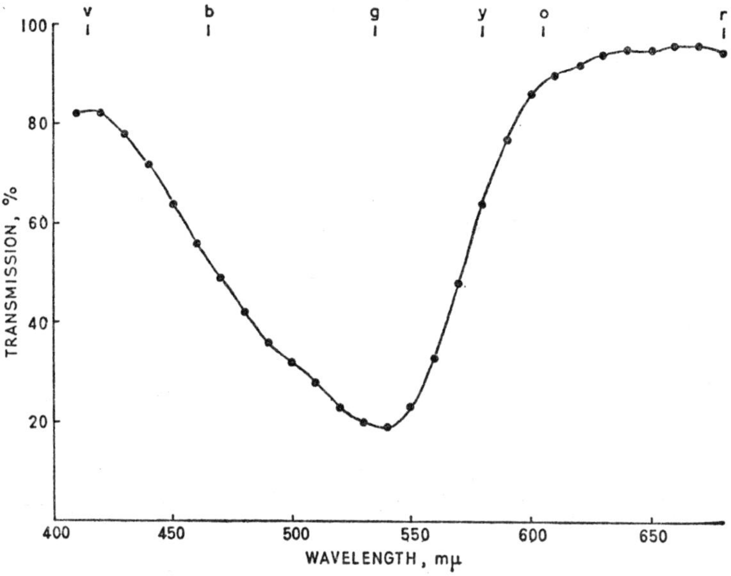

FIG. 17. Graph showing the transmission of light through a layer 1 cm thick of basic fuchsine, 0·00062% aqueous.[38] The following abbreviations are used in this and subsequent figures:—*v*, violet; *b*, blue; *g*, green; *y*, yellow; *o*, orange; *r*, red.

Since the active property of the dye is to absorb, not to transmit light, it is for some purposes desirable to express its absorptive capacity as the reciprocal of the transmission; for instance, instead of saying that the transmission at a particular wave-length is $\frac{20}{100}$ of the incident light, we may turn the fraction upside down and use the number 5 to represent the absorption. A curve of reciprocals for the same solution of basic fuchsine as before is shown in fig. 18. Alternatively, one may use the logarithms of the reciprocals. Such logarithms are called 'densities'. This method of expression is particularly useful with light-filters, because if one knows the density of each at any wave-length, it is only necessary to add the two densities together to obtain the density of two filters used one behind the other. As a result of this, the absorptive capacity of coloured substances is often given in the form of density-curves, such as that shown in fig. 19. It is necessary to understand this method of representation, because it is so com-

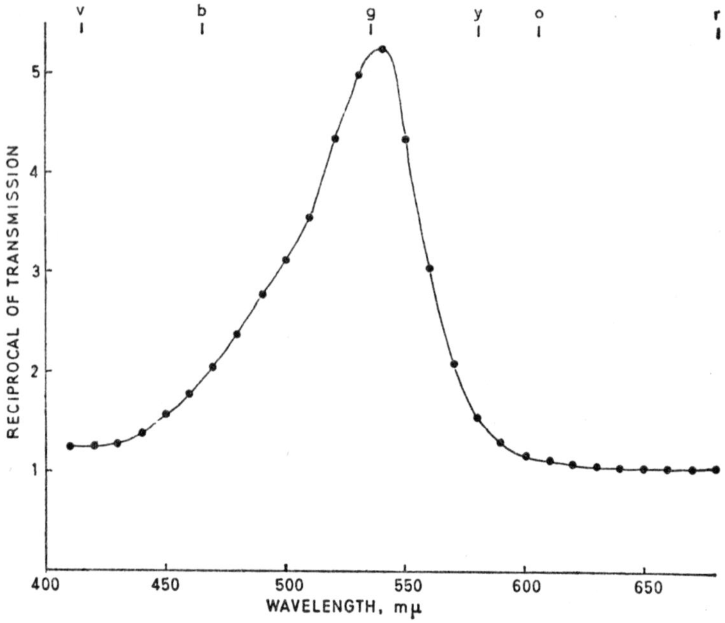

FIG. 18. Graph showing the reciprocals of the transmission of light through a layer 1 cm thick of basic fuchsine, 0·00062% aqueous.[38]

monly used; but in fact one seldom makes practical use of the ease of addition of densities, if one's main interest is in the use of dyes in microtechnique; and transmission gives a much more direct statement of what is actually the important point when one looks through the microscope—the colour of the dye.

Any change in structure that destroys the quinonoid linkage results in the loss of colour. Many dyes can be changed to colourless or *leuco*-compounds by the action of reducing agents such as sodium hydrosulphite ($Na_2S_2O_4$). The leuco-representative of pararosaniline may be regarded as a triaryl derivative of methane: that is to say, as methane in which three of the four hydrogen atoms have been replaced by aryl rings. (An aryl ring is either a simple phenyl ring or else a phenyl ring with a special group or groups (in this case $-NH_2$) replacing hydrogen in one or more places.) Dyes that are related to a leucobase of this kind are called *triarylmethane dyes*. Some of them are of great importance in microtechnique.

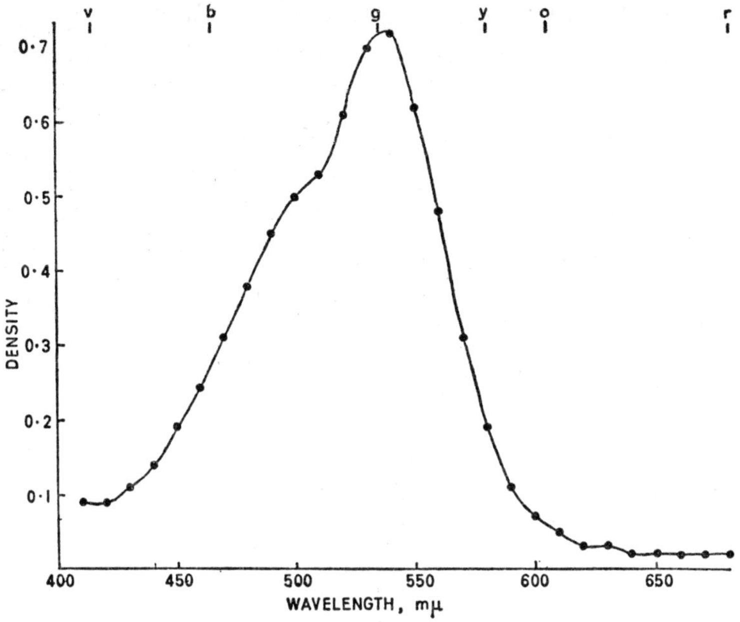

FIG. 19. Graph showing the optical density of a layer 1 cm thick of basic fuchsine, 0·00062% aqueous.[38]

Another colourless derivative of pararosaniline, in which $-OH$ replaces the hydrogen attached to the central carbon atom, may be regarded as a triaryl derivative of methanol. Compounds of this

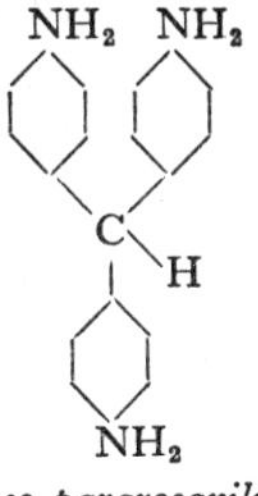

Leuco-pararosaniline

kind are called *leucobases*. This word has unfortunately been wrongly applied in recent years to Schiff's reagent, which is a colourless derivative of basic fuchsine but not a leucobase (p. 308).

Pararosaniline owes its colour to the cation. Dyes that have this character are called *basic dyes*. Since the anion is commonly chloride, they may be shortly represented by the formula R^+Cl^-, though other anions (sulphate, nitrate, acetate, or oxalate) are sometimes used instead in basic dyes, without influencing the colour. Many dyes, however, owe their colour to the anion. Since sodium is then usually the cation, these *acid dyes* may be represented by the short formula Na^+R^-. These dyes are often closely related to basic dyes, but they contain one or more acidic auxochromes, to which the negative charge on the dyeing ion is due. The auxochrome is commonly the sulphonic group, $-SO_3^-$.

To convert basic fuchsine (pararosaniline mixed with rosaniline) to its acidic counterpart, the dye is first treated with concentrated sulphuric acid at about 150° C. Water is then added and the solu-

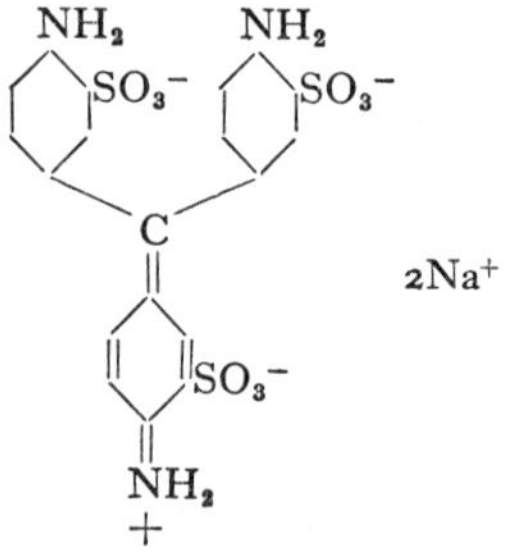

The acidic counterpart of pararosaniline

tion neutralized with calcium hydroxide. The very soluble calcium sulphonate of the dye is converted to the sodium salt by the addition of sodium carbonate. Calcium sulphate is precipitated and may be filtered off. The sodium salt is called *acid fuchsine*.

It will be noticed that the dyeing ion has a balance of two negative charges, which are equalized by the two sodium ions.

The transmission of light by acid fuchsine is shown in fig. 20. Comparison with fig. 17 (p. 161) will show that basic and acid fuchsine absorb light very similarly. Although their colour is so similar, their staining reactions are extremely different. (See pp. 192 to 196.)

The aesthetic desires of man have led to the elaboration of a huge variety of dyes capable of imparting almost every conceivable colour to textiles. An impression of the immense number can be gained by looking through Rowe's invaluable *Colour Index*,[434] which gives the chemical composition of each. Structural formulae for nearly all the dyes used in microtechnique are given in Conn's *Biological Stains*,[131] which is produced under the auspices of a commission set up in the United States to inspect commercial specimens of dyes used in microtechnique and to certify those that are suitable for use in particular techniques. It includes a number of dyes that are not used in the textile industry. The book gives the wavelength of maximum absorption of many dyes.

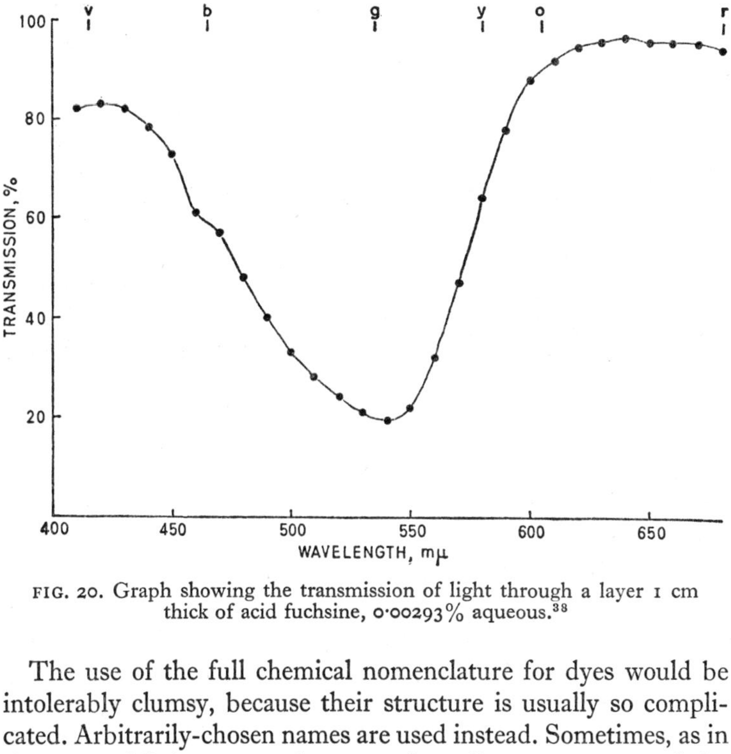

FIG. 20. Graph showing the transmission of light through a layer 1 cm thick of acid fuchsine, 0·00293% aqueous.[38]

The use of the full chemical nomenclature for dyes would be intolerably clumsy, because their structure is usually so complicated. Arbitrarily-chosen names are used instead. Sometimes, as in the case of light green, the name is a direct statement of the colour of the dye. The colours of the flowers of familiar plants give their names to gentian violet, dahlia, and fuchsine. Eosin, as its name suggests, has a colour reminiscent of the dawn. Sometimes a certain amount of chemical information is conveyed by a name, as for instance by naphthol yellow, which is in fact naphthol with various side-groups. In some cases, however, names that appear to be partly chemical are in fact misleading. Thus methyl blue contains no methyl group and methylene blue no methylene group; and azo-carmine is not an azo dye and is not related to carmine. Some dyes have names of fantastic, almost surrealist origin. It was discovered that a particular dye was especially convenient for the colouring of cotton. In the same year the European powers recognized the existence of the Congo Free State. These unconnected events led to the dye being called Congo red.[315]

Despite all that can be said about the lack of consistency in the naming of dyes, the fact remains that the words used are much more easily remembered than the numbers and strings of initial letters that are so often used nowadays in industry for purposes of this kind, without any thought for those who have to try to hold in mind which symbol refers to which object.

The names of many dyes are followed by letters or numbers. These generally serve to distinguish closely-related dyes. Thus 'B' usually means that the dye is more blueish than a related dye, and 'Y' or 'G' (*gelb*) that it is yellower. 'WS' conveys that it is water-soluble, while a related dye is not. The letters A, B, C are sometimes used as arbitrarily-chosen marks of distinction (for instance, with the azures, p. 268).

Although dyeing has been and is of such immense service to biology, the number of dyes that are really useful in microtechnique is not very great. There has been a tendency to try new dyes from mere whim and desire for novelty, and the introduction of many superfluous ones has been recorded from recipe-book to recipe-book as though it were the fruit of wisdom. Dabbling with dyes by persons ignorant of the chemistry of what they are doing has no counterpart in the rest of science and indeed cannot be regarded as a scientific activity. The stricture on this subject reproduced on p. 187 is as applicable today as when it was delivered by Gustav Mann more than half a century ago. More so, perhaps; for it would be a sobering experience to many present-day dabblers to note the real erudition brought to the subject by such men as Ehrlich and Paul Mayer, right back in the nineteenth century. The person who can show a particular dye to be superfluous in microtechnique usually deserves better of his colleagues than he who introduces a new one.

Dyes are classified by their chromophores, but the auxochromes and modifiers occur over and over again in dyes of different groups. It is proper to consider these first, for what can be said about them is of more general application than what can be said about chromophores.

With a few exceptions that will be mentioned later (p. 262), all dyes are basic or acidic. The fundamental basic group is $^{-}\overset{+}{N}H_2$. The usual acidic group is $-SO_3^-$, and many acid dyes are substituted sodium sulphonates. A sulphonic acid may be regarded as sulphuric acid that has lost an oxygen and a hydrogen atom and replaced them with an organic radicle (R).

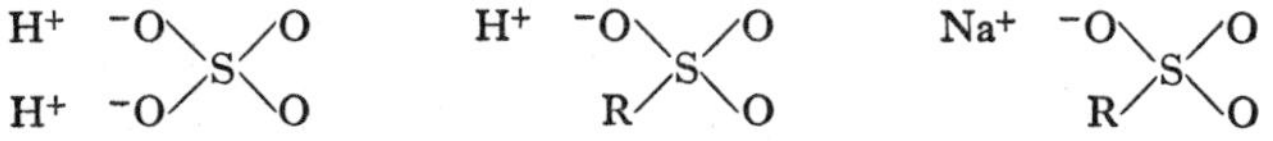

Sulphuric acid *Sulphonic acid* *Sulphonate*

Other acidic auxochromes are the carboxyl ($-C(=O)OH$) and hydroxyl (–OH) groups, ionized to produce negative charges ($-C(=O)O^-$ and $-O^-$). Many dyes possess hydroxyl as well as sulphonate groups, and a few (such as fast acid violet A2R) possess both carboxyl and sulphonate groups. Some dyes have both basic and acidic auxochromes. The acidic counterpart of pararosaniline (p. 164), for instance, has both $-\overset{+}{N}H_2$ and $-SO_3^-$, but the negative charges predominate in number.

In the textile industry it is usual to restrict the name of 'acid dyes' to those particular acid dyes, in the wide sense, that will dye wool only from a strongly acid bath, will not dye cotton directly, and are not used with intermediaries or 'mordants' (p. 207) to attach them to the fibre. Venkataraman, in his invaluable textbook,[523] adopts this nomenclature. It is desirable, however, to have a single name for all dyes in which the dyeing ion bears a negative charge, and the term 'acid dyes' will be used in this book to cover all such dyes.

A variety of different atoms and groups of atoms may replace hydrogen atoms attached to the aryl and other rings of dyes. Thus eosin Y contains four bromine atoms, and erythrosine B has exactly the same composition except that iodine replaces bromine. These atoms are held by covalent linkages, and there is no question of the formation of bromide or iodide. The particular atom or group of atoms attached in this way often affects the colour. The iodine of erythrosine B shifts the absorption-maximum somewhat towards the longer wave-lengths, and the dye is therefore bluer (less red) than eosin.

The hydrogens of the amino-groups are often replaced by methyl, ethyl, and phenyl groups. These particular modifiers tend to make a red or reddish dye blueish or blue. There are six such hydrogen atoms in pararosaniline. As more and more of these are replaced by methyl (in the various dyes that together constitute the mixtures called methyl and gentian violet), so the colour becomes bluer and bluer until the hexa-methyl compound, crystal violet, is reached; this is a very blueish violet dye. Ethyl groups have more blueing effect than methyl, and phenyl groups still more, till dyes are reached that are purely blue.

Crystal violet, an important dye for chromosomes, will serve to show the influence of resonance on colour. When a solution of the dye is acidified, a proton is added to one of the three dimethyl-

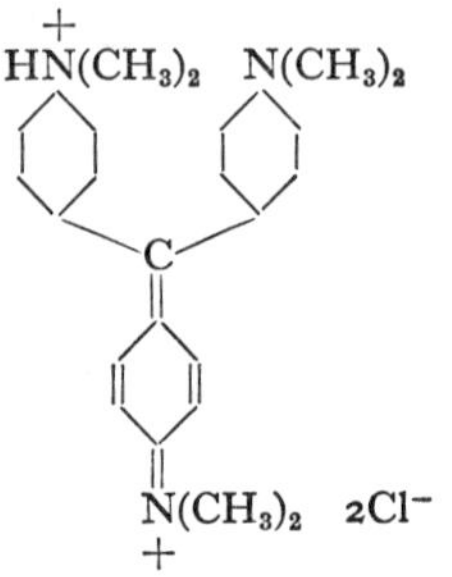

Acidified crystal violet

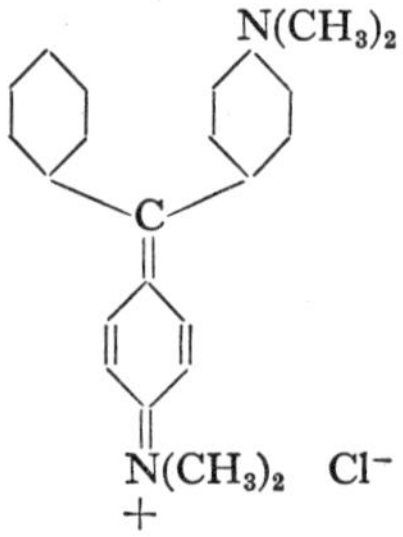

Malachite green

amino groups, and this now no longer participates in the resonance of the ion. The resonance is thus limited to that of the related dye, malachite green; and in correspondence with this, the colour changes from blue-violet to green.

The more central part of a dye ion also affects colour. If two benzene rings are put together to form naphthalene, there is more absorption of electro-magnetic waves than by two separate benzene rings, and the same principle applies again when a third ring is added to form anthracene. Many dyes have a double-ring component, and a whole group of dyes is based on anthracene (p. 175).

CHAPTER 9

The Classification of Dyes

Dyes are classified by their chromophores. In a logical classification the natural dyes such as haematein and carmine do not form a group apart, but divide themselves here and there among the various groups of dyes, interspersed among the synthetic ones. Haematein, for instance, owes its colour to a paraquinonoid ring, and is therefore much more closely related to the fuchsines than to a dye such as orange G, which owes its colour to an entirely different chromophore.

Nearly all the dyes used in microtechnique owe their colour either to a quinonoid ring, or to an azo linkage (–N=N–, p. 182), or to a nitro-group ($-NO_2$, p. 184). There are thus three main groups, of which the first is so diverse that it requires considerable subdivision, while the third (nitro dyes) is so small that it can be dealt with very shortly.

THE QUINONOID DYES

These mostly contain a paraquinonoid ring, but in some cases an orthoquinonoid ring is present instead (p. 181).

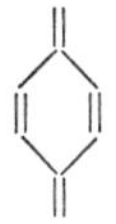

Paraquinonoid ring

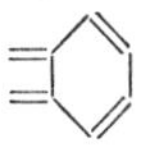

Orthoquinonoid ring

The chief subdivisions of the quinonoid dyes are represented here by skeleton formulae, from which the auxochromes and modifiers are omitted. The chromophore is indicated in each case in a particular resonance-position.

The 'X' in the formula for the azines and related dyes may represent nitrogen, oxygen, or sulphur. The double bonds of the quinonoid ring of this group of dyes cannot be represented in full in a skeleton-formula, because it is sometimes para- and sometimes orthoquinonoid.

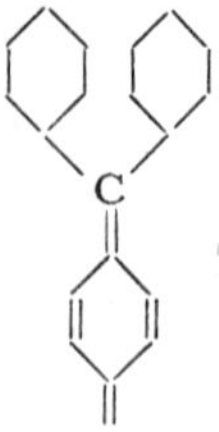

Triarylmethane

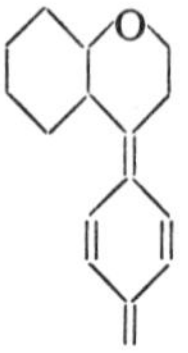

Haematein

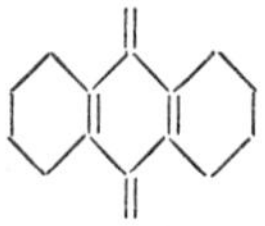

Anthraquinonoid

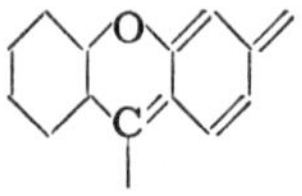

Xanthene

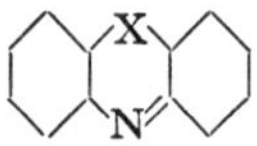

Azines and related dyes

Skeleton-formulae of the main groups of quinonoid dyes

The following table shows the classificatory positions of the quinonoid dyes mentioned in this book. The dyes named in heavy type are the ones that are particularly important in microtechnique. The table is intended for reference.

page

TRIARYLMETHANE 171

Basic. **Basic fuchsine** (pararosaniline with rosaniline), methyl violet, gentian violet, **crystal violet,** malachite green, **dahlia, methyl green, aniline blue** (spirit soluble)

Acid. **Acid fuchsine, methyl blue,** aniline blue WS, **light green**

HAEMATEIN 172

Acid. **Haematein**

ANTHRAQUINONOID 175

Acid. Alizarine, alizarine red S, purpurine, Kernechtrot, **carminic acid**

	page
XANTHENE	178
Basic. **Pyronine G**	
Acid. **Eosin Y,** erythrosine B, phloxine, fast acid violet A2R	
AZINE AND RELATED DYES (quinone-imine dyes)	179
Oxazine	180
Basic. Brilliant cresyl blue, **Nile blue A,** gallamine blue, **coelestine blue,** gallocyanine. (The three last-named are partly acidic.)	
Thiazine	180
Basic. **Thionine,** azure C, **azure A, azure B, methylene blue,** new methylene blue, methylene green, **toluidine blue**	
Azine	181
Basic. **Neutral red,** safranine O, mauveine, amethyst violet, **Janus green B** (contains also an azo chromophore)	
Acid. Azocarmine G, induline (nigrosine)	

THE TRIARYLMETHANE DYES

The basic triarylmethane dyes, like basic dyes in general, are used for colouring chromatin. Crystal violet is one of the best dyes for chromosomes (p. 225). Methyl green is particularly valuable in mixtures with the basic xanthene dye, pyronine G, because it is possible to arrange the proportions in such a way that DNA is coloured green (or blue), and RNA red (p. 230). Dahlia is a useful vital dye for various cytoplasmic inclusions.

Basic fuchsine can be converted by sulphurous acid to a colourless substance that becomes coloured in the presence of aldehydes. This forms the basis of certain important histochemical tests (p. 308).

Acid fuchsine is one of the best dyes for mitochondria (p. 241), and it can also be used for the differential colouring of collagen, though other acid dyes of the same group—methyl blue, aniline blue WS, and light green—are preferable for this purpose. Acid fuchsine and light green are also available as background dyes.

HAEMATEIN

This single acid dye has no relative that is commonly used in microtechnique, but it is so important that rather a full description is necessary. Indeed, it is probably the most useful of all dyes in microtechnique. It is used with intermediaries (mordants) between it and the tissues (p. 207). The importance of haematein derives from the variety of different objects that can be dyed by it, the ease with which dyeing can be controlled, the insolubility of the colour in neutral aqueous and alcoholic media after dyeing, and the possibility of obtaining a jet black that is permanent in Canada balsam (such a black being very convenient in photomicrography).

Haematoxylon campechianum Linn. is a small, spreading tree with crooked branches (fig. 21, A), thorny when young and remarkable for the gnarled appearance of the stem of old specimens (fig. 21, C). The stem attains a circumference of about two feet. It is a leguminous plant, belonging with the tamarind to the group Caesalpiniaceae, in which the sepals are nearly or quite separate, instead of being fused as in the pea and its relatives. The pods and heart-shaped leaflets are shown in fig. 21, B. For a fuller botanical description see Bentley and Trimen.[60]

The sap-wood is white, but the heart-wood red. The latter is

"... what Campeachy's disputable shore
Copious affords to tinge the thirsty web."
(From *Dyers' Fleece*, quoted by Bancroft.[40])

The dyeing property of the heart-wood was known to the natives of Campeche before the arrival of Europeans. The Spaniards brought the wood to Europe soon after the discovery of America. It seems to have been brought to England early in the reign of Queen Elizabeth. Strangely enough, the dye was thought to fade and a law prohibiting its use, under severe penalties, was in force for nearly a century.[40] Its virtues were eventually recognized, however, and the tree was introduced in 1715 into Jamaica, where it is still cultivated at the present day; for haematein still survives the severe competition of modern synthetic products in the dyeing of black on wool, silk, leather, and nylon. Extracts of logwood were first used in microtechnique in the eighteen-forties.[409]

FIG. 21. *Haematoxylon campechianum.*

A, a small, much-branched specimen in the Peradeniya Botanic Garden, Ceylon. B, leaves, flowers, and fruit. The side of the square photograph represents 12 inches. C, part of the trunk of a mature specimen in the Museum of the Royal Botanic Gardens, Kew. The white strip of cardboard is 10 cm long.

(A and B are photographs by Mr T. B. Worthington, 9th Jan. 1956; C, by the author, is reproduced by kind permission of the Director of the Gardens.

The tree is commonly felled at the age of about ten years. The bark and sap-wood are chipped off and the heart-wood exported as 'logwood' in pieces three feet long. These logs are reduced to chips for the extraction of the substance, *haematoxylin*, from which the dye originates. Haematoxylin itself is a colourless solid, but if

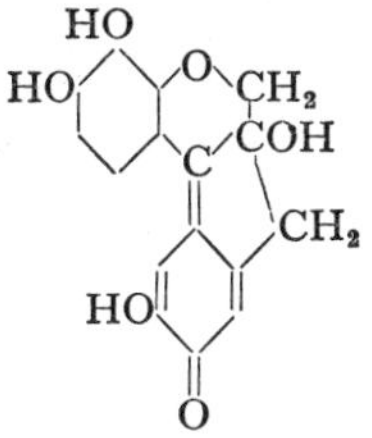

Haematein

damp becomes oxidized by atmospheric oxygen to the reddish dye, *haematein*. It is partial oxidation that gives colour to logwood. Products in which various proportions of the parent substance have been oxidized to the dye are available for use in the textile industry. The biological significance of the presence of haematoxylin in the wood is unknown.

The second *e* in the word haematein (four syllables) distinguishes this dye from an entirely unrelated coloured substance, *haematin*, the non-protein component of haemoglobin. The word hematine is used in the textile industry for partly or wholly oxidized haematoxylin.[76]

Haematoxylin is the leuco-counterpart of haematein. The ring represented in the formula as quinonoid in haematein is non-quinonoid in haematoxylin. The oxygen attached to the ring that is shown as quinonoid in haematein is replaced by –OH in haematoxylin. The carbon atom attached by a double bond to the upper end of this ring in haematein has one valency free in haematoxylin, and this is satisfied by hydrogen. In all other respects the formulae for the two substances are the same. Haematoxylin crystallizes with three molecules of water.[210]

Many different ways of oxidizing haematoxylin in the laboratory have been suggested.[513, 210, 272, 529] Sodium iodate is as convenient an oxidizer as any. It requires 0·187 g of this to oxidize 1 g of haematoxylin crystals fully.[210]

One might think it best always to buy haematein and make one's solutions from this, but in fact haematein solutions lose their strength rather quickly by flocculation of the products of further oxidation, and it is therefore generally best to have some haematoxylin present in addition to the haematein, to replenish the dye gradually through oxidation by atmospheric oxygen. Solutions of haematoxylin are often allowed to 'ripen' gradually. Long ago Unna [513] used to stabilize his solutions after a time by the addition of sulphur, which was presumably changed in part to hydrogen sulphide and thus acted as a reducing agent; when he wanted a strong haematein solution, he fully oxidized a part of the stabilized solution. A simpler plan is to oxidize a haematoxylin solution only partly, by the use of less sodium iodate than would be necessary for full oxidation.[39] One half or one quarter of the full amount of the oxidizer is suitable. The solution is ready for use directly it has been made up, and maintains its strength for a long time by gradual 'ripening' of the part of the haematoxylin that was not oxidized by iodate. It is desirable to start with wholly unoxidized (white) haematoxylin.

The products of further oxidation diffuse more slowly into tissues than haematein, work less well with mordants, and tend to precipitate.[455] It is doubtful whether definite stages of further oxidation can be distinguished, and such terms as *Trioxyhämatein* [210] are not acceptable without more evidence than we possess.

If one makes a simple aqueous solution of haematein, one has an acid dye of a dirty reddish colour. It will be guessed from the chemical formula that this substance is in fact of weakly acidic nature, like the related substance, catechol. Most acid dyes, as we

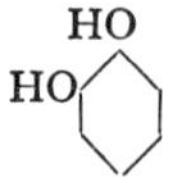

Catechol

have seen, are salts, usually of sodium, but in haematoxylin and a few others (carminic acid and picric acid among them) the metal is replaced by hydrogen. For clear contrast with other colours one requires a dye with rather a narrow absorption band, and there are so many suitable red acid dyes that haematein is never used in this form.

As we shall see (p. 193), certain acid dyes become basic when rather strongly acidified, but here again haematein is useless, for a different reason. At the pH at which it becomes a basic dye, it is orange, and for reasons that will be mentioned (p. 229), we seldom use yellowish basic dyes in microtechnique. Thus haematein is not used alone. In conjunction with mordants, however, it gives the blue, blue-black, and black colorations that make it pre-eminent among the dyes used in microtechnique. This subject of mordants, however, is so complex that a separate chapter must be devoted to it (p. 207).

THE ANTHRAQUINONOID DYES

The simplest dye in this group is alizarine, the chief coloured constituent of madder. This plant-product has been used for dyeing since ancient times, especially in the production of 'Turkey-red'. The synthesis of alizarine in the laboratory in 1869 gradually put an end to the cultivation of the plant and led to the synthesis of a number of related compounds, some of which are important in the textile industry. The 1 : 2 : 4-trihydroxy-compound is *purpurine*, which occurs also in madder. Alizarine is too insoluble

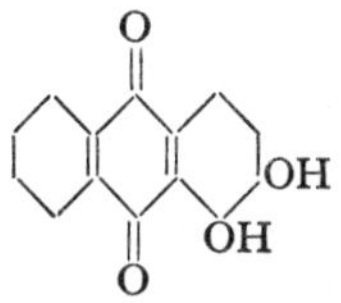

Alizarine

for convenient use in microtechnique, but purpurine, though only sparingly soluble in water and alcohol, is used in histochemical tests for calcium. *Kernechtrot* or 'calcium red' is an anthraquinonoid dye (or mixture of dyes) of unstated composition, soluble at about 0·25% in water; it can conveniently be used for the same purpose.[435] A soluble sodium sulphonate, alizarine red S, can be made from alizarine almost exactly as acid fuchsine is made from basic fuchsine; it is used for staining chromatin, with a mordant, in Benda's [55, 56] method for mitochondria.

Of far more general use in biology than any of these is *carminic acid*, a dye of quite special interest for several reasons. It is more protean than any other colouring agent, for it can be used as a direct basic dye, is of great value as an acid dye used with a mordant, and is also capable of being taken up directly as an acid dye and afterwards changed in the tissues to a basic one. The mordanted dye is excellent for chromatin, remarkably permanent in Canada balsam, and particularly suited to the colouring of whole mounts.

Carminic acid is the only dye used in microtechnique that is of animal origin. It is obtained from the scale-insect, *Dactylopius cacti*. The female of this animal is wingless and lives on the succulent plant, *Nopalea coccinellifera* (fig. 22), a native of Central America. Like other scale-insects, she produces a whitish waxy material from the surface of her body.[80] The male is winged and contains only a little of the colouring matter, but the latter is so abundant in the females that it constitutes about 10% of their dry weight.[332] The Central American Indians cultivated the animals for the sake of the dye from remote times. The plant on which it lives is nowadays grown in the Canary Islands, North Africa, and elsewhere, as food for the insect. The dye has never been synthesized. It is less used in the textile industry than formerly, on account of the competition of synthetic products, but it is still used for uniforms and hunting 'pink'.[75]

On opening the body-cavity of the fresh female, it can at once be seen that the colour is in the lobulated fat-body.[333] A red pigment is contained in globules situated near the periphery of the cells. These globules give reactions to metals that are characteristic of carminic acid. The male has as much fat-body as the female, but there are only a few red globules scattered here and there in it. The yolk of the ripe egg also contains red droplets, and the colour is carried over into the embryo and then develops anew in the fat-body. No other part of the animal contains the dye. The contents of the alimentary canal are not red. It is clear that the animal synthesizes the substance, but for what purpose is not known (conceivably to neutralize some poisonous constituent of the sap of the plant, which forms its only food).

The dried females constitute *cochineal*. To extract carminic acid from this, the material is powdered and boiled in water, and the fluid then filtered. On the addition of lead acetate, a dark red-violet precipitate is formed. This is the lead salt of the dye. It is dried and ground up with strong ethanol. On the addition of concentrated sulphuric acid, lead sulphate is formed and carminic acid dissolves in the alcohol. The yellowish red solution is evaporated at a moderate temperature. (Heat alters carminic acid. It bebecomes amorphous and eventually insoluble in alcohol.) The material is purified by repeated extraction with benzene or other suitable solvents of contaminating substances, and finally allowed to crystallize slowly from ethanol.[450] The pure acid consists of red, prismatic crystals.

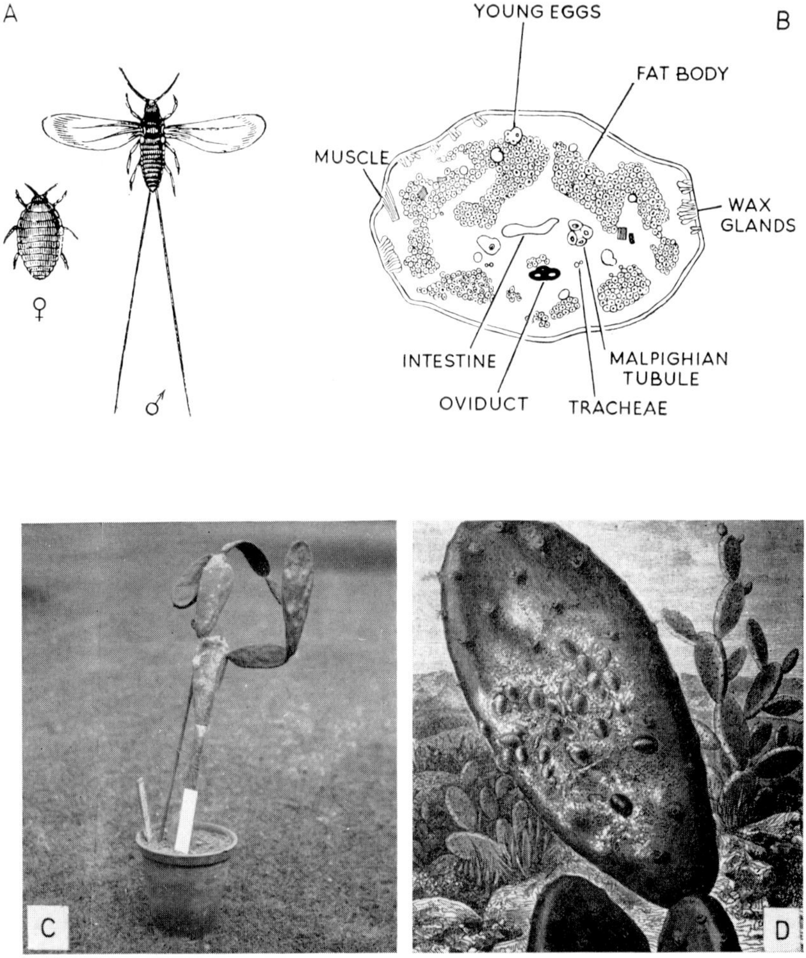

FIG. 22. The cochineal insect and its food-plant.

A, *Dactylopius cacti*, female and male. B, transverse section through the abdomen of a female. C, *Nopalea coccinellifera*, a specimen in the Oxford Botanic Gardens. The white strip of cardboard is 10 cm long. D, female *D. cacti* attached to *N. coccinellifera*.
(A from Hartwig;[217] B from Mayer;[333] C, photograph by the author, reproduced by permission of Prof. C. D. Darlington; D from Blanchard.[80])

Carminic acid is a complex substance. It was shown long ago [145] that treatment with hot 50% sulphuric acid splits off a side-group containing six carbon atoms; that subsequent heating to 170° C in a mixture of sulphuric acid with one-third its volume of water splits off a carboxyl group; and that a methyl-trioxyanthraquinone

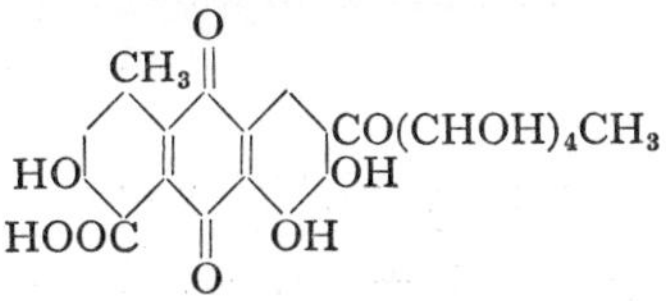

Carminic acid

is left. The six-carbon side-group is a methyl-pentose sugar, which, when free, has the formula $C_5H_9O_5.CH_3$.[351] This is an amorphous, bitter-sweet, honey-coloured substance. It is claimed that carminic acid should not be regarded as a glucoside, because the linkage of the sugar with the anthraquinone is through a carbon atom of the sugar, not an oxygen atom.[146]

A crude form of carminic acid is often obtained by extracting cochineal with water and precipitating with alum. The product, *carmine*, contains only about 56% of carminic acid, with a high proportion of protein; also some aluminium and calcium combined with a part of the carminic acid, and other substances in smaller amounts.[332] Carmine has the advantage of being considerably cheaper than pure carminic acid. The impurities render it almost insoluble in distilled water. It is suitable for use in ordinary microtechnique, but should not be used when one wants to know exactly what one is doing.

Alcoholic solutions of cochineal were used in the study of plant stems so early as 1770,[232] and a few years later Baron von Gleichen [194] was feeding ciliates on carmine, in an investigation of their method of nutrition. Carmine was first used as a dye in microtechnique in 1849, by Goeppert and Cohn.[195] In the following years its use was repeatedly rediscovered, and the general recognition of the value of dyeing in biological microtechnique was due to researches made with this substance.[29]

Pure carminic acid was first used in microtechnique in 1884,[144] but it was Mayer [332] who gave it popularity by introducing some excellent dye-mixtures containing it.

Carminic acid is soluble in both water and ethanol. It is a fairly strong acid, capable of setting free carbon dioxide from marble.

It forms blueish red or violet salts with metals. The salts with the heavy metals are insoluble, but those with the alkali metals are readily soluble, and indeed carmine itself is soluble in solutions of borax or ammonia.

The use of carminic acid (or carmine) as a basic dye, and as an acid dye converted in the tissues to a basic one, is described on p. 193. For its use with mordants, see chapter 11 (p. 207).

THE XANTHENE DYES

These are to be regarded as derivatives of xanthene, which itself exists in the form of colourless leaflets. A few dyes related to this substance occur in nature, including a yellow one extracted in

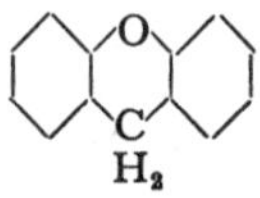

Xanthene

India from the urine of cows that have fed on mango leaves; but the xanthene dyes in the strict sense are synthetic products. A para-quinonoid ring gives colour; an oxygen atom participates in the

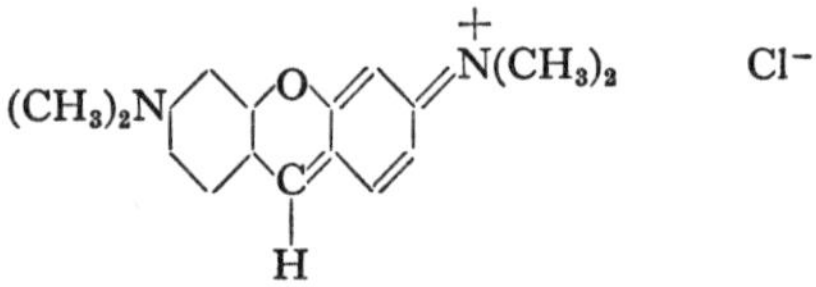

Pyronine G

linking of two rings. The simplest of those that are useful in microtechnique is the basic dye pyronine G, which, as has already been mentioned (p. 171), is used with methyl green in histochemical tests designed to distinguish DNA from RNA.

A far more familiar dye in the laboratory, and indeed one of the best known of all that are used in microtechnique, is eosin Y. This is unusual among synthetic acid dyes in not being a sulphonate. The auxochromes, as the formula shows, are $-C\genfrac{}{}{0pt}{}{\nearrow O}{\searrow O^{-}}$ and $-O^-$. It is used chiefly as a background dye. Its yellowish red colour and small capacity to overstain fit it rather well for this function, especially when blue dyes such as aluminium-haematein (p. 215) are used for chromatin. A special advantage of eosin is that the dye is able to penetrate red blood-corpuscles and other close-textured components of tissues. This is so despite the large size of the dyeing ion. The capacity to penetrate well must be ascribed to the tend-

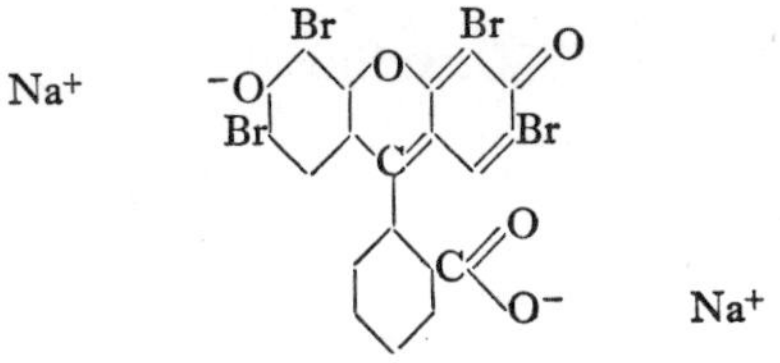

Eosin Y

ency of the ions to remain separate instead of aggregating (p. 236). Eosin Y can be made to stain red blood-corpuscles most powerfully, and show them up so vividly that the course of the capillaries can sometimes be traced as easily as in an injected specimen. Mann's methyl blue/eosin (the long method) is particularly well adapted to this purpose.[322]

Phloxine and erythrosine B are background dyes closely resembling eosin Y. The former has chlorine atoms as well as bromine; erythrosine B is exactly the same as eosin Y except that four iodine atoms are substituted for the four bromines. In both of these, especially phloxine, the greatest absorption of light is shifted towards the longer wavelength, with a resulting bluer (less yellow) shade.

AZINE AND RELATED DYES
(quinone-imines)

These dyes include some of the most valuable in microtechnique. The chromophore is a quinonoid ring associated with a substituted imino-group (HN=). They may be regarded as derived from quinone-imine by the substitution of an aryl ring for the H of the imino-group, and they are therefore sometimes called quinone-imine dyes. In the simplest of these the two rings are connected

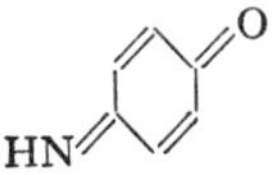

Quinone-imine

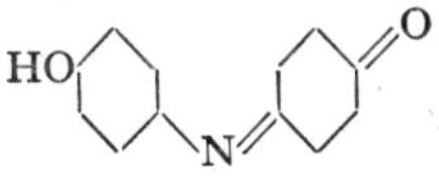

Indophenol

by a single link only (–N=). The indophenol dyes, which are briefly mentioned on p. 288, are of this nature. In all the quinone-imine dyes that are important in microtechnique, however, there is a second link between the two rings, through an atom that may be oxygen, sulphur, or nitrogen. The three groups of dyes are respectively the oxazines, thiazines, and azines.

Oxazines

The link through oxygen will be noted in the structural formula for the basic vital dye, brilliant cresyl blue. Among the small group of oxazine dyes used in microtechnique, a high proportion

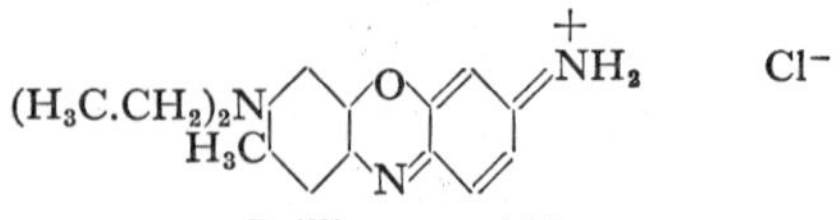

Brilliant cresyl blue

can be used vitally. This applies to Nile blue A, which is also used in a roundabout way in histochemistry for the distinction between certain groups of lipids (p. 301).

Several of the oxazine dyes, coelestine blue B among them, can be used with a mordant (p. 215).

Thiazines

In these dyes, several of which are among the most important used in microtechnique, the second part of the link between the

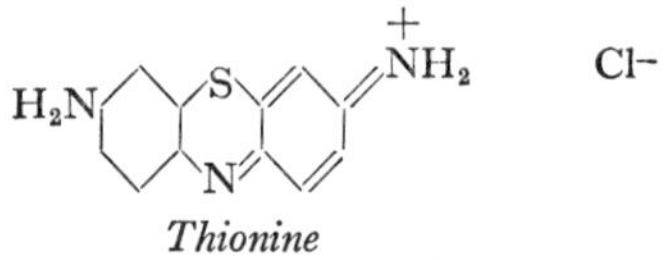

Thionine

two rings is formed by an atom of sulphur. The simplest is thionine, a violet dye. The thiazine dyes used in microtechnique are all basic.

The four hydrogens attached to nitrogen can be replaced by methyl groups. One is so replaced in azure C, two in azure A, three in azure B, all four in methylene blue (see p. 268). As the replacement occurs, so the absorption maximum shifts towards the longer wave-lengths, and methylene blue is a pure blue dye. Toluidine blue is closely related to azure A.

The thiazine dyes are useful for staining chromatin. Their chief virtue, however, is that they have a particularly strong tendency towards metachromasy: that is to say, towards the dyeing of different tissue-constituents in different colours. This subject is discussed elsewhere (p. 243). It is their metachromatic property that makes these dyes so important in the staining of blood-corpuscles (p. 268).

Methylene blue is a valuable vital dye (p. 287).

Azines

The structural formulae for all the diverse dyes that have been mentioned so far can be written down so as to include a para-quinonoid ring, and this has been done. In some dyes, however, there may in fact be resonance between the paraquinonoid and orthoquinonoid configurations, and for the azine dyes it becomes

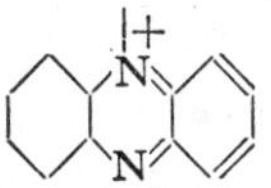

Skeleton-formula for azines

necessary to write orthoquinonoid formulae. In this group an imino-nitrogen once more forms one of the links between two rings, but, in contrast to the oxazines and thiazines, the second link is also formed by a nitrogen atom. The skeleton-formula shown here is satisfactory for most of the azine dyes. Most of the ones

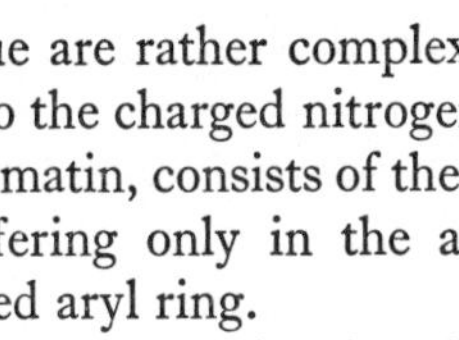

Safranine

used in microtechnique are rather complex. The safranines have an aryl ring attached to the charged nitrogen atom. Safranine O, a useful red dye for chromatin, consists of the dye shown here mixed with another dye differing only in the absence of the methyl group from the attached aryl ring.

Two other basic azines, neutral red and Janus green B, are of outstanding importance as vital dyes. The former, a dull red, is an exceptionally innocuous dye that colours certain kinds of globules in the cytoplasm. It turns yellow on the alkaline side of neutrality, and one can thus learn something of the acidity or alkalinity of cytoplasmic inclusions by using it. Janus green B, a very complex dye that contains also an azo chromophore, has been much used in the study of mitochondria (p. 292).

A few acid dyes of this group are important for the production of navy blue on wool. Only two or three are used in microtechnique. Azocarmine G, red, is exceptional among acid dyes in being used for colouring chromatin (p. 205). Induline WS, blue-black, is useful for collagen. This is an acid dye related to safranine, but with the hydrogens of the amino-groups replaced by phenyl rings. Usable specimens, blended with other dyes to produce a blacker colour, are sometimes sold as nigrosine, and this name occurs frequently in the older literature of microtechnique. There is unfortunately considerable variation among dyes sold as nigrosine, and some of them are unsatisfactory as collagen-dyes.

THE AZO DYES

We now leave quinonoid rings behind us and pass to an entirely different chromophore, –N=N–. The azo dyes nowadays occupy a predominant place in the textile industry. A huge variety of reds, oranges, and yellows is available, as well as other colours. The number of acid azo dyes suitable for the direct colouring of wool

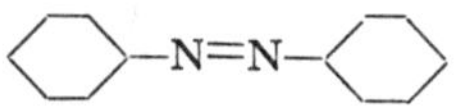

Skeleton formula for azo dyes

exceeds that of all such acid dyes of other groups put together. Many azo dyes can be used with mordants, and about three-quarters of all the dyes that are used with mordants in the textile industry are of the azo-group. The number of azo-dyes that are useful in microtechnique is, however, rather limited.

It is a remarkable fact that no living organism produces an azo dye. They are made synthetically by the action of nitrous acid or nitrites on amino-derivatives of benzene. If aniline hydrochloride is treated with sodium nitrite, a nitrogen atom leaves the sodium

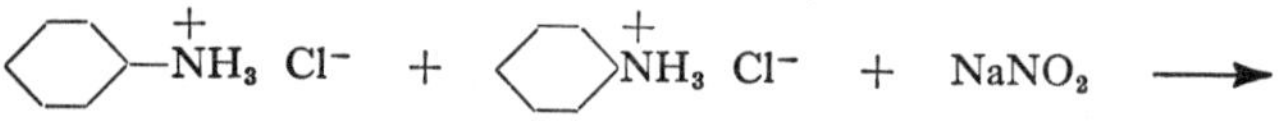

Aniline hydrochloride *Aniline hydrochloride* *Sodium nitrite*

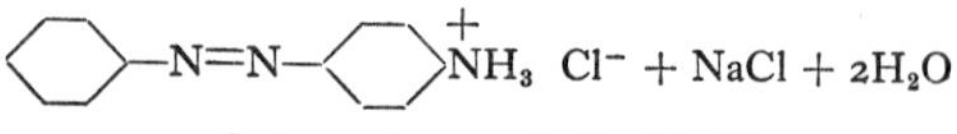

Aminoazobenzene hydrochloride

nitrite and associates through a double bond with a nitrogen atom of an amino-group, thus introducing itself as a link between two rings. The resulting ion gives a yellow basic azo dye.

Dyes that contain a single –N=N– group are called mono-azo; two, disazo; and three, trisazo. The azo dyes mentioned in this book are listed below, for ease of reference. The ones that are particularly important in microtechnique are named in heavy type.

MONO-AZO

Basic. **Janus green B** (contains also an azine chromophore)
Acid. **Orange G,** chromotrope 2R, ponceau 2R (= xylidine red), Bordeaux red, metachrome yellow RA, tropaeolin, methyl orange

DISAZO

Basic. Bismarck brown Y
Acid. Congo red, Congo rubin, **trypan blue**

TRISAZO

Acid. Chlorazol black E

MONO-AZO DYES

Several acid mono-azo dyes, of the yellow, orange, or red colour that is so usual among azo compounds, are useful for background coloration. Examples are orange G, chromotrope 2R, and ponceau 2R (= xylidine red). These three dyes are closely related. Orange

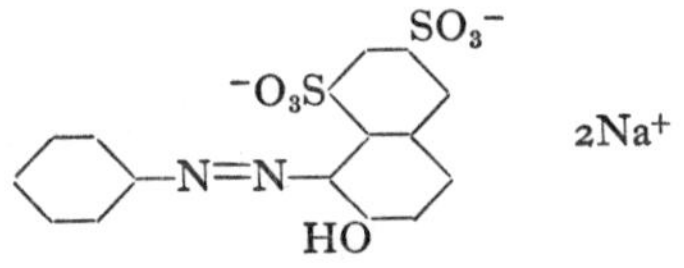

Orange G

G is of a colour near the middle of the visible spectrum that contrasts well with the familiar blue dyes for chromatin, and has the further advantage that it is very unobtrusive: that is to say, it has little tendency to overstain or to apply itself to the objects that are coloured by basic dyes.

Metachrome yellow RA is a simple azo dye, remarkable in more than one respect. It is an acid dye that has carboxyl and hydroxyl groups for its auxochromes, and also possesses a nitro-group; it is scarcely soluble in water, and therefore lends itself, unlike other acid dyes, to the background staining of specimens that will be mounted in aqueous media.

The only important basic mono-azo dye that is important in microtechnique is Janus green B, which, since it contains an azine as well as an azo chromophore, has already been mentioned (p. 181).

DISAZO DYES

Bismarck brown has the distinction of being the first synthetic vital dye that was ever used (p. 274). The colour is distinctly unusual in microtechnique. For most purposes it is desirable to use dyes that have either a rather sharply marked absorption-band, so that strong contrasts with other colours can be arranged, or else a general absorption throughout the visible spectrum to give a black that will contrast with an unstained or lightly stained background. Thus browns and yellowish browns are seldom chosen. Bismarck brown Y is a basic dye, convenient for colouring lipid cytoplasmic inclusions during life.

Trypan blue and its relatives are acid dyes that are used in a very special way in a particular kind of vital work (p. 276).

Several coloured disazo compounds that are not dyes, because they do not ionize, are very useful in microtechnique for colouring lipids (p. 299).

TRISAZO DYES

The only trisazo dye that has found favour in microtechnique is chlorazol black.[102] It is an acid dye with some basic tendency.

THE NITRO DYES

The third and last main chromophore with which we shall be concerned in this book is the nitro-group, $-NO_2$. Colour is due to resonance between two possible positions of the negative electric charge. Nitrobenzene, like other substances containing $-NO_2$, is

Nitrobenzene in two resonance positions

coloured. It is a yellow liquid, but lacks an auxochrome and so is not a dye. Trinitrobenzene possesses three chromophores but is not a dye. If phenol be mixed with concentrated nitric acid, however, a yellow crystalline substance is formed, which, because it contains both $-NO_2$ (three times) and the auxochrome $-OH$, is a dye. It is trinitrophenol or picric acid, the only substance that is used both as a fixative (p. 96) and a dye. It is one of those acid dyes that are actually acids, like carminic acid, for instance. Its

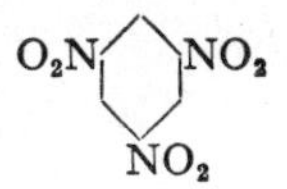

Trinitrobenzene

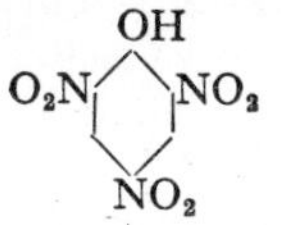

Picric acid

salts are seldom used in microtechnique. It is a far stronger acid than phenol, and this to some extent militates against its use as a background dye, for which its pale yellow colour otherwise fits it; for it has a tendency to remove basic dyes. Its acidity, however, is not harmful to the action of other acid dyes, and it is useful in such mixtures as picro-nigrosine (p. 236).

The only other nitro-dyes that are at all frequently used in microtechnique are naphthol yellow and aurantia. The latter is the background dye in Kull's [275] method for mitochondria.

Orcein is a dye of considerable interest. It has long been used for showing elastin (p. 233) and latterly has come into favour in techniques for showing chromosomes in smeared preparations. It cannot be included in the classification of dyes, because its structural formula is unknown.

Since ancient times dyes have been prepared from lichens. The modern usage, however, dates from the fourteenth century, when a Florentine merchant began to make them on a large scale. *Rocella*, the genus most commonly used in the preparation of orcein, commemorates the name of his family. The lichens of this genus, commonly called orchil- or archil-weed, grow chiefly on rocks near the sea-shore in the warmer parts of the world. They form tufts of bluish-grey or whitish strap-shaped fronds, up to 6 inches long. Several species are usable as raw materials for the preparation of orcein. In Scandinavia and other cool climates another lichen, *Lecanora tartarea*, replaces *Rocella*, and indeed many other species can be used.[477] *Lecanora* is often called cudbear, but the familiar names are loosely used.

These various lichens contain lecanoric acid, which splits to produce orcinol when the plants are boiled with water.[391] It will be noticed that orcinol is resorcinol with a methyl group attached. It crystallizes as colourless, hexagonal prisms, freely soluble in water. The substance can also be prepared synthetically.[128]

In the presence of ammonia and atmospheric oxygen, orcinol becomes transformed into orcein, which precipitates. Stale urine was formerly used as a source of ammonia. Orcein is sold as a fawn-coloured powder, soluble in alkaline or alcoholic solution or in

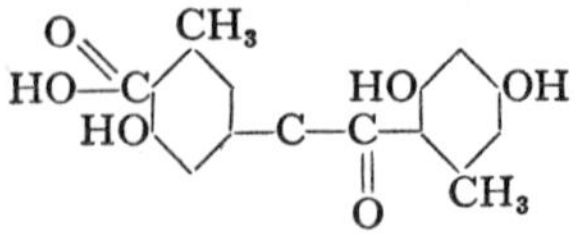

Lecanoric acid

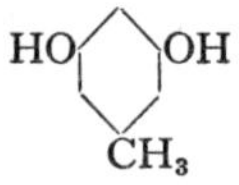

Orcinol

acetic acid. It is used for dyeing wool and silk directly (that is, without the use of mordants). It gives dull magenta shades, but is often used with other dyes to give browns.[270]

The formula for orcein is probably $C_{28}H_{24}N_2O_7$. It has been suggested that the two nitrogen atoms may form the chromophore of an azine dye.[559]

Litmus is produced from the same lichens as orcein by a closely similar process, but potassium carbonate is required in addition to atmospheric oxygen and ammonia.[391, 420]

The indigo-dyes are briefly mentioned on p. 307, the acridine dyes on p. 310.

CHAPTER 10

The Direct Attachment of Dyes to Tissues

An insoluble pigment requires the addition of an adhesive if it is to remain in position when applied to any object. A dye is presented in solution without the addition of any adhesive; it is not itself sticky or obviously adhesive; yet it adheres. It is the purpose of this chapter to explain the reason for this.

It has already been mentioned (p. 172) that certain dyes are used with intermediaries between themselves and the tissues. These intermediaries or mordants are not adhesive in any ordinary sense. Chapter 11 (p. 207) deals with the use of these substances.

The nature of the process of dyeing has been studied more elaborately by the research-workers of the textile industry than by those who use dyes in microtechnique. This is due partly to the fact that larger funds are available in industry and partly to the relative homogeneity of the textile fibres—especially cotton—in comparison with the kinds of tissues usually studied by biologists. It must be confessed also that the practical dyer has generally adopted a more scientific approach to his work than the histologist and cytologist. As Mann [322] unkindly remarked:—

> 'The method of staining, once having taken root in the animal histologist, grew and grew, till to be an histologist became practically synonymous with being a dyer, with this difference, that the professional dyer knew what he was about, while the histologist with few exceptions did not know, nor does he to the present day.'

The realization of this truth is helpful to the biologist, but it is necessary to point out another truth that has been overlooked, namely, that there are many very important differences between the dyeing of textile fibres and the dyeing of tissues in biological microtechnique. This is true although some of the most important textile fibres are those of plant and animal origin. The chief differences are tabulated here.

Textile dyeing	*Dyeing in microtechnique*
Cotton has been especially investig ted, because it is so homogeneous chemically. It is extremely peculiar, because it is a negatively charged object ordinarily dyed by acid dyes.	The dyeing of cotton is of little interest except to those studying the cell-walls of plants. The biologist does not ordinarily dye negatively charged objects with acid dyes.
Whether, in particular circumstances, a dye is acting as a basic or an acid dye may not be known (p. 209).	A glance down the microscope at a dyed preparation usually shows whether the dye used was basic or acid.
Dyes are generally used at or near the boiling point of water.	Dyes are usually used at room temperature.
The dye-bath is usually exhausted or nearly so.	The tissue takes up only a minute part of the dye in the dye-bath.
Almost perfect fastness to water is generally required.	Fastness to water is unnecessary, as one can quickly transfer the tissue to some other medium in which the dye is insoluble.
The fibres—of cellulose or special proteins, or synthetic—are non-living and unfixed.	The objects dyed are generally either fixed or alive (p. 274).
No differential dyeing at the microscopical level is required, and there is no process of differential extraction of the dye.	The whole purpose is differential dyeing at the microscopical level; it is often achieved by differential extraction.
Anionic chromium is used to mordant for azo-dyes; other mordants and mordant-dyes are seldom used nowadays.	Iron, aluminium, and cationic chromium are used to mordant for haematein, carmine, and certain oxazine dyes (see p. 207).

In this book every effort will be made to profit from the valuable researches of the textile chemists, and it is necessary to make special acknowledgement of the admirable presentations of this subject by Vickerstaff,[524] Bird,[75] and Venkataraman; [523] but the object throughout will be to concentrate attention on the use of dyes in microtechnique.

In microtechnique we are primarily concerned with the electric charges on the dye-ions and on the objects dyed.

The electric charges on dye-ions are investigated by subjecting dye-solutions to the action of an electric current. Cataphoretic

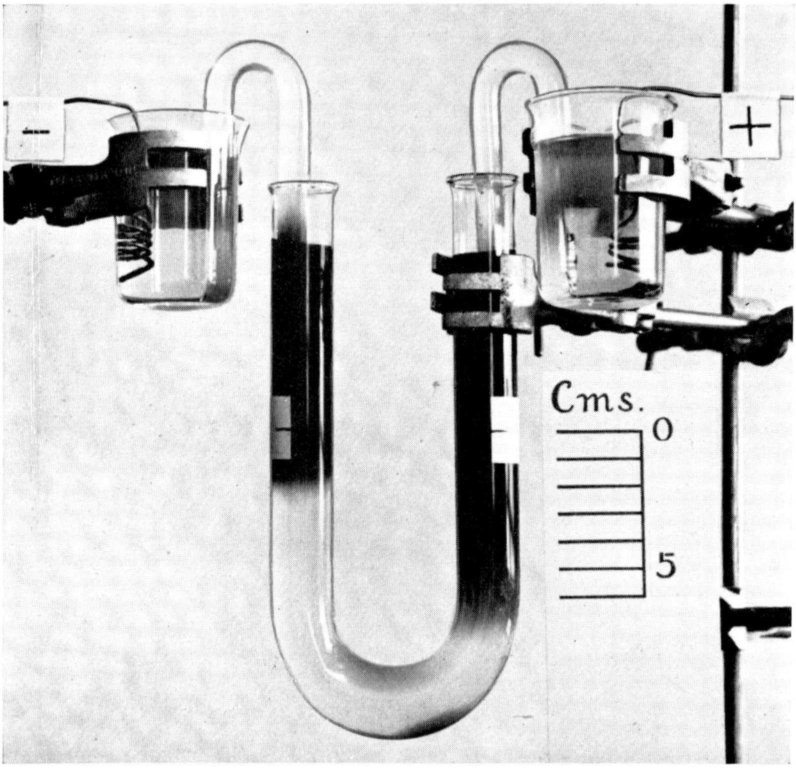

FIG. 23. Apparatus for cataphoretic experiments with dyes.[38] The wires from a 12-volt accumulator are marked − and +. Methylene green (a basic dye) is being tested. The current has been passing for 42 hours. For full description see accompanying text and Appendix (p. 321).

(Photograph by Mr P. L. Small)

experiments of this sort have been carried out especially by Seki,[452] a Japanese investigator who has played a particularly important rôle in the scientific study of microtechnical dyeing. An apparatus resembling Seki's is shown in fig. 23; practical instructions for setting it up are given in the Appendix (p. 321).

The main part of the apparatus is the large U-tube in the middle of the photograph. This contains an aqueous agar gel. It fills the tube up to the level marked 'O' on the cardboard scale fixed beside the tube: the level is also marked by a line on a label stuck on to each limb of the tube. The gel is buffered at a known pH. The dye-solution, buffered at the same pH, is poured into both limbs of the U-tube to the same height. In principle, one might now put the positive wire from an accumulator into the dye solution in one limb and the negative wire into the dye solution in the other, and the experiment would begin. If this were done, however, the electrolysis of water would take place, gas would be formed at the electrodes, and the latter would become depolarized. The rest of the apparatus exists solely to prevent this. It consists of two beakers into which dip electric wires (marked + and –), and two small U-tubes, each of which dips on one side into a beaker and on the other into the dye in the big U-tube. The current passes through agar gel in the small U-tubes. The contents of the beakers and U-tubes are given in the Appendix.

The electric current should be switched on as soon as the dye has been poured into both sides of the big U-tube. The current now tends to make the dye move towards either the negative or the positive pole: that is to say, to make it descend in one or the other of the two limbs of the U-tube. Simple diffusion also occurs, however, and this causes some descent on both sides. When basic dyes are used, there is also an attraction between the dye and the agar, which causes some descent on both sides. The much greater descent on one side than on the other shows that the dye moves in response to the current. In the figure the dye in question (methylene green) has moved in 42 hours about 6 cm towards the negative pole and only about 2 cm towards the positive. The dye ions are clearly positively charged or cationic, and indeed methylene green is a cationic or basic dye.

This apparatus enables us to find whether any dye is basic or acid at any particular pH. In general, any dye that is shown by its chemical formula to be basic will behave like methylene green, while any acid dye will move in the opposite direction. The speed with which the various dyes move varies considerably. Thus methylene green is one of the fastest, dahlia one of the slowest of the basic dyes.[452] This depends partly on the electric charge, partly on diffusibility.

Most dyes remain cationic or anionic, as the case may be, throughout the range of acidity and alkalinity within which dyeing ordinarily takes place in microtechnique: that is from about pH3 to 9. Examples of basic dyes that are typical in this respect are crystal violet and safranine, while orange G and picric acid are typical acid dyes.[458] Some dyes, however, are amphoteric, being cationic below a certain pH (the iso-electric point) and anionic above. Examples are lithium carminate and haematein, with iso-electric points about pH 4·5 and 6·6 respectively.[458] These facts facts can be expressed in a simple diagram.

	pH 3 4 5 6 7 8 9
Typical basic dye .	+ + + + + + + + + + + + + + + +
An amphoteric dye.	+ + + – – – – – – – – – – –
Typical acid dye .	– – – – – – – – – – – – – – – –

The majority of dyes fall within this scheme, though some are bleached by acidity and more by alkalinity, so that their behaviour cannot be studied cataphoretically at the ends of the pH range.

Electrically-charged groups occur also in the tissues, especially in the proteins and certain lipids. The most obvious sources of electric charges in the proteins are the $-\mathrm{C}\genfrac{}{}{0pt}{}{\diagup\!\!\diagup \mathrm{O}}{\diagdown \mathrm{OH}}$ and $-NH_2$ groups of certain amino-acid residues. A part of a protein chain at the iso-electric point is here represented by an example.

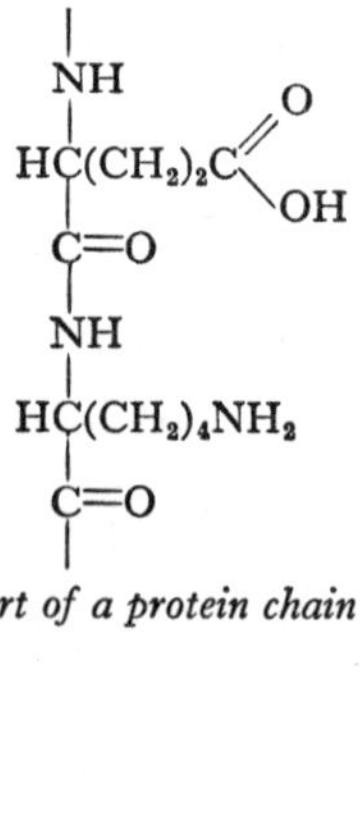

glutamic acid

lysine

Part of a protein chain at the iso-electric point

The degree of ionization of the carboxyl and amino-groups depends on the pH. When this is pushed to the acid side of the iso-electric point, the dissociation of the acid groups is suppressed and that of the amino-groups increased. The protein thus becomes progressively more positively charged as more and more of the amino-groups become ionized. They are all ionized when pH 2 is reached, or thereabouts.

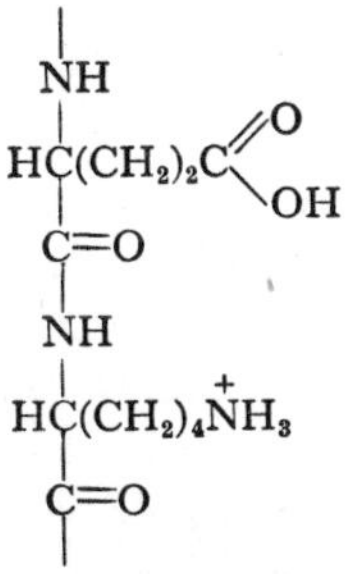

ionized amino-group

Part of a protein chain on the acid side of the iso-electric point

When the pH is on the less acid or more alkaline side of the iso-electric point, the dissociation of the amino-groups is suppressed, while that of the carboxyl groups is increased. Thus the protein becomes more negatively charged as the pH increases, until all the carboxyl groups are ionized at about pH 11.

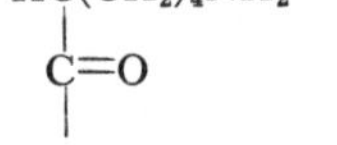

ionized carboxyl group

Part of a protein chain on the less acid or more alkaline side of the iso-electric point

Although the protein as a whole is neutral at the iso-electric point, yet a few ionized carboxyl and amino-groups still exist at this pH. This is important for dyeing.

The chief amino-acid side-groups that can give negative charges are aspartic, glutamic, and hydroxy-glutamic. The –OH group of tyrosine and serine can also be ionized, and so can the terminal carboxyl group of a protein chain.

The amino-acid side-groups that can give positive charges are lysine, arginine, and histidine, and the terminal amino-group of the chain is also available.

The majority of the amino-acids (glycine, alanine, leucine, phenylalanine, etc.) cannot provide electric charges unless they happen to lie at the end of a chain.

Since different proteins contain different proportions of acidic and basic amino-acids, the position of the iso-electric point varies.

The tissues provide a number of other charged groups beyond the amino-acid residues of proteins. Negative charges can occur on the phosphoric acid groups of nucleic acids and phospholipids, and on the uronic and sulphuric groups of mucopolysaccharides.

The various tissue-constituents that are visible under the microscope vary in their content of these positively and negatively charged groups. Some are preponderatingly positive or basic, others negative or acidic, others again amphoteric or easily swayed by changes in pH. Some characteristic examples are these:—

Acidic . .	DNA and chromatin
	RNA and ribonucleoprotein
	matrix of cartilage
	many mucous secretions
	most lipids other than triglycerides
Amphoteric .	cytoplasm of most cells
	contractile substance of muscle
Basic . .	collagen
	cytoplasm of red blood-corpuscles
	granules of eosinophil leucocytes
	nuclei of the spermatozoa of certain fishes.

The substances here listed as acidic and basic act as such within the range of pH at which dyeing usually takes place in microtechnique, but if the dye-solution be made sufficiently acid or alkaline it will be found that they are in fact amphoteric.

Since both dyes and tissue-constituents are electrically charged, it is natural that they react with one another. The acidic constituents have an affinity for basic dye-ions and are therefore called *basiphil* (see p. 329), while the basic attract acid dye-ions and are called *acidophil*. The fact that basic dyes show a 'most striking conformity with one another' in their reactions with tissue-constitu-

FIG. 24. Ehrlich at the age of 24. At about this age he discovered the fundamental differences between basic and acid dyes in their reactions with the constituent parts of cells, and also showed how the varying permeability of these parts aided differential dyeing.

(From Marquardt,[324] by kind permission of Messrs William Heinemann Medical Books, Ltd.)

ents was remarked by Ehrlich [152] in 1879. In the same year he noted [153] the affinity of acid dyes in general for the granules of what he had named 'eosinophil' leucocytes. In the following year he gave a short general statement of the differences between basic and acid dyes in their reactions with cell-constituents, illustrating his remarks by reference to leucocytes.[154] He noted that the granules of eosinophils have an affinity for acid dyes, of *Mastzellen* for basic dyes, and of polymorphs for both constituent parts of 'neutral' dyes (p. 262). These papers, published when Ehrlich was about 25 years old, mark a turning-point in the history of scientific microtechnique.

In preliminary studies of the reactions of dyes with electrically-charged objects it is convenient to use simple, homogeneous models to represent the tissue-constituents. The acidic and amphoteric constituents are the most worthy of attention, because every cell contains them. Seki [458] chose collodion as a model for the acidic components, gelatine gel for the amphoteric. These substances are very convenient, because they can so easily be cut in slices of uniform thickness, but collodion is more acidic than most tissue-constituents. The depth of colour may be recorded on an arbitrary scale, or determined by photometry. Practical instructions for some experiments resembling Seki's are given in the Appendix (p. 323).

The results with collodion (fig. 25) are simpler than those with gelatine because the former, when in water or aqueous solutions, maintains its negative charge throughout the relevant range of pH. It is strongly dyed by crystal violet or any other typical basic dye throughout the range. Typical acid dyes, however, scarcely tinge collodion, except in strongly acid solutions, in which the charge on the collodion is somewhat lessened.

An amphoteric dye behaves as one would expect on theoretical grounds. On the acid side of its iso-electric point it is positively charged and therefore dyes collodion strongly, but on the less acid side it becomes negatively charged and its tendency to colour collodion disappears about neutrality. These facts are well exemplified in the practical use of carminic acid, the iso-electric point of which is about pH 4·2. It is commonly used for dyeing chromatin in the form of aceto-carmine.[444] In this strongly acid solution it acts as a basic dye.[454, 458] In alkaline solution, however, in the form of borax-carmine,[199] it acts as an acid dye, but is usually converted subsequently into a basic one by treatment with acid.[454]

Orcein is another amphoteric dye,[457] much used in its basic state—that is, in strongly acid solution—for the dyeing of chromosomes.[279] Haematein is yet another amphoteric dye,[455] but it is not used as a direct dye in either its basic or acid character, as the colours it produces are feeble and indefinite. The way in which this important dye is used in practice will be explained in chapter 11 (p. 207).

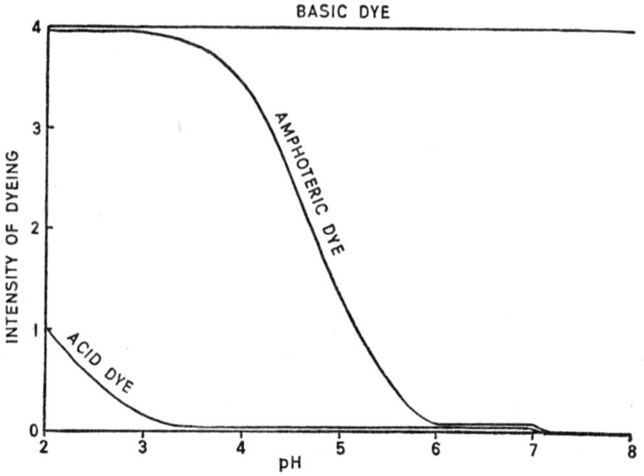

FIG. 25. Diagrammatic representation of the dyeing of collodion by typical basic, amphoteric, and acid dyes. The ordinate is divided into arbitrary units. In general conformity with the data of Seki.[458]

It was mentioned above that most tissue-constituents are shown to be amphoteric if tested towards the extremes of pH, and their reactions to dyes are therefore more complicated than those of collodion, which remains negatively charged even in strongly acid solution. More than half a century ago Bethe [72] tried the effects of adding varying amounts of sodium hydroxide or sulphuric acid to solutions of toluidine blue. He tried these solutions on various mammalian tissues fixed in alcohol. Whereas everything was colourable in alkaline solutions or at neutrality, there was a tendency for the various constituents to fail to take the dye as more and more acid solutions were used: one constituent after another dyed more feebly or failed to dye at all. Cytoplasm fell off rapidly in capacity to be coloured, chromatin much less rapidly and mucus less rapidly still, while the ground substance of cartilage remained as deeply dyed in the most acid solution tried as at neutrality.

Pischinger [401] developed Bethe's study by using both basic and acid dyes, instead of basic only. He showed that the effect of both basic and acid dyes tended to fall off sharply towards a particular pH, which he regarded as the iso-electric point of the object dyed. Thus an acid dye would colour a particular object in the tissues in strongly acid solutions, but would scarcely act above a certain pH. A basic dye would act strongly on the same object in alkaline solution, but the affinity between dye and object would fall and

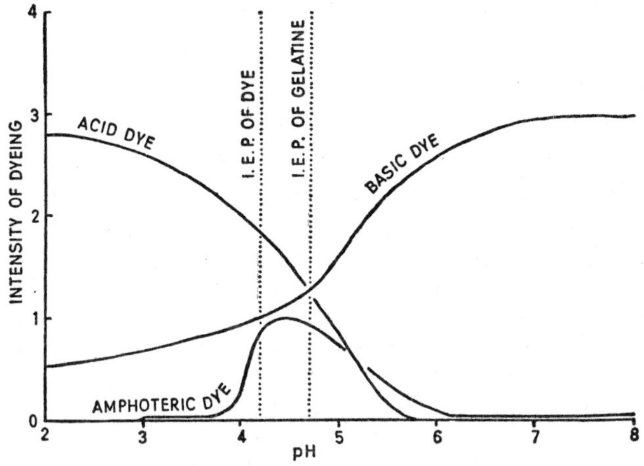

FIG. 26. Diagrammatic representation of the dyeing of gelatine by typical basic, amphoteric, and acid dyes. The ordinate is divided into arbitrary units. In general conformity with the data of Seki.[458] I.E.P., iso-electric point.

nearly disappear at about the same pH as that at which the acid dye failed to act. The position of the iso-electric points of tissue-constituents may indeed be roughly estimated by Pischinger's method, but different pairs of dyes do not give exactly the same results, and Lison [310] prefers to speak of the 'apparent iso-electric point' when referring to information gained in this way.

Instead of using a basic and an acid dye separately, one may mix them together and judge the position of the iso-electric point of any particular object in a microscopical preparation by noting the pH at which the mixed colour is given.[505] Methylene blue and eosin make a good pair for this purpose.

Methods of this sort help in the histochemical identification of certain tissue-constituents. Carboxyl groups soon lose their negative charge as the pH is lowered, then phosphoric groups, and the sulphuric groups of certain mucosubstances later still. These facts provide the explanation of Bethe's results.

Seki [458] chose sheet gelatine, untreated by any fixative, as a model for amphoteric tissue-constituents. The dyeing of gelatine is represented diagrammatically in fig. 26. The iso-electric point of this substance is about pH 4·7. Since gelatine is positively charged at lower pH than this, it has a strong affinity for ordinary acid dyes ('levelling' dyes, see p. 235) in strongly acid solution. In the same way wool has a strong affinity for these dyes at low pH. The affinity of gelatine or wool for ordinary acid dyes falls off rapidly as the pH rises above the iso-electric point, but now, as the charge has become negative, there is increasing affinity for typical basic dyes.

The curves representing the uptake of acid and basic dyes by proteins cross somewhere, and the place of crossing is at or near the iso-electric point of the protein. It might be thought that dyeing would be impossible beyond the iso-electric point, so that the curves for acid and basic dyes would reach the base-line here, instead of crossing. It must be remembered, however, that proteins are not without electric charges at the iso-electric point: such charges still exist, but the positive and negative ones balance one another.

Accurate figures for the amounts of a basic and an acid dye taken up by thin fibrin films have been obtained photometrically by Singer and Morrison.[473] Their results confirm the general correctness of the data provided by Seki, who used an arbitrary scale (+, ++, etc.) to represent the intensity of dyeing as judged visually.

The facts just recorded explain the common custom of using basic dyes in weakly acid solution. The pH is low enough to prevent the amphoteric cytoplasm from taking the dye, but not low enough to prevent the chromatin from being coloured strongly.

The complication is considerable when an amphoteric substance is treated with an amphoteric dye. If the iso-electric points of both were the same, there would be little affinity between them. If, however, the iso-electric points are different, there will be a narrow range in which the charge on the dye will be opposite to that on the protein; dyeing will occur within and to some extent also beyond this range (fig. 26).

After basic dyes have acted, there is generally little tendency to extraction by water at neutrality. Many acid dyes are less fast to water. (Metachrome yellow, an acid dye, is exceptional in being nearly insoluble in water, so that it could not easily escape even if the bond with the tissue were loosened.) Acids and alkalis release respectively basic and acid dyes, and once again the iso-electric point of a particular tissue-constituent will determine whether or how quickly it will release a dye. These facts form the basis of regressive dyeing. Instead of allowing the dye to colour the tissues progressively until the desired effect is obtained, one may over-stain and then extract the excess of dye differentially from the various tissue-constituents. The hydronium and hydroxide ions diffuse through the tissue much more rapidly than any dye-ion and therefore give more even results, especially if the piece of tissue be thick.

Although the basic dyes show more resistance to washing out by distilled water than acid ones, many of them are very quickly removed by the alcohols used in dehydration, which remove the acid dyes more slowly. Indeed, the dehydrating alcohols act like very weak acids.[486] Their powers of extracting basic dyes diminish in the series methanol, ethanol, propanol, butanol, pentanol.[268] There appears not to have been any full investigation of the various isomers of the three last-mentioned alcohols, but tertiary butanol is said not to remove any toluidine blue that has combined with nucleic acids.[297] It might be thought that basic substances, such as aniline and pyridine, would help to hold basic dyes in the tissue, but in fact they appear to compete with the dye for the acidic components of the tissues.[486]

A simple experiment, described in the Appendix (p. 324), shows that acidified alcohol rapidly removes all colouring of tissues by a basic dye, but leaves intact the colouring by an acid dye.

Möllendorff[355] considered that a sharp distinction was to be drawn between the action of acid dyes and certain basic ones on one hand, and of other basic dyes on the other. The latter, in his view, were especially liable to flocculation and tended to be precipitated on the surface of the objects for which they had an affinity. This process he called precipitation-dyeing (*Niederschlagsfärbung*). Other dyes, not subject to flocculation, penetrated into the interstices of objects and dyed them uniformly throughout. He called this permeation-dyeing (*Durchtränkungsfärbung*). This distinction is perhaps valid. One sometimes has the impression that certain basic dyes and dye-lakes (p. 207) have a tendency to thicken the objects for which they have an affinity, which would be impossible if they only acted by permeation.

It is also easy to gain the impression that basic dyes show a considerable degree of specificity, while acid ones act diffusely. This apparent contrast is not quite genuine. The difference lies in the objects dyed rather than in the dyes. If every cell contained a characteristic object as basic as chromatin is acidic, we should not be so much struck by the diffuseness of acid dyes, for we could arrange to colour the basic object differentially with them. The eosinophil granules of certain leucocytes can indeed be dyed sharply in this way. Often, however, we deliberately use the acid dyes to colour the amphoteric cytoplasm feebly, in a colour contrasting with that of the acidic objects, which have been strongly coloured by a basic dye

Despite this, there is reason to believe that acid dyes really are somewhat more diffuse in their action on tissues than basic ones, and it is reasonable to look for a component of the protein chain with which they might react, other than the specifically basic residues of lysine, arginine, and histidine. The dyeing of the synthetic fibre, nylon, is instructive here.[524] This substance does not provide much opportunity for the kind of reaction with acid dyes that we have been considering, because an amino-group only occurs at one end of the long chain-molecule. When this has reacted by salt-formation with the anion of an acid dye, an increase in acidity will result in a further, sudden uptake of dye, which is

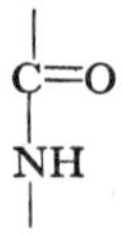

Peptide group in nylon or protein

attributed to the activation of the peptide groups that occur repeatedly in the nylon chain.[396] It is supposed that hydrogen ions are taken up by the peptide groups, which thus become positively charged and therefore attract dye-anions. If nylon can behave in this way, there would not seem to be any reason why acid dyes should not attach themselves similarly to the protein chain, in acid solution. It is a striking fact that wool becomes softened ('tendered') when it has taken up more than a certain amount of an acid dye. This suggests an alteration in the main chain of the protein. A reaction of this sort would be unrelated to the preponderance of basic or acidic side-groups, and the appearance under the microscope would therefore necessarily be diffuse.

The electrostatic forces between oppositely-charged ions act over much greater distances than the 'short-range' forces concerned in covalent and hydrogen bonds. That such short-range forces may be important in textile-dyeing has been especially stressed by Neale.[365] The process we have been considering will not account for the dyeing of cotton, for this is a negatively-charged substance that is generally dyed by acid dyes. The energy necessary to bring the similarly-charged bodies together is provided by thermal agitation. It is for this reason that high temperatures are used.

In microtechnique we are not faced with exactly this situation, because we do not ordinarily dye negatively-charged objects with the negatively-charged ions of ordinary acid dyes. Still, if dye-ions can be attached to cotton by forces other than those we have been considering, the possibility exists that in microtechnical dyeing long-range electrostatic forces merely play a part similar to that played by thermal agitation in the dyeing of cotton. The close, final attachment may then be achieved through hydrogen bonds. A useful short summary of the various parts of dye-molecules that could serve for hydrogen bonding to the hydroxyl groups of cellulose is given by Evans.[161] Either the hydrogen or the oxygen of these hydroxyl groups can participate in a hydrogen bond, the former making a link (for instance) with the nitrogen atom of an amino-group or one of the nitrogens of an azo-group of the dye, the latter with a hydrogen of a $-\underset{H}{C}=\underset{H}{C}-$ group. These reactions only occur so long as the hydroxyl groups of the cellulose are intact. Bonding of this sort presumably takes place in microtechnique when we use a direct cotton dye such as Congo red to colour cellulose cell-walls.

It is not only cellulose, however, that can bind itself to dyes in this way. The dyes commonly used to colour nylon (which are not the ones mentioned above (p. 198) as colouring it from acid solution) appear to form hydrogen bonds with the peptide groups of the fibre. If so, the amido-groups of the protein chain may also be available for the attachment of dyes. A link could be formed connecting the hydrogen of a peptide group forming part of a protein chain to the nitrogen of an amino- or azo-group in a dye.

General intermolecular attraction (van der Waals forces) are thought sometimes to aid the indiscriminate anchoring of dyes, provided that close enough approximation to the substrate can somehow be attained.

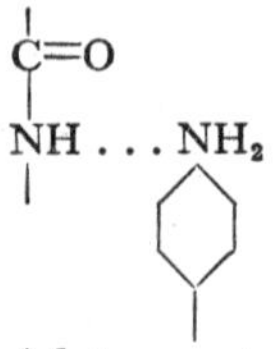

A possible hydrogen bond (. . .) between a peptide group of a protein chain and an amino-group of a dye

Although these kinds of bonding must be kept in mind, yet there is at present no strong evidence that they play a dominant rôle in microtechnical dyeing. The biologist has one advantage over the textile chemist in judging the forces that bind a dye to its substrate. The tissues of plants and animals, as studied in the biological laboratory, provide us with very obvious visible indications as to whether a dye is acting in conformity with the electric charge on its ions or not. It is true that the cortex of wool is somewhat acidophil and the medulla somewhat basiphil; [524] but the contrast is not very sharp, and anyhow textile chemists do not devote a very great deal of study to the appearance of dyed wool under the microscope. When a biologist looks at a microscopical preparation of the most ordinary kind, every cell suggests the supremacy of ionic forces, for the acidic chromatin is coloured by the basic dye and the ground cytoplasm in a contrasting colour by the acid dye.

If reactions between oppositely charged groups play a major part in dyeing, one would expect a stoichiometric relation between the amount of protein dyed and the maximum amount of dye that could be taken up. This is a subject on which there is conflicting evidence. One can measure the amount of inorganic acid with which a particular protein will combine, and then compare this with the amount of acid dye that can be taken up. It is claimed that the number of arginine, lysine, and histidine groups in the protein account for the amount of acid or acid dye that will combine, and that there is therefore no reason to suppose that the peptide groups of the main protein chain participate in the process of dyeing.[119] Singer and Morrison,[473] however, found that at pH 2 and 11 fibrin films took up far less acid and basic dyes respectively than would be expected if all the charged groups of the protein in fact reacted with dyes. The subject, however, is complicated. Dye ions often aggregate (p. 238), especially at the low temperatures used in the biological laboratory, and this would make stoichiometric proportions unlikely. Dyes ions with two or more charged groups might not be able to use all of them to make attachment to oppositely charged groups on the protein, because the latter might be too far apart.

The effect of inorganic salts takes up a good deal of space in works on textile dyeing, but is of less direct interest to the biologist. In dyeing cotton and other cellulose fibres with certain direct cotton dyes, it is usual to add sodium chloride or sulphate to the dye-bath, as this greatly aids the uptake of dye. The dye-ions of these direct cotton dyes carry the same (negative) electric charge as the cellulose. In general, inorganic salts help dyeing in those particular cases in which the electric charges oppose it, apparently by favouring the near approach that is necessary for dyeing by close-range bonds.[472] On the contrary, they interfere with the linkage between a dye-ion and an oppositely-charged group in the object to be dyed, and also lessen the activity of many dyes by increasing their tendency to flocculation and thus lowering their capacity to diffuse. Salts should not be used as buffers for dyes without regard for these facts.

Temperature affects dyeing in several ways.[524, 472] High temperatures increase the rate of diffusion of dye-ions and also reduce any tendency they may have to aggregate into larger particles (p. 241), which would move more slowly and penetrate less easily. High temperatures also loosen the covalent bonds that hold protein chains together, and dissolve the disulphide links: thus the protein becomes more easily permeable. At 100° C 1 hour may suffice for wool to take up as much of a dye as it can hold; but if the temperature be kept down to 20° C, 5 months may elapse before equilibrium between wool and the same dye is reached. Temperature does not have much effect, however, on the amount of dye eventually taken up: rather *more* is taken up from cold solutions, except in those cases in which a dye cannot enter at all in the cold (p. 241). Most biological material is very much more easily penetrated by dyes than wool is, and the temperature is only raised above that of the laboratory when a dye has a special tendency to flocculation (e.g. azocarmine) or when a tissue-constituent is particularly close-textured and therefore difficult to penetrate (e.g. mitochondria).

The effects of fixation on dyeing have already been mentioned in chapters 5 and 6, under the headings of the eight primary fixatives selected for separate description. A general review of the subject will be given here.

These effects can be studied by experiments on proteins. Seki [459] fixed small pieces of egg-white in various ways; he either dyed them whole or made paraffin sections. Fibrin films are very suitable for work of this sort; they are chemically uniform and require no sectioning.[473] The effects of fixation on the dyeing of tissues has been investigated by several workers.[505, 552, 459, 472, 38] Whether proteins or tissues be used, the material may be exposed to a mixture of a basic and an acid dye,[505] or to basic and acid dyes separately; the pH of the dye-solution may be controlled.[505] Instructions for carrying out simple experiments on the effects of fixatives on dyeing are given in the Appendix (p. 325).

For studies of this kind it is important to avoid cells (such as nerve-cells) that have much RNA in the cytoplasm, for this would colour strongly with most basic dyes and therefore mask the effect of fixatives on the reactions of dyes with the cytoplasmic proteins. Strongly basic protoplasm, such as that of mammalian red blood-corpuscles, is also unsuitable. One needs an example of typical cytoplasm and typical chromatin. Seki [459] chose the skin of the frog and mammalian kidney. The convoluted tubules of the latter organ and the spermatogonia and spermatocytes of mammals are particularly suitable.[38] The late spermatids and spermatozoa themselves are unsuitable not only because of the absence of unspecialized cytoplasm, but also because of the basic nature of the nuclear material.

Certain possible sources of error should be noted. The intensity of dyeing is often different at different depths in a piece of tissue. It is best to use small pieces of nearly uniform size and to compare cells in the middle of each piece. Some fixatives shrink the cytoplasm strongly. In its shrunken condition it may give a false impression of taking up a lot of dye, when in fact the same amount of dye, spread over the cytoplasm of an unshrunken cell, would look quite pale. It is desirable that quantitative methods should be introduced into work of this kind. This has already been done in studies made with fibrin films.[473]

Unfixed proteins are generally neither strongly basiphil nor strongly acidophil. Chemical fixation generally increases their colourability by dyes. This is well shown by fibrin films.[473, 472] Fixation by heat has the same effect.[473] Certain fixatives favour the action of basic dyes, others that of acid dyes; others again allow easy coloration by both.

Formaldehyde favours basic dyes more than any other fixative does. Mercuric chloride has the same tendency, but in a less marked form. These facts are related to what is known of the reactions of these two fixatives with proteins, for in both cases there is blocking of $-NH_2$ groups (pp. 53 and 59). Kelley [264] considers that mercuric chloride favours acid dyes; he thinks the main effect of the salt is to block carboxyl groups, while amino-groups are left free for linkage with acid dyes. This is contrary to what is known of the chemistry of fixation by mercuric chloride, and also to the observable facts of dyeing.

From some of the remarks in the literature one might imagine that osmium tetroxide almost abolishes the capacity of tissues to be dyed. Even Seki,[459] so reliable as a general rule, gives this impression. Actually this fixative leaves cytoplasm readily colourable by basic dyes (after bleaching), though scarcely at all by acid ones. Anyone can prove this for himself by carrying out the experiment described in the Appendix (p. 325). Basic fuchsine gives particularly intense coloration of the convoluted tubules of the mammalian kidney. These facts cannot be correlated with what is known of the reaction of osmium tetroxide with proteins (p. 62). From the effects of dyes one would suppose that amino-groups were almost eliminated, but carboxyl groups left free for linkage with basic dyes. It must be mentioned that basic dye-lakes (p. 207) act less strongly than basic dyes on tissues fixed by osmium tetroxide.

Ethanol is intermediate between such fixatives as formaldehyde, mercuric chloride, and osmium tetroxide, which make cytoplasm basiphil, and the ones to be mentioned below, which have the reverse effect. It allows fairly easy dyeing of proteins and cytoplasm by both basic and acid dyes. This must be correlated with the fact that it is a non-additive fixative, which would not be expected to have much effect on the proportions of the acidic and basic groups available for linkage with dyes.

In sections of tissues fixed with acetic acid, the cytoplasm shows a general resemblance in its reaction to dyes to that fixed by ethanol, for both basic and acid dyes are taken up fairly readily.

The fixatives that particularly favour acid dyes are trichloracetic acid,[461] picric acid, and chromium trioxide. Seki [461] would add potassium dichromate to these, but after this substance has been allowed to act on the mammalian kidney, the cytoplasm of the convoluted tubules is readily coloured by basic dyes.[38] The facts suggest that trichloracetic acid, picric acid, and chromium trioxide tend to block carboxyl groups and leave amino-groups free for linkage with acid dyes. Seki, however, considers that basic proteins are immobilized by the alkaloidal reagents and remain in the tissues to react with acid dyes, while acidic proteins simply dissolve away. Unfortunately we have not nearly so much knowledge about the reactions of these substances with proteins as we have about the reactions of formaldehyde and mercuric chloride.

The effects of fixatives in blocking carboxyl or amino-groups are reflected in changes in the iso-electric points of proteins. Such changes are most readily observable with fibrin films, which can be subjected to direct cataphoresis.[473] The iso-electric point of unfixed fibrin is pH 6·0; denaturation by heating (1 min. at 100° C) lowers this figure to 5·7; fixation by formaldehyde (10%, 16 hr.) to 5·2. Estimations of the shift of the iso-electric point of tissue-constituents can be made by experiments with dyes used at particular pHs.[505, 552] It is found that various objects in the tissues can be coloured by basic dyes at a lower pH after fixation by formaldehyde than after any other fixative. Ethanol has been found to have little effect on the position of the iso-electric points of tissue-constituents.

The dyeing of chromatin is more complicated than that of the cytoplasm. It is likely that fixatives generally split DNA from protein, and the colouring is then mainly that of DNA by the basic dye.[264] A virtue of acetic acid is that it leaves chromatin scarcely colourable by acid dyes, so that basic dyes are not masked. Many other fixatives allow strong coloration of chromatin by basic dyes: for instance, mercuric chloride, formaldehyde, ethanol, and potassium dichromate. The two latter, however, do not coagulate nucleoproteins, and there is therefore no immobilization of chromatin in its original sites. When it is contained in the nucleus, however, it is prevented from escape by the nuclear membrane, and it distributes itself almost at random within the nucleus. This is particularly obvious when one examines the first meiotic prophase in a gonad fixed in ethanol or potassium dichromate. No one would choose either of these fixatives for work on chromosomes, except in mixtures.

Despite statements to the contrary, osmium tetroxide allows quite strong colouring of chromatin by basic dyes; but it often makes the nuclear sap basiphil, and this clouds the structure of the nucleus. One would not use this fixative alone in studies of prophase or telophase changes, or in any other investigation requiring a glassy nuclear sap.

The reactions of picric acid with chromatin are very curious.[38] The protein part is coagulated, but the DNA is set free and dissolves (compare Levene [298]). The chromatin is now represented only by protein, and the nature of the latter determines the reaction with dyes. Since the protein component of chromatin is seldom strongly acidic, there is little affinity for basic dyes; but if it chances to be definitely basic (as, for instance, in the spermatocytes of mammals), acid dyes will be taken up strongly. We thus have the curious spectacle of meiotic chromosomes strongly dyed by acid dyes, but feebly by basic ones.

It is not unusual for the protein of chromatin to be colourable by acid dyes, after DNA has been separated off by the action of fixatives. Acid dyes are used to colour chromatin in certain techniques (acid fuchsine in Mallory's tricolour method [320] and azocarmine in Heidenhain's 'Azan').[220] After most fixatives the DNA will still be present, but will not mask the colouring of the protein by an acid dye unless a basic dye is used as well. The acid dye works more strongly, however, if DNA-ase is allowed to act first.

In the chromatin of the spermatozoa of certain fishes the DNA is combined with protamine. The strongly basic nature of this simple protein, caused largely by its high arginine content, predominates to such an extent that the substance as a whole, with its DNA, is highly acidophil and scarcely takes basic dyes, whatever fixative may have been used.

If a particular fixative leaves some of the phosphoric groups of DNA still combined with protein, this will necessarily result in a low affinity for basic dyes.

The reason why formaldehyde interferes with the dyeing of chromatin by borax-carmine follows from the foregoing considerations. In alkaline solution carminic acid is an acid dye, without affinity for DNA. It can combine, however, with the basic groups of the protein of nucleoprotein, and then, when the tissue is subsequently acidified and it becomes a basic dye, it can fix itself to the DNA. If, however, formaldehyde was used as fixative, the protein of the chromatin will have been rendered acidic and therefore will have taken up little carmine from alkaline solution; there will consequently be little dye present in the chromatin to link up with DNA on subsequent acidification.

In routine preparations we usually want to dye the chromatin in one colour with a basic dye (or basic dye-lake, p. 207) and the cytoplasm in a contrasting colour with an acid dye. What does this involve? Ideally, the DNA must be struck off by the fixative from the protein with which it was combined, and precipitated instantly without change of position in a form in which it will combine readily with basic dyes; the protein of the chromatin must have a low affinity for acid dyes, so that the colour given by the basic dye shall not be masked; and the cytoplasm must have little affinity for basic dyes, but link strongly with acid ones. The fixative that leaves tissues most nearly in this condition is chromium trioxide, though chromatin fixed by this substance has not so strong an affinity for basic dyes as one would wish, and retains some affinity for acid ones. Acidified potassium dichromate acts similarly to chromium trioxide. (See p. 132.)

Many fixative mixtures probably owe their continued popularity largely to the fact that they happen to leave the iso-electric points of the tissue-constituents in convenient positions; that is to say, in positions in which it is easy, without troubling to use buffers, to colour the chromatin with a basic dye and the ground cytoplasm with an acid dye of contrasting colour. Some shift of the iso-electric point towards higher pH would be helpful towards this end, which is realized by such fluids as that of Zenker.[554] If this shift did not occur, ground cytoplasm would generally be too basiphil to give good colour-contrast with chromatin.

It is a special property of mercuric chloride that it leaves tissues in a state in which they are readily coloured by dyes of all sorts (but particularly by basic dyes and dye-lakes). No adequate explanation of this familiar fact has been provided. One must suppose that the tissues are fixed in such a form that they are readily penetrated by large ions, and that many acidic and basic groups (especially the former) are available for linkage with dyes.

CHAPTER 11

The Indirect Attachment of Dyes to Tissues

It is characteristic of dyes that when they are dissolved in water or a mixture of water and alcohol, they attach themselves directly to certain tissue-constituents in one or more of the ways described in the last chapter. Some of them, however, have an alternative method of attachment, involving the presence of another substance besides dye and solvent; and when this substance is present, their behaviour is quite different. Great advantage can sometimes be taken of this difference in behaviour, both in the textile industry and in microtechnique. A dye that is almost useless in simple solution may become of major importance.

The salts of certain metals are the chief substances that radically alter the behaviour of particular dyes. These salts were called *mordants* because they were thought to bite into certain textile-fibres and thus give attachment to dyes that would not work satisfactorily by themselves. The word was taken into our language from French.

A mordant is capable of entering into chemical combination with a dye. The resulting substance is called a *lake*. This word had a curious origin. It derives from *lac*, a Hindustani word meaning the waxy material produced by the scale-insect, *Tachardia lacca*. The females of this species attach themselves in great numbers to the twigs of certain Indian trees and produce an abundance of waxy material, which fills up all the spaces between them. The waxy material is sold as lac, or, if in flat plates, as shellac. The insects themselves contain a large quantity of lac-dye, which is chemically related to carminic acid; [330] this is said to be stored chiefly in the ovaries.[80] The dye is dissolved out and then precipitated by potassium alum. The name *lac* became attached not only to the wax but also to the precipitated dye, and afterwards to the product of the cochineal insect precipitated in the same way ('crimson lake'). The alum plays a double rôle. It precipitates the protein fraction of the crude dye, and also combines with at least part of the true dyestuff.

In a valuable work on dyeing published in 1813,[40] it was suggested that dyes used with mordants should be called 'adjective' dyes; those used without mordants, 'substantive'. These words are still used, but it must be remembered that there are not really two classes of dyes. It is true that many dyes cannot be used adjectively; but dyes that can be so used will also colour biological material substantively.

The great advantage of the use of mordants in microtechnique is that when once the tissue/mordant/dye complex has been formed, it is insoluble in all the neutral fluids ordinarily used, so that subsequent colouring with other dyes is easy and there is no hurry in dehydration. The lakes are basic in action. Their fastness to alcohol is a very great advantage over ordinary basic dyes and renders them particularly suitable for the colouring of whole mounts, which cannot be dehydrated quickly. Certain dyes, particularly haematein, produce weak, indefinite, or unsuitable colours when used substantively, but give brilliant or intense results when used with a mordant.

In the textile industry the mordant is sometimes used first and the dye afterwards (two-bath method); sometimes the two are used together (single-bath method); sometimes the dye is used first and the mordant afterwards ('afterchrome' method).[75] In microtechnique the mordant is very seldom used after the dye, though two examples can be quoted.[175, 455] The other methods are both in common use.

Two problems at once present themselves. How does the mordant attach itself to the tissue? How does the dye attach itself to the mordant (or, in other words, what is the lake)?

In trying to answer these questions one gets far less help from the textile chemists than might be expected. There are several reasons for this. In microtechnique the most important dyes used adjectively are haematein and carmine, but in the textile industry the great majority of mordant dyes belong to the azo-group. No azo dye is used with a mordant in ordinary microtechnique. A far more important reason, however, concerns the nature of the mordants used. The industrial dyer uses chromium mordants almost exclusively nowadays, but these are much less used in microtechnique than iron and aluminium. This would perhaps not matter very much in itself, for knowledge gained about the action of chromium might be applicable to the other metals; but unfortunately the ordinary use of chromium in industry is radically different, for it is applied as sodium or potassium dichromate, that is to say in an anionic complex, whereas in microtechnique we use the metals as cations. It is an extraordinary misfortune that in fixation we cannot profit from industrial chemistry because we use *anionic* chromium while the tanner uses cationic, and in dyeing we cannot profit much because we use *cationic* chromium while the professional dyer generally uses anionic.

In short, in microtechnique we generally use iron, aluminium, and cationic chromium to mordant for haematein, carmine, and certain oxazine dyes, while the industrial dyer generally uses anionic chromium to mordant for azo dyes.

The chemistry of mordanting in industry has been thoroughly investigated and must be mentioned very shortly here.[192, 410, 182, 75] The acid azo dye commonly has –OH groups *ortho* to the azo-group. On reaction with the mordant at high temperature, the

Skeleton formulae of an azo dye before and after linkage with chromium

chromium atom makes covalent links with the oxygens of the hydroxyl groups and obtains a dative covalency from one of the nitrogens, at the same time showing a negative electric charge. It follows that the mordant/dye complex has the character of an *acid* dye, which can react with wool by making salt-linkages with amino-groups of the protein. The dye generally possesses sulphonic groups (not shown in the skeleton formula), which again are acidic and can react in the same way.[182] Thus the mordant/dye complex reacts with positively-charged groups in the wool, or indeed may even be held in place by the mere fact that it forms too large a particle to escape from the pores of the wool.[410] There is strong evidence that one chromium atom reacts with two molecules of the dye,[410] but this is not represented in the simplified formula shown.

Long and complicated researches have led to the conclusion that the mordant/dye complex of the textile dyer is to be regarded as acting as though it were an acid dye, but in microtechnique we can generally tell whether any colouring agent is acting as an acid or a basic dye by a momentary glance down the microscope. The fact is that most tissues of plants and animals are extremely unlike wool. An acidic lake would be useless to us (see p. 223): our mordant colours act as basic dyes, and are indispensable.

The chemistry of mordanting in microtechnique is perhaps the most perplexing of all the subjects with which this book deals. The literature is scattered among journals seldom read by any one person, and no serious attempt has ever previously been made to integrate our knowledge of the subject into a comprehensible whole. The most convenient arrangement for the reader would be to start with a detailed description of the whole process of attaching one particular dye with one particular mordant, and a facile story of this kind could indeed be written. In fact, however, we lack the knowledge necessary for a consecutive account of the whole process. Ionization of mordants has been best described with the salts of chromium, lake-formation with those of aluminium, and the attachment of metal to tissue with ferric salts: different dyes must be chosen to illustrate particular points. In what follows no attempt will be made to push generalization too far. It is better at the present time to disclose the gaps in our knowledge than to put forward a consistent theory too confidently. Certain important facts and the rough outline of a synthesis will emerge.

The mordants commonly used in microtechnique are salts, especially sulphates, of iron, aluminium, and chromium. Double sulphates or alums are generally used. The use of alums rather than simple sulphates is perhaps due in part to historical causes. Alums are easily crystallized and therefore easily purified from the rocks that contain them. Their use in dyeing has been known from rather remote times.[41] Until about the middle of the fifteenth century the only rocks known to contain a suitable mordant were in Turkey; but when Constantinople had been captured by the Turks in 1453, a fragment of the knowledge that escaped with the refugees to the western world concerned the recognition of suitable rocks and the mode of preparation of the mordant. Alum was first mined in Great Britain about the end of the reign of Queen Elizabeth I.

The alums generally used are these:—

potassium alum, $Al_2(SO_4)_3.K_2SO_4.24H_2O$;
ammonium alum, $Al_2(SO_4)_3.(NH_4)_2SO_4.24H_2O$;
iron alum, $Fe_2(SO_4)_3.(NH_4)_2SO_4.24H_2O$;
chrome alum, $Cr_2(SO_4)_3.K_2SO_4.24H_2O$.

The formulae can also be written as $AlK(SO_4)_2.12H_2O$, etc., but this method shows the composition less well. There is no purpose in constructing the formula in a way that will make the molecule appear as small as possible, since there are no molecules in the crystal, and only ions in solution.

Aluminium, ferric, and chromic sulphate act similarly to the alums. There is no compelling reason why sulphates should be used in preference to other salts. Ferric chloride, for instance, has been used as a mordant for haematein.[207]

The cation of the mordant salt, then, is the part that interests us in microtechnique. The metal is in each case tervalent in these compounds, but the ions are complex. The water of crystallization is partly bound up with the metal. For instance, the ferric cation that gives the violet colour to a crystal of iron alum is $[Fe(OH_2)_6]^{+++}$. The chromic cation makes a crystal of chrome alum almost black. When a brilliant light is shone through a small crystal, the colour is seen to be violet. The red component is easily seen when a small crystal is held against an electric light bulb. The cation responsible for the deep violet colour is $[Cr(OH_2)_6]^{+++}$. The water molecules are co-ordinated to the iron or chromium atoms by dative covalent bonds from their oxygen atoms. The metal may be described as sexi-covalent.

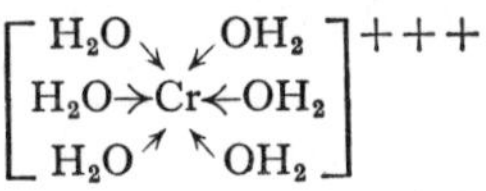

The chromic cation in a crystal of chrome alum

When iron alum is dissolved in water, the solution is not violet but yellow or brownish-yellow. When chrome alum is dissolved in hot water, the solution is not violet but green. In both cases the solutions are quite strongly acid. The changes that occur in the cation have been carefully studied in the case of chromium. A

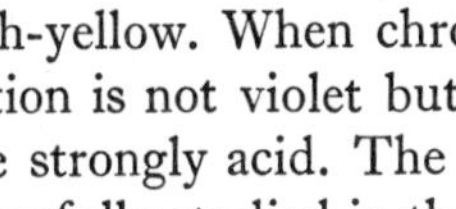

The chromic cation in a solution of chrome alum

hydrogen ion is lost from the cationic complex and goes off with one of the positive electric charges, thus acidifying the solution, and a second may follow suit.[192] Similar changes occur when iron alum dissolves, though the yellowish cationic complexes produced are usually represented as $[FeOH]^{++}$ and $[Fe(OH)_2]^{+}$.

Aluminium behaves similarly, but both the simple sulphate and potassium alum are colourless, and there is no production of colour to call attention to the change in chemical composition of the cation on solution. The cationic complexes in solution are usually represented as $[AlOH]^{++}$ and $[Al(OH)_2]^{+}$.

The production of hydrogen ions in the course of these changes results in varying degrees of acidity. The following figures relate to solutions of the crystalline salts (ordinary laboratory chemicals, as used in microtechnique):—

	pH
chrome alum, 5%	1·8
iron alum, 5%	1·9
,, 2·5%	2·1
potassium alum, 5%	3·2

The complex cations that we have been discussing, in solutions rendered acid by the act of dissolving, are the essential part of the mordant—the part that must react with the dye and with the tissue, so as to leave a link between them. As we have seen, the attachment with the dye can occur either before or after the mordant metal has fixed itself to the tissue. It will be convenient to consider first the nature of a soluble lake as used in the single-bath process; that is to say, we shall forget the tissue for the moment and think only of the reaction of the dye with the mordant.

Dyes that form lakes possess a phenolic –OH group, which plays an important part in lake-formation. Sodium hydroxide will react with phenol to form sodium phenolate and water, and ferric

Sodium phenolate

iron will react in a comparable way. A much firmer bond can be made between certain phenols and iron, however, because they can provide a second link with the metal. For instance, salicylic acid reacts with iron in the way just described, the metal replacing the hydrogen of the phenolic –OH group.[212] This happens when salicylic acid is treated with ferric chloride. This, however, is not all that happens.[469] The organic acid has an oxygen atom conveniently situated to donate electrons to an iron atom that has replaced the hydrogen of the –OH group. Thus a dative covalency is formed, as is shown in the formula by an arrow, and the iron atom

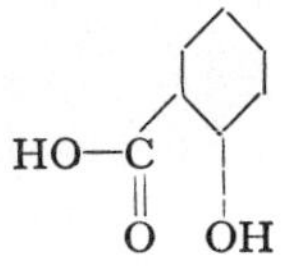

Salicylic acid

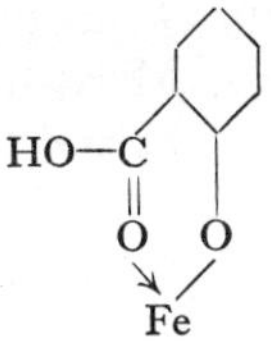

Salicylic acid combined with iron

is now gripped by a pair of pincers. It will be noticed that the iron atom now forms part of a six-membered ring, the atoms being in this order:—iron, oxygen, carbon, carbon, carbon, oxygen. Actually the iron atom is gripped by three pairs of pincers (though this fact is not shown in the formula), for the iron atom can replace hydrogen in three molecules of salicylic acid, each of which can form a second bond with the iron through an oxygen of its carboxyl group.

To achieve this result, one must have a bi- or multivalent metal that is capable of forming covalent bonds, and an organic compound that possesses both a hydrogen atom replaceable by the metal and also a donor atom (here oxygen), capable of donating electrons to the metal to form a second link. These are exactly the conditions that are fulfilled when a lake is made. This fact was first realized nearly half a century ago by the celebrated Swiss chemist, Werner, in 1908,[533] though naturally he did not express himself in terms of the electronic theory of valency. He realized that the mordants were salts of metals that could form *innere Komplexsalze*, and he quoted $[Cr(OH_2)_6]X_3$ (X standing for the anion). The fact that chromium is capable of receiving dative bonds is implied by the formula he wrote, for, as we have seen, six molecules of water are co-ordinated to the metal through their oxygen atoms. Werner expressed himself thus:—'From the reported results of the experiments it can no longer be doubtful that the property of chemical compounds to connect with mordants depends upon the property of the latter to form internal metal-complex salts.' [533] Of the other partner to the transaction he wrote, 'Dyes that connect with mordants are hereafter constitutionally characterized by the fact that a salt-forming group and a group capable of producing a co-ordinate bond with the metal atom are so arranged that an internal metal-complex salt can originate'.[533]

It is fitting that the man who was afterwards to receive the Nobel Prize for his co-ordination theory of valency was the first person to grasp the essentials of lake-formation.

The chemistry of this process was subsequently studied by two British chemists, Morgan and Smith,[359] to whom we owe the expressive term 'chelate' for the pincer-like grip in which a dye holds the mordant metal. They took the word from the analogy with the chela of a lobster.

It now remains to look for the phenolic –OH group and nearby donor oxygen in the dyes that form lakes. Alizarine provides a very simple example. A phenolic –OH is close beside a suitable oxygen atom, and a six-membered ring (aluminium, oxygen, carbon, car-

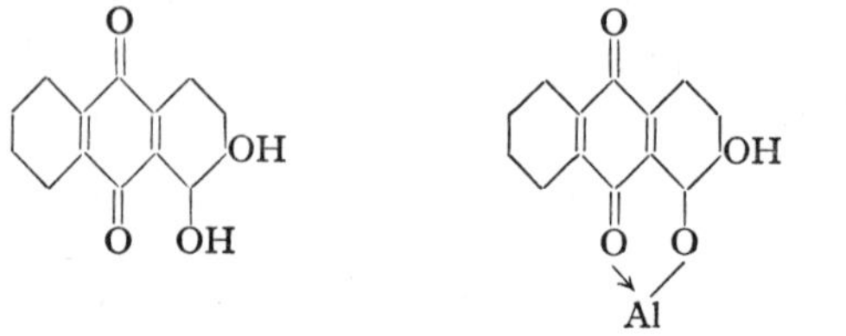

Alizarine *Alizarine aluminium lake*

bon, carbon, oxygen) is readily formed. (The aluminium atom is capable of linking with three alizarine molecules.) One has only to look at the formula for carminic acid on p. 177 to see that it is capable of acting in exactly the same way.

It is important to notice that the metal atom (whether aluminium, chromium, or ferric iron) makes two different kinds of links with the dye. These links may for convenience be called primary and secondary. The primary link is made by the substitution of the metal for a hydrogen atom in an acidic –OH group. It is reasonable to suppose that this primary linkage is initially electrovalent, though it may be replaced by a somewhat polar covalency. The secondary link is the dative covalency with the electron-donor oxygen atom. The cation could make room for such a covalency by losing one of the $-OH_2$ or –OH groups held to the metal by 'subsidiary' valencies. (See the formulae for chromic cations on p. 211.)

It was pointed out nearly 70 years ago[304] that many anthraquinone dyes that can be used with mordants have two phenolic –OH groups in *ortho*-relationship to one another, and it was at first thought that these gripped the metal. The action of the second

–OH group is not clear,[359] and indeed it is absent in lac-dye, which works with mordants.

In the absence of a nearby donor oxygen atom, two –OH groups can form chelate compounds with iron. Certain derivatives of catechol provide examples of this.[469] It is therefore interesting to

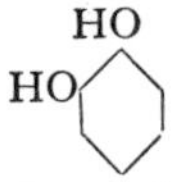

Catechol

find two –OH groups in this position in haematein (p. 173), which, however, also possesses a phenolic –OH in close proximity to a donor oxygen, in another part of the molecule.

The oxazine dyes that work with mordants also possess the necessary groupings for the formation of chelate bonds with

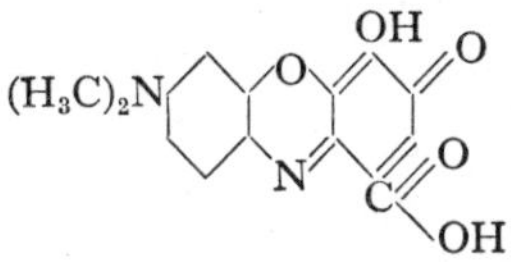

Gallocyanine

metals, though perhaps in a somewhat disguised form. The formula for gallocyanine is printed here in such a way as to emphasize the relationship with other mordant dyes.

Haematein, carminic acid, and the mordant oxazines are all amphoteric dyes. Indeed, one would suspect this from a glance at their chemical formulae. Gallamine blue, for instance, presents two basic groups ($-NH_2$ and $-N(CH_3)_2$) and two acidic (–OH). The iso-electric point of haematein is about pH 6·5, of carminic acid about pH 4·2, of gallocyanine and gallamine blue about pH 4·1. The lakes are in all cases basic throughout the pH range in which they are used, and they act like basic dyes apart from their complete insolubility in neutral fluids after they have once attached themselves to the tissues. The positive charge on the lakes can be proved by cataphoretic experiments.[454, 455]

Sometimes a dye-lake has a colour that is different from that of its parent dye. Thus haematein is yellowish at about pH 1 and changes its colour towards orange and dirty red as acidity lessens towards neutrality; but the aluminium lake is blue and the characteristic iron lake dark blue or blue-black. Carminic acid is somewhat less affected, for the dye by itself, whether in acid or alkaline solution, is red, and the aluminium lake crimson. The oxazine dyes are still less affected in colour by combination with mordants.

The oxazine dyes do not generally work well if applied in simple solution after mordanting [407] and the single-bath method is commonly used with them Carminic acid forms very stable soluble lakes, and here again the single-bath method is appropriate. Haematein can be used as a soluble lake, but the two-bath method is generally used when iron is the mordant for this dye. In Hansen's so-called *Trioxyhämatein* [210] the dye is mixed with the iron mordant, but precipitation is liable to occur unless precautions are taken. It was Ehrlich [155] who first overcame the instability of the aluminium haematein lake, by the addition of acetic acid. The effect is to keep the dye separate from the mordant. The acidified solution shows the reddish colour of the dye, not the blue of the lake. How the acid acts is not perfectly clear. The mordants themselves, as we have seen (p. 212), are themselves strongly acid. It must be presumed that the additional hydrogen ions stop the loss of hydrogen from the phenolic –OH group of the dye, and thus prevent its replacement by the metal of the mordant. (See p. 221.)

Strong acid is needed to prevent the formation of an iron lake or break it when once formed. Sulphuric acid may be added at 2% to prevent the precipitation of the iron lake of coelestine blue.[294] Since this lake has the same colour as the dye, the course of events is not so easy to follow as when aluminium and haematein are kept apart by acidity and then allowed to join by neutralization. With iron and coelestine blue the tissue/mordant/dye complex can be formed without removal of the acid. The presence of tissue seems to favour the formation of the lake.

It is a curious fact that some lakes have greater powers of penetration than their parent dyes. Thus gallamine blue diffuses into gelatine more slowly than its lake with chrome alum.[356] It is evident that the dye by itself is somewhat aggregated, and that the mordant disperses it.

We turn now to the attachment of the metal to the tissue.

It is desirable to say at the outset that the metal will eventually be held to the tissues by bonds that are very similar to those that hold it to the dye. The primary linkage is with acidic groups in the tissues, and this is the reason why dye-lakes act as though they were basic dyes.

The attachment of metal to biological material has been specially studied by Wigglesworth.[540] Sections of tissues fixed in various routine fixatives were placed in solutions of iron alum and then washed in water; the places where the metal had become attached were made visible by treating the sections with ammonium sulphide and thus producing dark sulphide of iron. A careful survey of the resulting slides suggests strongly that the metal is taken up by the acidic groups in the tissue, notably the phosphoric groups of the nucleic acids and the $-C \begin{matrix} /\!\!/ O \\ \diagdown OH \end{matrix}$ groups of proteins. The chromatin (especially that of chromosomes) is darkened. This is partly the result of their nucleic acid content; but when the nucleic acids have been removed by treatment with hot trichloracetic acid, the colour still develops by reaction of the iron with the protein component of the nucleoprotein.

The reaction with different proteins is illuminating. Thus myosine, which contains many acidic groups, is strongly darkened, while salmine, which lacks such groups, is scarcely touched; other substances, acidic in varying degrees, show, with one or two exceptions, a degree of darkening proportional to their acidity.

If the carboxyl groups in the proteins are methylated by prolonged immersion in acidified methyl alcohol, the uptake of iron is much reduced. Deamination by formaldehyde has, on the contrary, no such effect, and it is thus clear that the $-NH_2$ groups are not concerned.[540]

Iron is taken up fairly strongly by the proteins of ground cytoplasm, but not nearly so strongly as by chromatin; also, in certain circumstances, by elastin (p. 233); also, though feebly, by collagen. Iron haematein colours these tissue-constituents brown, in contrast to the black or blue-black of chromatin. Möllendorff[356] claimed that iron haematein gave black when it acted as a precipitant dye, and brown when it permeated an object evenly (p. 197). The two contrasting colours resemble those produced when ammonium sulphide is substituted for haematein.[540] A completely satisfactory explanation of these facts has not yet been given.

There can be little doubt that iron is also bound by the phosphoric groups of phospholipids, if the fixation and after-treatment have been adapted to the retention of these substances. It has been shown that after all the RNA has been removed from the tissues by ribonuclease, the mitochondria are still colourable by iron haematein; [522] this has been attributed to their lipoprotein content.[521] Other acidic lipids besides phospholipids may perhaps also bind iron, but the idea that iron haematein attaches itself to nothing in cells except acidic lipids [204] is untenable.

Presumably the cationic iron complex first makes a salt-linkage with the available acidic groups, but the characteristic reactions of ferric iron are not given and it is thought that a non-ionizing complex is formed with adjacent carboxyl and hydroxyl groups.[540]

When iron has been taken up by the tissues, the compound into which it has entered is readily broken down by ammonium sulphide and it appears once more in an ionic form. Everything is at first coloured blue or blue-grey; but when the section is washed in water and thus exposed to atmospheric oxygen, an interesting differentiation occurs. Certain parts, such as the chromatin, remain blue or bluish, while others, such as collagen and the contractile substance of muscle, become brown. It is thought that the iron is ferrous in the one case, ferric in the other. At first the iron appears everywhere as ferrous sulphide, and remains as such wherever there are reducers (such as sulphydryl groups) in the proteins or other cellular constituents; but where there are not, oxidation can occur, with development of a brown colour.[540]

There is a contrast between the uptake of iron on one hand, and of chromium and aluminium on the other. The two latter metals attach themselves only slightly to the proteins of the cytoplasm, and scarcely at all to elastin.[356] They also have much less tendency than iron to attach themselves to lipids (though *anionic* chromium makes bonds with these; see pp. 107, 128). It is for these reasons that we choose iron haematein to show mitochondria and certain other cytoplasmic inclusions, but aluminium haematein or chrome oxazine when we want to colour chromatin and little else.

The attachment of the chromic cation to tissue-constituents has not been studied in detail. We have some information about its behaviour in the dyeing of textiles, although, as was mentioned above, most of the research on the chemistry of mordanting has been done with anionic chromium. It is thought that the attachment is by covalent linkages. Various groups in the protein of wool are thought to displace water or OH or both from their attachment to chromium in the cationic chrome complex, and to co-ordinate with the metal in their place. The hydroxyl, amino-, and amido-groups of the protein are mentioned in this connexion.[192] It is also claimed that where two protein chains are held together by an –S–S– bond, this may be split apart with production of two –SH groups, and these, reacting with the chromium atom, may cause the latter to act as a new link between the protein chains.[192] It will be remembered, however, that the dyeing of wool is carried out at high temperatures.

In biological preparations it is certain that the phosphoric groups of the nucleic acids and the acidic groups of certain mucosubstances make attachment with the chromium mordant more readily than proteins, especially at low pH,[437] and this also applies to aluminium.

It will be noticed that the attachment of mordants to tissues is a very complicated process, of which we have as yet only an imperfect understanding. The statement is commonly made in chemical textbooks that the mordant metals are deposited in textile fibres as gelatinous, insoluble hydroxides. This is not only an extreme oversimplification, but quite untrue. Neither in textiles nor in ordinary biological material are the metals deposited in this way. Had the statement been true, there could be no question of that delicately differential tying up of the metals with particular tissue-constituents that makes mordants so invaluable in microtechnique. There is no gelatinous mass pervading our sections. If there had been, its combination with a dye would have rendered microscopical study impossible.

Lakes can be used progressively. Hansen's iron haematein and the oxazine lakes are ordinarily used in this way. It is, however, almost invariable to work regressively when mordant and dye are applied separately.

There are three separate methods by which differentiation can be carried out when mordant-dyes are used regressively. The agents used are mordants, acids, and oxidizing agents.

The tissue/mordant/dye complex is broken by excess of the mordant. The dye distributes itself partly as a soluble lake with the free mordant and partly as a component of the insoluble complex; and since the amount of mordant in the complex is small in comparison with the amount in the differentiating fluid, nearly all of the dye will eventually associate itself with the latter, if enough time be allowed. Since there will be much dye in some tissue constituents (the chromosomes, for instance) and less in others (such as the cytoplasm), a particular degree of differentiation will leave no visible dye in certain parts though others remain strongly coloured.

Since a mordant will continue to extract dye even when it has already taken up a certain amount of it, but will colour tissues powerfully if it is heavily laden with dye, there must be an intermediate amount of dye that will cause strong colouring of basiphil objects but not of those that have less affinity for the lake. This explains the facility with which certain lakes can be used. Grenacher's alum carmine [199] and Mayer's carmalum [332] may be quoted as familiar examples. Chromatin is coloured, but cytoplasm has little tendency to take up the lake. As a result the exact period of dyeing is unimportant. These dyes are particularly suitable for whole mounts. Since too strong coloration does not occur easily, the outer parts of a piece of tissue are not dyed very differently from the interior.

It is interesting to notice the relative amounts of metal and dye in such fluids as these. The weight of potassium alum crystals that contains one atomic weight in grams of aluminium is almost exactly equal to one gram molecular weight of carminic acid. Therefore, if one wished to have a solution with 3 times as many dye molecules as aluminium atoms, so as to combine as much dye as is theoretically possible with the metal, one would take 3 g of carminic acid to 1 g of alum crystals. Now Mayer's carmalum only contains 0·1 g of carminic acid to 1 g of alum crystals.

If the amount of potassium alum in Mayer's carmalum is reduced by one-half, precipitation eventually occurs. It is evident that a dye-lake only remains in permanent solution if the mordant is present in great excess.

When a piece of tissue is mordanted with iron alum and then treated with haematein, elastic fibres are not coloured. When, however, the dyed tissue is put once more into iron alum, these fibres take up the colour.[356] This shows that during differentiation by the mordant a soluble lake is formed, which is capable in certain circumstances of dyeing, and also that the dye-lake penetrates the elastic fibres more easily than its two components separately.

In the differentiation of mordant-dyes by acids, either the bond between the tissue and the mordant or that between the mordant and the dye might be attacked (or indeed there might be a ing of both bonds). Evidence obtained since the first publicatio of this book has proved conclusively, in the case of aluminium-haematein, that the acid attacks the tissue/mordant bond. This follows from the fact that the acid exerts exactly the same effect whether it is used before or after the dye (but after the mordant). Also, when the acid has acted after the dye, a second application of the dye does not result in an increase in the intensity of colour, which would be a necessary consequence if the acid had removed part or all of the dye but had left the tissue/mordant bond intact. The experiments on which these findings are based are described in a paper by myself in the *Quarterly Journal of Microscopical Science*, Vol. 103, No. 4 (1962).

The hydrogen ions of the differentiating acid presumably tend to reconstitute, in unionized form, those acidic groups in the tissue (phosphoric, carboxyl, hydroxyl, etc.) that were ionized when they made their first contact with the cationic metal complex. The reactions, however, must be more complex than these few words suggest. The bonds of the metal with the dye and with the tissue are not, in their final forms, electrovalent. Further, it must be remembered that the mordant solutions themselves, in the absence of added acid, have already quite a low pH (p. 212), and dye-lakes are nevertheless taken up by the tissues from such solutions. The addition of weak acids to mordant solutions does not have a big effect on pH. For instance, 5% potassium alum shows a pH of 3·2; if 3% of glacial acetic acid is added, the figure only falls to 2·45; if 5%, to 2·3. Now 5% chrome alum and 5% iron alum by themselves (p. 212) are considerably more acid than 5% potassium alum acidified in this way; yet their low pH does not prevent dyeing.

Another problem is presented by the peculiar resistance of the tissue/iron/haematein complex to acidity. Quite strong acid must be used to break this bond. This fact is very useful in microtechnique, for we can use weak acids in subsequent procedures without removing the lake from chromatin; but the reason for the difference from other lakes is not obvious. We have not yet a full explanation of the action of acids in loosening the tissue/mordant and mordant/dye bonds that are formed in the ordinary processes of microtechnique. It is probable that we should have had a better insight into it if there had been any process corresponding to differentiation in the dyeing of textile fibres. In industry, however, there would be no purpose in trying to colour one part of one cell (in wool, for instance) with a dye, while leaving another part colourless; and that is the whole purpose of differentiation in ordinary microtechnique.

Benda [55] long ago distinguished *oxydirenden* from *einfach lösenden* differentiators of mordant dyes. *Einfach* is not a happily chosen word, yet the distinction is on the whole a useful one. Potassium permanganate has been used as an oxidizing differentiator, and ordinary bleaching agents are also available for the purpose. The dye is presumably oxidized to a colourless substance. The effect is one of differentiation, because certain objects contain so much dyed matter that they still hold plenty of it when the background has lost all visible trace. It has also been suggested [322] that oxidizing differentiators may act on the metal of the mordant. Chromium trioxide has been used as a differentiator of iron haematein.[54] Here one cannot be certain of the relative rôles played by oxidation and acidity. The same applies to picric acid, which is a moderate oxidizer and a weak acid. It differentiates iron haematein slowly.[507]

Another oxidizing differentiator is potassium ferricyanide. This is used after tissues have been mordanted with potassium dichromate and then dyed with haematein. The technique was introduced by Weigert [530] for the colouring of myelin, and subsequently perfected as a histochemical test for phospholipids.[143, 31] The mordant is anionic chromium, which acts very differently from the cationic complexes that we have been considering in this chapter. The attachment of the metal to phospholipids has been discussed in the part of the book that deals with fixation (p. 130). The oxidizer seems to act, partly at least, upon the fraction of the haematein that has been taken up by the tissues directly, not in the form of a lake. This technique does not provide a typical example of the action of an oxidizing differentiator on a lake.

From what has been said it will be clear that a mordant is a salt, the metal of which can combine with certain tissue-constituents and can also be held in the chela-grip of certain acid dyes. It will have suggested itself to the reader that a converse to a mordant might exist, an acidic substance that could be taken up by certain tissue-constituents and could also be linked to basic dyes. Such 'converse mordants' do in fact exist. Many basic dyes that are familiar in microtechnique are used in this way in the textile industry. Tannic acid is often used as intermediary between fibre and basic dye. Cotton and linen have a remarkable power of taking up this acid from aqueous solutions, and they can subsequently be coloured by basic dyes, for which they have no direct affinity. The compound between tannic acid and a basic dye has not, however, the chemical complexity of a lake. The word 'mordant' loses some of its meaning if used to include tannic acid and similar substances. There is a suggestion of similarities that do not exist, even in converse form.

Tannic acid and similar intermediaries are only occasionally used in this way in microtechnique. The reason is this. The tissues consist as a general rule of an amphoteric background with acidic (basiphil) substances (especially chromatin) distributed in it in the form of separate objects. One first colours these separate objects with a basic dye or lake, and then the background with an acid dye. The value of a lake is that it is insoluble and does not dissolve while the acid dye is acting. A converse to a lake would only be useful if the separate objects in cells were basic (acidophil) and we wished to dye them with an insoluble substance before colouring the background with a basic dye.

Tannic acid is occasionally used in quite a different way, to prevent the escape of a basic dye that has already attached itself to an acidic object. Any dye that began to dissolve out would at once be precipitated. A blood-smear dyed with methylene blue/eosin, for instance, may be treated with a solution of tannic acid,[288] and this will retard or prevent the subsequent loss of methylene blue from chromatin. The dye attaches itself in accordance with its own electric charge and those of the tissue-constituents, and the tannic acid merely traps it. There is here a strong contrast to the action of mordants. When a mordant is used, the distribution of the colour in the finished preparation depends on the affinity of the various tissue-constituents for the metal.

The use of iodine by Gram [198] to hold gentian violet in certain bacteria is the most celebrated example of the use of trapping agents in microtechnique. The method was invented by the Danish pathologist in 1884, by accident. It was his intention to introduce a double-colouring technique for diseased kidneys containing casts (*Harncylindern*) in the tubules. He intended that chromatin should be blue with gentian violet and the casts brown with iodine dissolved in potassium iodide solution. Gram's hope was not realized, for the dye disappeared quickly from the sections on subsequent treatment with alcohol. Luckily for the cause of bacteriology, however, he decided to find whether the dye would again be quickly lost if the same method were applied to other organs, and he chose some that were infected with bacteria. The result was startling, for the bacteria were intensely dyed by the gentian violet, while all the tissue-constituents of the host organism lost every trace of blue in the alcohol used for differentiation. Thus the bacteria were rendered more easily visible than had previously been possible.

Gram found that only certain particular kinds of bacteria lost their blue colour in the alcohol, and this fact became the basis of an important technique for distinguishing bacteria as Gram-positive and Gram-negative. The method is still used in a slightly modified form to the present day. Crystal violet is usually substituted for gentian violet, which is a variable mixture of the former with related dyes. Another dye, of a contrasting colour, is often used subsequently. This tends to disguise or displace the dye trapped by iodine, and the term 'true Gram-positive' is sometimes restricted to those bacteria or other objects that retain the colour of the first dye when the second has been applied, and when 95% alcohol has been allowed to act subsequently for a certain period. There is, however, a large subjective element in all this, for the periods in the various fluids are arbitrarily chosen.

Most basic dyes can be substituted for gentian violet in Gram's technique, provided that a second dye is not used subsequently. Only eight are known, however, that give a satisfactory 'true Gram-positive' reaction, and these are all triarylmethanes.[46, 47] It was formerly held that pararosanilines were suitable, while rosanilines were not.[511] It may be recollected that rosaniline has a methyl group attached to one of the three aryl rings, while pararosaniline lacks this (p. 159). Dahlia, however, is one of the eight dyes that give a satisfactory 'true Gram-positive': yet this is one of the rosanilines, for it possesses the methyl group.

Gram-positive and Gram-negative bacteria can be distinguished by the use of crystal violet alone, without any trapping agent.[46] The differentiation in alcohol is so difficult, however, that the technique is not suitable for routine use in the bacteriological laboratory. It seems that iodine is used because it is extremely convenient rather than because it is theoretically necessary.

There is very strong evidence that the substance that retains crystal violet in Gram-positive bacteria is a ribonucleoprotein.[225, 226, 48] It may be mentioned, however, that this is not undisputed.[126] It is certain that a positive Gram reaction does not always denote the presence of RNA. Indeed, one would not expect this to be so, for crystal violet will behave like other basic dyes and is likely to be held by iodine wherever its affinity for acidic objects has caused it to be present in particularly large amount. In the spermatozoon of *Ascaris*, for instance, there is a large cytoplasmic inclusion, the 'refringent cone', which is an object consisting of highly acidic (basiphil) proteins. This naturally takes up a lot of crystal violet and is strongly Gram-positive.[382] It contains no RNA whatever.

The use of iodine to trap gentian violet in chromatin and thus allow slower differentiation in alcohol was introduced by Hermann,[228] whose technique involved the use of another dye as well. Gentian and crystal violets, trapped by iodine, are much used in modern chromosome studies.[367, 247, 278, 39] The method is valuable, for the cytoplasm is of glassy transparency, while the delicate chromosomal threads of early meiosis retain the dye.

Iodine is the most familiar trapping agent. It is seldom used in its blue, molecular form, but is nearly always dissolved in aqueous potassium iodide solution and thus presents itself as brown potassium tri-iodide, KI_3. Various other trapping agents are available, such as bromine, mercuric iodide, mercuric chloride, potassium permanganate, and picric acid.[350, 47] In vital studies, as we shall see (p. 294), ammonium molybdate is used to trap methylene blue. It is characteristic of all trapping agents that they precipitate certain dyes from aqueous solutions, and that the precipitated material has a low solubility in alcohol. Thus crystal violet is freely soluble in 95% alcohol, but the crystal violet/potassium tri-iodide precipitate will only dissolve at concentrations up to 0·07%. The crystal violet/potassium permanganate precipitate is even less soluble (up to 0·02%). All basic dyes that have been tested are precipitated by iodine, but a few of the precipitates are fairly soluble in 95% alcohol. Thus the neutral red precipitate dissolves at 0·43% and the rhodamine B at 0·61%. These dyes would not be selected for work with trapping agents. Very few acid dyes are precipitated by iodine. Acid fuchsine and aniline blue WS are exceptional in this respect.[46, 47]

Potassium permanganate traps crystal violet so effectively that it cannot take the place of iodine in the Gram technique, for it holds the dye even in Gram-negative bacteria.[47]

The way in which trapping agents work has not been fully established. Two possibilities present themselves. On one hand it may be that the basic dye attaches itself as usual to acidic objects and remains there subsequently in the presence of an extracting agent (ethanol) simply because iodine is also present, which instantly precipitates any dye that begins to be extracted. On the other hand it is possible that an object/dye/iodine complex is formed, which is not easily split by alcohol.

If the first possibility is true, the continued presence of iodine all round the object is obviously necessary. The Gram method depends upon the fact that iodine can enter Gram-positive bacteria in the form of potassium tri-iodide, but its outward diffusion in the form of iodine/ethanol (which is also brown) may be hindered by factors that did not prevent its entrance. It may be that Gram-positive bacteria have an external part, perhaps a special membrane or cell-wall,[57] that hinders the escape of iodine/ethanol. The density and permeability of the Gram-positive substance itself may also affect the escape of iodine. If iodine can escape, the dye will be extracted by alcohol.

On this view a bacterium could be Gram-negative because the dye could not enter it, or because there was little or no acidic material in it to hold a basic dye, or again because there was no special cell-wall or other feature having the property of hindering the escape of iodine. It is noteworthy in this connexion that if Gram-positive bacteria be crushed, they appear Gram-negative,[350] presumably because the iodine is free to escape into the alcohol. If 0·25% of iodine be added to the alcohol, the extraction of the dye from Gram-negative bacteria is prevented and they therefore appear to be Gram-positive.[257]

If this first possibility is true, it follows that iodine does not influence the affinity of the dye for the objects it colours, but only prevents its subsequent escape.

If the second possibility is true, the object/dye/iodine complex may have special properties of its own; that is, it may not only be less easily split by ethanol than the object/dye compound, but may also have special affinities, so that the objects coloured will not be exactly the same as those coloured by the dye alone. This appears to be the opinion of Panijel,[382] who has made a careful study of crystal violet trapped by iodine. We have, however, no knowledge of the chemical reactions involved, if indeed the triple complex is actually formed.

CHAPTER 12

The Differential Action of Dyes

The purpose of dyeing in microtechnique is nearly always to obtain *contrast* between the constituent parts of an object. If a dye were perfectly diffuse in action, the whole of a section or other microscopical preparation would be uniformly coloured by it. This result is not produced by any dye, for certain parts of the specimen are always somewhat more strongly coloured than others, even by the most diffuse acid dyes. The mere production of contrast between the specimen as a whole and its surroundings is seldom useful, though plankton organisms (for instance) may sometimes be dyed with no other intention than this, when the desired end is recognition by external characters rather than study of internal structure.

We have seen in chapter 10 that the different objects in a preparation may take up different amounts of the same dye, and that different dyes (a typical basic and a typical acid dye, for instance) may attach themselves differently to the same object; that is to say, one dye may dye it deeply, another slightly or not at all.

It is now necessary to consider in greater detail the way in which dyes may be used to give striking contrasts and thus exhibit clearly the diversity of the parts of a microscopical preparation.

There may be chemical or physical reasons for the stronger coloration of a particular object. It may be dyed strongly *either* because it possesses many chemical groups capable of reacting with the particular dye used; *or* because there is a lot of colourable matter in it per unit volume (that is to say, because it is very dense in the physical sense); *or* because it is easily permeable by the dye used, while other tissue-constituents are more difficult to penetrate. In short, depth of coloration is affected by *chemical affinity*, *density*, and *permeability*.

The matter is complicated, because these three factors may either act together or antagonize one another. Thus chromatin is chemically reactive (basiphil), dense, and permeable, and therefore easily dyed; red blood corpuscles are chemically reactive (acidophil) and dense, but relatively impermeable and therefore not easily dyed except by easily diffusing dyes. Also, different dyes are differently affected by the chemical affinity and permeability of the various tissue-constituents. The matter will be analysed by a separate consideration of each of the three factors affecting intensity of colouring.

Chemical Affinity

The simplest way of getting sharp colour contrasts is to take advantage of different chemical affinities by using a basic and an acid dye in succession. Not every pair of colours is suitable, however. It is an interesting fact that yellow basic dyes are scarcely ever used. This has come about by a process of natural selection, and no one seems to have mentioned the subject. The reason is curious. Basic dyes are above all dyes for chromatin, and chromatin exists in the form of separate objects in cells and never forms a background against which other cellular constituents are viewed. It is therefore desirable to stain it *darkly*, and to use a *light* dye for the background. Now when the colour-receptors in our eyes receive light near the middle of the spectrum, in the region of the yellow and greenish yellow, they are stimulated in such a way that the colour appears highly 'unsaturated'; that is to say, in this region of the spectrum there is an appearance of the adulteration of the light by whiteness and thus the colours appear pale, while the regions towards the ends of the spectrum are not diluted in this way. It would never enter anyone's head, therefore, to stain chromosomes yellow and surround these objects on all sides with cytoplasm stained blue, for the blue would make it difficult to see the chromosomes. We therefore choose our basic dyes from the regions towards the ends of the spectrum, while we usually avoid the ends when choosing our background dyes.

When a black dye is used to colour chromatin (iron haematein, for instance), any acid dye used merely to colour the background must somewhat reduce the contrast.

It might be supposed that we only needed one basic and one acid dye. It is true that we could dispense with many that are used, without detriment; nevertheless we could not limit ourselves to two. Differences of chemical affinity among acid dyes and among basic dyes account in part for this fact, though differences in capacity to penetrate are more important (p. 234).

A good example of a difference in chemical affinity among dyes is seen in the colouring of cellulose cell-walls. Most acid dyes have no affinity for these: they act chiefly by the linking of their anions to the amino- and other basic groups of proteins. As we have seen, however (p. 199), Congo red and some other dyes are able to act in quite a different way by forming hydrogen bonds with the hydroxyl groups of cellulose.

Certain basic dyes used in pairs exhibit small differences in affinity that can result in very noticeable differences in effect. The most familiar example is the mixture of methyl green (triaryl-methane) and pyronine G (xanthene). This was introduced by Pappenheim,[384, 385] who showed that the basiphil cytoplasm of certain leucocytes could be coloured red with pyronine, while chromatin became green or bluish green. Thus two basic dyes coloured two basiphil substances differently. The method was improved by Unna,[516] whose formula included phenol. There is no full and universally accepted explanation of the results obtained with Unna's mixture and its modern variants, but the following facts are relevant.

It will be remembered that if a basic dye be used at various levels of pH, certain tissue-constituents show themselves capable of combination with it even in very acid solution, while others are more sensitive to acidity and react with the dye only at somewhat higher pH (p. 194). Now if two basic dyes differed somewhat in their capacity to bind themselves to objects at a particular pH, it should be possible for a mixture of them to give differential colouring.

If the usual mixture of the two dyes be used at pH 1·5, the pyronine predominates everywhere over the methyl green; the converse is true at pH 9·3. At intermediate pH both dyes act, but not equally on all tissue-constituents.[184, 185] One sees now how Unna hit on the use of phenol by empirical experiments, for it gives the weak acidity that allows both dyes to act, and to act differentially. It is usual nowadays to buffer at pH 4·7 or 4·8.[86, 253]

At the appropriate pH, chromatin is coloured mainly by the methyl green and appears green, blue-green, or blue, while nucleoli and basiphil cytoplasm are red with pyronine. It is usually RNA that binds the red dye, but this must be confirmed by failure to colour after the use of ribonuclease. (For a very convenient source of this enzyme, see Bradbury.[87]) Similarly, reliance must not be placed on methyl green as an indicator of the presence of DNA, for as a basic dye it has general affinity for acidic tissue-constituents. Some objects that do not contain DNA have a very marked affinity for this dye. Certain globules in the 'vitelline' glands of the liver-fluke, *Fasciola*, are an example.[484] Attempts to raise this dye to the status of a histochemical reagent are misplaced. For a full discussion of this subject, see Sandritter.[437]

It is claimed [276, 277] that pyronine has a strong affinity for nucleic acids that are depolymerized and methyl green for those that are highly polymerized, and that the distinctive reactions are due to the fact that RNA occurs in the tissues in a feebly polymerized and DNA in a highly polymerized form. It is not known why the two dyes should differ in this respect, if in fact they do. It is to be remarked that in the absence of pyronine, methyl green will colour the chromatin in tissues that have been subjected to a sufficient degree of acidity to depolymerize DNA.[495]

Certain dyes are metachromatic; that is to say, they are capable of imparting one colour to certain objects and another to others. This forms an important distinction between one dye and another; but the difference is not exactly one of affinity, and metachromasy will therefore be considered separately (p. 243).

Density

Authors often say that the cytoplasm of a particular cell is 'dense', but in fact it is very difficult to find out how much matter there is in a microscopical object. We can tell that a nucleolus is denser than the nuclear sap if we see it fall under the influence of gravity in a living cell,[215] but this kind of opportunity seldom presents itself. Phase-contrast may help us, but there are plenty of traps for the unwary.[45] A study of the Becke line effect followed by the use of the interference microscope is the surest guide, but the actual measurement of the refractive index of microscopical objects is not in any circumstances easy. No information on the subject of density can be obtained by simply noticing the depth of colouring with a dye.

In life, the greater part of the cell is made up of protein chains intimately associated with water through $-\mathrm{C}\begin{smallmatrix}\diagup\!\!\!\diagup \mathrm{O}\\ \diagdown \mathrm{OH}\end{smallmatrix}$, $-\mathrm{OH}$, $-\mathrm{C{=}O}$, $-\mathrm{NH_2}$, and other hydrophil groups.[180] The substance resulting from this association usually has a refractive index not very far above that of water. Ground cytoplasm commonly gives figures in the neighbourhood of 1·353, but a less aqueous object like the mitochondrial *Nebenkern* of an insect spermatid gives 1·376.[429] When fixation takes place, the water-relations of the protein chains are entirely changed, and in a balsam preparation there is no water left. The protein has now a refractive index of about 1·52 to 1·54. It is precisely for this reason that we use Canada balsam as a mounting medium: it has almost the same refractive index as the 'dry' protein, and the latter is therefore transparent. The spaces between the protein fibres are now filled with balsam. If there was a lot of water associated with the protein in life, these fibres will be far apart; if there was little water, they will be close, so that there will be more fixed tissue per unit volume. In a word, the tissue will be *dense* in the strict physical sense.

Let us now imagine the dyeing of two fixed tissue-constituents, the one consisting of a greater length of protein chain per unit volume than the other (and therefore *denser*), but both exactly equal in the number of acidic and basic groups per unit length of protein chain. Any dye, whether basic or acid, will necessarily be taken up in greater quantity by the former, which will appear darker in the finished preparation. There will, however, be no obvious indication of the cause of the uptake of more dye. It might equally well have been due to an entirely different cause: not to any difference in density, but to the fact that the protein was (for instance) particularly acidic, and therefore bound to a lot of basic dye.

Metaphase chromosomes are genuinely denser than the surrounding cytoplasm, and the depth of their colouring is partly due to this. One would expect them to take up basic dyes strongly, but the protein constituent of the nucleoprotein can also be much more strongly coloured than the cytoplasm by *acid* dyes. In the same way the chromatin of the interphase nucleus can be much more deeply coloured than the nuclear sap by acid dyes. Indeed, deliberate use is made of acid dyes to show chromatin in certain techniques. It has already been mentioned (p. 205) that acid fuchsine was used for this purpose by Mallory,[320] and azocarmine by Heidenhain.[220] The colouring of the protein of chromatin by acid dyes is made easier if the nucleic acids are first eliminated by enzyme action. Ribonuclease and DNA-ase both help the subsequent action of acid dyes in colouring the protein of nucleoprotein.[259]

An interesting though complicated example of reliance on density for differential colouring is provided by Taenzer's method for showing elastic fibres. (The method is usually called Unna's, but Unna himself [512] attributed it to his pupil.) The dye used is orcein (p. 185), dissolved in strongly acid alcohol. Differentiation is carried out in alcohol, usually strongly acidified. The brownish-red colour is better retained by elastic fibres than by other tissue-constituents. The explanation appears to be as follows.[457]

Orcein is used in alcoholic solution because it is scarcely soluble in water. It has no particular chemical affinity for elastin. It is an amphoteric dye (p. 190). In neutral solution (40% ethanol), being now itself acidic, it colours collagen more strongly than elastin. On the more acid side of its iso-electric point (about pH 5·7), it is a feebly basic dye. It now no longer colours collagen strongly, on account of the mutual repulsion of the positive charges, but is taken up by various negatively charged objects. It diffuses easily if dissolved in 70% ethanol, and is thus able to enter tissue-constituents that would have been impermeable to it in weaker ethanol. Elastic fibres, the matrix of cartilage, acidic mucus, and chromatin are coloured rather strongly, other tissue-constituents feebly. Orcein is not sufficiently basic to have a particularly strong affinity for the three last-mentioned substances, which are more electro-negative than elastin. When the section is subsequently put in acidified ethanol, the dye is easily extracted because it is extremely soluble, and it would eventually be washed out everywhere. Density rather than electric charge now controls events. Since there is a lot of matter in the elastic fibres, a lot of dye is held by them; and when the somewhat less dense constituents (chromatin, etc.) have lost all visible remnants of it, enough still remains in the elastic fibres to show them clearly.

If this explanation be correct, the result is achieved partly because elastin is even denser than chromatin, partly because orcein is too feeble a basic dye (at the pH at which it is used) to allow the final result to be controlled primarily by electric charges.

The dyeing of elastic fibres by orcein is probably not controlled only by their density: permeability is likely to play a part.

Carminic acid, another amphoteric dye, behaves rather similarly to orcein; but much more of it is taken up by the chromatin, presumably because the dye is more strongly basic at the pH of the solution. During the differentiation in acid alcohol the dye is therefore retained as long by chromatin as by elastin.

Permeability

The permeability of the various tissue-constituents plays a more important rôle than chemical affinity in determining the differential action of acid dyes in microtechnique. This fact was clear to the genius of Ehrlich right back in 1879.

It has already been mentioned (p. 193) that Ehrlich had discovered a striking fact about the granules of certain leucocytes (eosinophils), namely, that they have a special affinity for acid dyes. These were what he called the 'α' granules. He observed that the 'β' granules of certain leucocytes of the rabbit were also acidophil. Leucocytes of this second kind do not occur in man. Ehrlich drew a distinction between those acid dyes (eosin among them) that diffused rapidly, and those (such as nigrosine and induline) that diffused slowly. The eosinophil leucocyte could most easily be distinguished from that containing the β granules by using eosin in a single solution with nigrosine or induline. He found that the mixture showed the α granules in the colour of the rapidly diffusing dye, and the β granules in that of the slowly diffusing. It was for this reason that he named the former kind eosinophil.

> 'A consequence of this consideration', he wrote, 'is the hypothesis that the α granulations are of closer texture (*dichter*) than the β granulations; that is, that in the former the groups of molecules (*Micellen* of Naegeli, *Syntagmen* of Pfeffer) are larger and the intermicellar spaces smaller than in the latter. . . . The molecules of the easily-diffusing eosin penetrate much more quickly into the narrow osmo-regulatory spaces of the α granulations than those of nigrosine, which diffuses with difficulty; and so the micelles of the granulations are already saturated with eosin before the second dyestuff can enter them at all. In contrast to this, the molecules of nigrosine can enter the wider intermicellar spaces of the β granulations and so achieve an important colour-effect.' [153]

Thus Ehrlich showed that while differential dyeing was in some cases caused by the chemical differences between basic and acid dyes, in others it was due to physical factors in dyes and objects. At the time he was 24 or 25 years old (fig. 24, opposite p. 193).

The differential action of the rapidly and slowly diffusing dyes is reflected in their industrial use. The professional dyer classifies the acid dyes according to the way in which they can best be used to colour textile fibres.[524,75,523,315] It is desirable to mention this classification, partly because one can only follow the textile literature if it is understood, partly because it has direct significance for microtechnique. The classification is not based on chemical relationship: dyes in a single group may possess quite different chromophores. The limits of some of the groups are not sharp, and the textile authorities do not all use exactly the same classification. Three groups will be defined here.

The 'levelling' dyes are acid dyes that are used to colour wool and other protein fibres from a strongly acid bath. They have not a very high affinity for such fibres and will not dye them at neutrality, and they have no affinity for cellulose fibres. The word 'levelling' (or 'equalizing') means that they dye very evenly. They are often called simply 'acid dyes',[523,315] in reference to the pH of the bath in which they are dissolved, but this confusing name will not be used here.

The 'milling' dyes colour protein fibres very strongly at low pH, but their action is so uneven that they are not used in this way in practice. They are dissolved, on the contrary, in neutral or weakly acid solution and are therefore sometimes called 'neutral-dyeing' dyes. Their special character is that they are not decolorized nor extracted by 'milling', which is a felting process applied to wet wool after dyeing, often in the presence of soap. It is characteristic of milling dyes that they are of greater molecular weight than the levelling ones, and form colloidal solutions.[524]

The 'direct cotton' dyes colour cellulose fibres without the use of any mordant.

The relevance to microtechnique of this classification of acid dyes will appear shortly.

Acid dyes are very often used in pairs. Sometimes the dyes are mixed together, sometimes one is used after the other. If the affinities of the two were exactly the same, no advantage would be secured, for the appearance given would be exactly the same as though only one dye had been used (apart from the colour being mixed). In fact, however, the dyes are selected in such a way that certain objects are coloured by one of them, others by the other, and some (usually) by both. Sometimes three acid dyes are used.

One component of the pair (or trio) colours collagen fibres, another the ground cytoplasm. Other tissue-constituents are coloured by the one or the other or by both, but for our present purpose it will suffice to concentrate on collagen and cytoplasm. The dyes that colour the cytoplasm from these pairs are mostly the ordinary background dyes used for contrast with the basic dyes that colour chromatin. The most typical of these are eosin (xanthene), orange G (azo), ponceau 2R (azo), and especially picric acid (nitro). It is interesting to notice that most of the dyes used in microtechnique for colouring the background are levelling dyes.

The most typical of the dyes for collagen are aniline blue WS and the closely related methyl blue (triarylmethane), induline WS and nigrosine W (azine), diamine blue 2B and naphthol black B (azo), and indigo-carmine (indigo dye). The striking fact about this apparently random set of dyes is that none of them belongs to the 'levelling' group. Methyl blue, for instance, can be used for cotton, and is indeed sometimes called 'cotton blue'; diamine blue 2B is a direct dye for cotton; and others in this group, to a greater or lesser extent, have the characters of milling dyes.

In brief, then, the feeble levelling dyes colour the cytoplasm, the vigorous milling and cotton dyes colour collagen. Why?

The first clue was obtained in the twenties by a Frenchman, Collin,[129, 130] who made a special study of Mann's methyl blue/eosin.[321, 322] He dissolved gelatine at various concentrations in warm water, dipped microscopical glass slides in these solutions, dried the films thus produced, and fixed them in a mixture of formaldehyde and alcohol. After washing and again drying them he put them in Mann's mixture. The films made from concentrated solutions of gelatine took up the red colour of eosin, those from weak solutions the blue of methyl blue.

These results suggested that methyl blue was not able to enter concentrated gelatine, but that wherever it could enter and compete with the eosin, it dominated the latter. In other words, the blue dye was more vigorous in action, but penetrated with greater difficulty than the red.

Collin now showed that eosin penetrated much more rapidly than methyl blue into gelatine gel contained in a test-tube, and also that if Mann's mixture was put in a collodion sack, and the sack in water, the water was coloured by eosin before the methyl blue escaped. A solution of methyl blue, filtered, contains particles that are visible under the microscope: a solution of eosin does not. Various objects (blotting-paper, animal charcoal, etc.) take up much more methyl blue than eosin from solutions at the same concentration. The blue dye is well retained, the red easily washed off.

Möllendorff[355] reached essentially the same conclusions from independent studies. As he put it, from mixtures of two acid dyes, the more diffusible goes into the more compact structures, the more colloid into the more pervious.

This subject was subsequently investigated in detail by Seki, who noted the rate of penetration of dyes into agar gels contained in test-tubes. In his experiments, of which full particulars are given in his paper,[452] the distances penetrated by certain acid dyes in 15 hours were these:—

picric acid	25 mm
acid fuchsine	10 ,,
nigrosine W	7 ,,
methyl blue.	5 ,,
aniline blue WS	4 ,,

Details of a similar experiment are given in the Appendix (p. 322). The distances penetrated in 48 hours were as follows:—

orange G	37 mm
methyl blue.	16 ,,
aniline blue WS	12 ,,

It will be noticed that the dyes that penetrate quickly are the ones that colour the cytoplasm and those that penetrate slowly are the ones that colour collagen, when suitable pairs are used.

The fact that we are not concerned here with special chemical affinities, but rather with the factors that have been mentioned, is shown very clearly by the behaviour of acid fuchsine. This is an intermediate kind of dye, for it is level-dyeing, yet moderately fast to milling,[270] and it penetrates at moderate speed. In Mallory's technique[320] it is used with aniline blue, and here it colours the cytoplasm; but in various techniques, of which Hansen's[209a] is an example, it is mixed with picric acid, and now it colours the collagen. It is clear that we are not concerned here with specific affinities, but with the relative positions of the various dyes in a scale of characters. The extremes in this scale are aniline blue WS and picric acid.

The dyes used for collagen mostly have higher molecular weights than those used for cytoplasm (e.g. methyl blue 800, picric acid 229). This, however, is not the direct cause of the difference in rate of penetration. Many dyes are dispersed in particles that are larger than single anions. Estimates of the size of the particles can be obtained by measurement of osmotic pressure or electrical conductivity or by ultra-centrifuging or ultrafiltration through cellophane or collodion film.[461, 75] It is clear that the anions of a dye often aggregate together. Sometimes the particles are formed by the aggregation of several anions only, sometimes by the aggregation of several anions with a smaller number of sodium or other cations, the remaining cations being free. The resultant charge on the dye particles of an acid dye therefore varies in different cases, though it is always negative. It is probable that ion-aggregates break up below the temperatures at which textile dyeing takes place, but they are present at room temperature and play an important part in microtechnique. The dyes used for collagen are the ones that form large aggregates, while the typical background or levelling dyes are dispersed as single ions.

The size of the spaces or pores in the constituent parts of fixed microscopical preparations is not known, but the structure of wool may give us some impression of what to expect. The protein is here in the form of long, nearly parallel chains. Here and there the chains become quite parallel and closely bound together, so that a submicroscopic crystal or micelle is formed. The chains emerging from the end of a micelle wander loosely for a bit before they enter and form part of several other micelles. Thus there are minute crystalline and non-crystalline regions in the fibre, the former too compact to be entered by any dye, the latter loose and containing spaces between the somewhat irregularly arranged threads. The available evidence suggests that the diameter of these spaces is about $3\frac{1}{2}$ or 4 mμ.[523] Dyes must enter these if they are to permeate and colour the fibre. The cellulose fibres of cotton are arranged in a very similar way, the spaces being about 2 to 10 mμ in diameter.

It is probable that the various constituents of a fixed microscopical preparation vary greatly in the size of the spaces within them.[453, 461] Collagen is an example of a substance of very loose texture, readily entered by any dye; cytoplasm has a tighter consistency and is more selective towards dyes; the contractile substance of muscle is somewhat tighter still; while the red blood-corpuscles of mammals are among the least pervious of all tissue-constituents. The dyes that colour red blood-corpuscles are those that diffuse particularly easily through a fine collodion membrane. With three selected acid dyes one can colour collagen, ordinary cytoplasm, and red blood corpuscles in three different colours; for instance, with aniline blue, acid fuchsine, and orange G respectively in Mallory's technique.

The fact that it is so particularly easy to colour collagen and red blood corpuscles differently is interesting. Both of these are examples of strongly basic substances, for collagen contains a high proportion of arginine and lysine, and the globin of haemoglobin is rich in these and in histidine. One would therefore expect them both to be strongly acidophil. So indeed they are; but red blood corpuscles are so impermeable that the most powerful milling dyes can scarcely enter them. Thus in the differential action of acid dyes, physical or mechanical factors predominate over chemical affinities.

Some of the powerful but slowly diffusing dyes used for collagen colour collodion in acid solution.[458] Thus aniline blue and methyl blue dye it strongly from pH2 to pH5. This is unusual behaviour for acid dyes. One would expect the negatively charged anion to be repelled by the similarly charged substrate. These two dyes have some capacity to act as though they were basic, and indeed aniline blue possesses an amino-group. Once again acid fuchsine is intermediate between these and the levelling dyes, for it colours collodion moderately from pH2 to pH6.

Induline colours collodion less strongly at low pH, because it tends to flocculate.

When it is desired to colour collagen differently from other tissue constituents, use is often made of phosphomolybdic acid. The techniques employed are variants of the procedure introduced in 1900 by the American histopathologist, Mallory.[320] In his technique sections were treated with an aqueous solution of phosphomolybdic acid and then with a mixture of aniline blue, orange G, and oxalic acid in water. The oxalic acid served simply to lower the pH and thus help the action of the levelling dye, orange G. (The dyeing of the chromatin is irrelevant and will not be considered here.)

Molybdenum is a metal related to chromium. The yellowish white oxide, MoO_3, insoluble in water, dissolves in ammonia solution to produce ammonium molybdate (p. 294), which reacts with orthophosphoric acid, to produce ammonium phosphomolybdate; this, when dissolved in *aqua regia*, deposits pale yellow crystals of phosphomolybdic acid. The composition of these crystals is not quite constant, but approximates to $H_3PO_4(MoO_3)_{12}$, with water of crystallization.

The function of phosphomolybdic acid in Mallory's and similar techniques was explained by the researches of Möllendorff [355] and Seki.[461]

If an Irish bull be permissible, one may say shortly that phosphomolybdic acid acts as a colourless acid dye (for it scarcely colours the tissues). Luckily one can convert it into a coloured substance by exposure to bright light. A blue lower oxide of molybdenum, of indeterminate composition, is produced. If a microscopical section be soaked in a solution of phosphomolybdic acid and then exposed to light, the blue colour reveals that it was present chiefly in the collagen; much less in the cytoplasm; less again in muscle; and least of all in red blood corpuscles. This is exactly the same distribution as is shown by methyl blue, and phosphomolybdic acid thus acts as though it were a very slowly diffusing acid dye. The anion is large, and its size is increased by hydration.

If a section be treated with phosphomolybdic acid and then with one of the background or levelling dyes at low concentration in the presence of the same acid, the background dye will colour nothing except the red blood-corpuscles. Thus the phosphomolybdic acid acts as a dye-excluder towards the background dye. One is reminded of the use in the textile industry of 'resists', or substances that prevent the subsequent action of dyes.[75] A mixture can be made of normal and 'resisted' wool, and this gives a variegated effect when dyed. Some of the substances used are colourless sodium sulphonates, which act very much as though they were colourless dyes.

When a section is first treated with a typical background dye, or with some other dye (such as acid fuchsine or azocarmine) that diffuses more easily than aniline blue, and then with phosphomolybdic (or phosphotungstic) acid, the latter competes with the dye wherever it can enter. It enters the collagen most easily. If the treatment be stopped at the right moment, the dye is turned out of the collagen, but left in the cytoplasm, muscle, and red blood-corpuscles. If now the section be rinsed and treated with aniline blue or a similar dye, the collagen will be coloured exclusively by this.

The treatment with phosphomolybdic acid also helps differential colouring in another way. In so far as it enters the cytoplasm and muscle and is taken up by them, it helps to exclude the aniline blue, and thus appears to favour the background dye. All these processes, however, have to be carefully controlled. If the aniline blue or similar dye is allowed to act for too long, it will spread to the cytoplasm and muscle and eventually replace the background dye.

It is particularly to be noticed that the phosphomolybdic acid *opposes* the action of the aniline blue. One sometimes sees statements to the effect that it 'mordants' the tissues for the aniline blue. Not only is it impossible for such a substance to mordant for an acid dye, but in fact the aniline blue colours everywhere more powerfully if the treatment with phosphomolybdic acid be omitted. To prove this it is only necessary to take two sections from the same block, put one of them in a solution of phosphomolybdic acid, and then colour both sections with aniline blue for the same period.

An object that is difficult to penetrate will resist the escape of a dye that has succeeded in entering it. One may take a dye that does not diffuse readily, heat it until the ion-aggregates have dispersed, allow it to enter the tissues in this form, and then cool the dyed object: the dye is unable to escape. This process is much more applicable to textiles than to microscopical preparations, for the former are nearly always dyed at high temperatures. Polar yellow R, for instance, will not enter wool at all below 40° C, because the ion-aggregates are too large.[524] Similarly one can scarcely colour mitochondria strongly with cold acid fuchsine solution, but the dye enters them readily when the temperature is raised to near boiling point. If subsequently the section be treated at room-temperature with another acid dye, even a readily-diffusing one, the acid fuchsine will be replaced in the cytoplasm before it leaves the mitochondria. This is probably the basis of Metzner's [341] and several other methods for mitochondria, in which acid fuchsine is used hot and another dye (or dyes) at room-temperature subsequently. In Altmann's original method,[5] hot acid fuchsine was followed by warm picric acid. He himself admitted that the differentiation was difficult. It is far easier to use cold picric acid solution, as Metzner [341] did. (See also Meves [342] for details of Metzner's method.)

That this is the correct explanation of the usual mitochondrial methods is suggested by the fact that red blood-corpuscles generally retain the acid fuchsine in such preparations. Mitochondria can also be coloured by the basic dye, crystal violet, used hot.[55, 56]

Chemical affinity, density, and permeability play their allied or antagonistic parts in the colouring of tissue-constituents, and it is difficult to disentangle their effects. If, in any particular case, we can be sure that there is no obstacle to penetration, and if we know the chemical constitution of the substance that reacts with the dye, and if further the substance is chemically homogeneous or nearly so (unlike most proteins as they occur in the cell), we may be able to estimate the amount of the substance present by measuring the optical density of the dye taken up by it.[361] Thus a basic dye may be used, at a pH too low to colour protein, to obtain an approximation to the amount of DNA in a nucleus or of RNA in a nucleolus or in the cytoplasm. A mordant-dye, chrome alum/gallocyanine (p. 215), is particularly recommended for this purpose.[473]

CHAPTER 13

Metachromasy

The words metachromatic and metachromatism were introduced in 1876 [1,2] in reference to the changes in colour undergone by certain substances when heated. The adjective, however, is used in biology in a different and very special way. The corresponding noun is metachromasy or metachromasia.

If a section is dyed with toluidine blue, many tissue-constituents will be coloured blue, but if there is any cartilage in the preparation its matrix will be dyed purple or red. It might be thought that the dye was impure, owing to faults in manufacture or subsequent changes in chemical composition, but the effect is observed equally well with pure specimens of the dye. This is a typical example of metachromasy. Toluidine blue is said to be a metachromatic dye. The matrix of cartilage is called a chromotrope: that is to say, a substance capable of altering the colour of a metachromatic dye. The corresponding adjective is chromotropic. The word orthochromatic is used to mean non-metachromatic. Thus whatever is coloured blue by toluidine blue is said to be dyed orthochromatically, and dyes that do not give metachromatic effects are called orthochromatic.

Certain dyes are not stable in solution, but gradually give rise to other colouring agents, which are then present as impurities and can be separated by suitable means. The presence of such impurities of spontaneous origin is called allochromasy. This has no necessary connexion with metachromasy, but an orthochromatic dye may give rise to metachromatic impurities by allochromasy. Solutions of Nile blue are allochromatic, for they contain not only the ions that one would expect, but also another substance, an oxazone (p. 301); in addition to this, however, the cation of the dye is feebly metachromatic.

Metachromasy was discovered independently three or perhaps four times during the year 1875. The evidence suggests that Jürgens [255] was the first to report his findings publicly. He used dahlia in a study of amyloid degeneration and was surprised to find that this violet dye coloured the amyloid corpuscles a brilliant red. The celebrated histologist, Ranvier,[412] dyed cartilage metachromatically with cyanine, while another Frenchman, Cornil,[132] used methyl violet on the same two chromotropes that were observed by Jürgens and Ranvier. It is possible that the Austrian pathologist, Heschl,[231] also saw a metachromatic effect in 1872 and published it in 1875, but this is not certain. He accidentally dyed the skin of his fingers with some violet ink and then tried it on various other tissues, including liver and kidney in amyloid degeneration. The degenerate parts were coloured dark rose-red, everything else blue. This effect may indeed have been due in part at least to the metachromatic dye, aniline blue (spirit soluble), which was present in the ink; but since this also contained basic fuchsine, one cannot be sure. (In this paragraph the modern names of the dyes have been used throughout.)

It has been stated more than once [308, 310, 347] that the word metachromasy was introduced and defined by Ehrlich in his paper of 1877.[151] Very remarkable things are believed about this paper, by persons who have not read it. It has been said that he here introduced the idea of classifying dyes as acid and basic,[312] and objects in tissues as basiphil and acidophil.[310] In this paper, written when he was still a medical student, Ehrlich gives a competent account of the form, distribution, and reaction to dyes of what were obviously the basiphil cells (*Mastzellen*) of connective tissue, though he refers to them throughout as *Plasmazellen* (see Westphal [534]). He notes the colour-change of the dye, but makes no attempt to define metachromasy and does not use the word. Ehrlich's first scientific paper foreshadows rather faintly his subsequent contributions to our understanding of the action of dyes in biological microtechnique.

In a later paper [152] Ehrlich remarks that certain dyes colour the granulated cells (*Mastzellen*) of connective tissue 'metachromatically, that is, in a tint differing from the colour of the dye used'. It is more accurate to define metachromasy as the colouring of different tissue-constituents in different colours by a single dye. The words 'single dye' must here be taken to mean that the substance that dyes the different tissue-constituents in different colours can be extracted from the dye-solution in dry form as one pure chemical compound, not as two or more.

In this sense there are both basic and acid metachromatic dyes, but whereas the basic ones are important in microtechnique and all act on the same chromotropes, apparently in essentially the same way, the acid metachromatic dyes are relatively unimportant and act quite differently from the basic ones.

The metachromasy of basic dyes will be dealt with first. Unnecessary repetition of the words 'of basic dyes' will be avoided. All general remarks on metachromasy are to be understood as referring to the metachromasy of basic dyes, unless the contrary is clearly indicated.

Metachromasy is of importance in histochemistry, because very simple techniques give striking results that help towards the chemical identification of tissue-constituents.

The most obviously chromotropic tissue-constituents are the following:—

- the matrix of cartilage;
- the secretions of certain mucous glands;
- the granules of the basiphil cells (*Mastzellen*) of connective tissue;
- the corpuscles of amyloid degeneration;
- the 'volutin' granules that occur in yeast and in certain diatoms and bacteria.

Certain substances prepared from the cell-walls of various red algae are strongly chromotropic. Agar is an example.

Various tissue-constituents other than these are also chromotropic, but less strikingly so. Chromatin is an example of a weakly chromotropic substance. All chromotropes that occur in microscopical preparations are basiphil, though not all basiphil objects are chromotropic.

The first person to study metachromasy in detail by the examination of pure substances in glass vessels was Lison,[308] whose work has been extended by Sylvén.[494]

A large number of substances that are very familiar components of organisms are not chromotropes. The following are examples:— *d*-ribose, sucrose, maltose, cellobiose, lactose, dextrins, glycogen, starch, hemi-cellulose, cellulose, inulin, gums, mucilages, pectic substances; serum-albumin, serum globulin, fibrin, collagen, keratin, myosin, silk. No lipid is obviously chromotropic, though cerebrosides may perhaps be feebly so. It is characteristic of chromotropes that they are acidic, and this is of course related to the fact that the chromotropic objects seen in microscopical preparations are basiphil.

The acidic groups present in chromotropes are sulphuric, phosphoric, and carboxyl. As Lison [308] showed, most of the familiar, strikingly chromotropic objects owe their character to the presence of sulphuric esters of polysaccharides of high molecular weight. It will be recollected that a sulphuric ester differs from a sulphonic acid by the possession of an extra oxygen atom linking the organic radicle to the sulphur. Several such com-

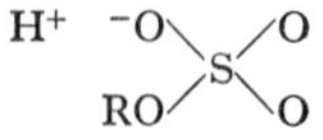
Sulphuric ester

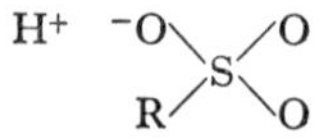
Sulphonic acid

pounds are mucosubstances. Chondroitic acid, for instance, makes the matrix of cartilage chromotropic; mucoitic acid does the same for certain mucous secretions; heparin for the granules of *Mastzellen.* Chromotropic sulphuric esters need not, however, be mucosubstances. Agar, for instance, is the calcium salt of a sulphuric ester of a pentose polysaccharide that lacks any amino-group.

The phosphate group does not generally confer such a strongly chromotropic character as the sulphuric. Adenosine triphosphate is not chromotropic, RNA only very slightly so, DNA rather weakly (see p. 258). Some of the metaphosphates, however, are rather strongly chromotropic. These substances tend to polymerize. The general formula for potassium metaphosphates is $(KPO_3)_n$. When n is very large, the substance is strongly chromotropic; when moderate, weakly; sodium trimetaphosphate is not chromotropic at all. A potassium metaphosphate of rather high molecular weight can be extracted from the mould, *Aspergillus niger*, and this is strongly metachromatic. It has been shown [538] that the metachromatic particles present in yeast (*Saccharomyces cerevisiae*) and certain bacteria contain a metaphosphate associated with protein. There seems to be no doubt that this substance corresponds to the 'volutin' of earlier authors.

The carboxyl group has less chromotropic effect than the phosphate. It occurs in uronic acids as a component of many mucosubstances. Some of these are sulphuric esters, and the sulphuric component then overshadows the carboxylic in chromotropic effect. When, as in hyaluronic acid, there is no acidic group other than the carboxylic, the substance is only feebly chromotropic.

Other acidic radicles than sulphuric, phosphoric, and carboxylic do not confer the chromotropic property. Thus nitrocellulose is acidic and therefore basiphil, but not a chromotrope. (For the reaction to dyes of nitrocellulose in the form of collodion, see p. 193.)

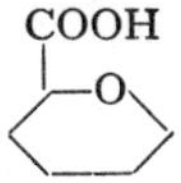

Skeleton formula of glucuronic, galacturonic, and mannuronic acids

There is a strong correlation between degree of polymerization and chromotropic effect. Thus glucuronic acids by themselves and hyaluronic acid are scarcely chromotropic, while sodium alginates (high polymers of mannuronic acid) and concentrated gels and solid films of hyaluronic acid show a definite colour-shift.[494] There appear to be two separate factors affecting the degree of chromotropy achieved: the nature of the acid groups, and their degree of separation in space. With any particular acid group, the greatest colour-shift will be shown by a highly polymeric gel or film crowded with sites of negative charge.[494] This crowding will naturally result also in strong basiphilia towards orthochromatic dyes.

It was shown by Lison [308] that various non-chromotropic high polymers occurring in the tissues of organisms, such as glycogen, starch, cellulose, gum arabic, and chitin, can be rendered strongly chromotropic by their artificial conversion into sulphuric esters. Sylvén [494] has extended this work by observing the gradual increase in degree of chromotropy as more and more carboxymethyl groups are introduced into cellulose.

Similar effects can be observed in microscopical preparations. If a section be treated briefly with concentrated sulphuric acid, any neutral polysaccharides will be transformed to sulphuric esters and will therefore become basiphil and metachromatic.[74] Glycogen and neutral mucopolysaccharides (such as the 'mucoid' of the cells lining the mammalian stomach) give this reaction. The same result may be achieved in a different way, by simply placing a section in a solution of chromium trioxide.[73, 125] Lison [310] considers that a $>C<^{OH}_{H}$ group of the saccharide component is oxidized through aldehyde to $-C^{\nearrow O}_{\searrow OH}$, presumably with breakage of the ring.

Certain inorganic substances with complex anions are markedly chromotropic.[43, 494] Among these anions are ferricyanide, thiocyanate, and especially phosphotungstate. The subject has not been fully investigated, presumably because it is not of much interest to biologists, who are the chief people concerned with metachromasy.

More than two dozen basic dyes are known to be definitely metachromatic, most of them being triarylmethanes and azines. No azo-dye is metachromatic except Janus green, which owes this character to the fact that it is also an azine. Among the lake-dyes certain oxazines are remarkable for giving strongly metachromatic effects.[234, 407, 296] The most useful metachromatic dyes are probably these:—

methyl violet (triarylmethane)
brilliant cresyl blue (oxazine)
cupric, ferric, and aluminium lakes of coelestine blue (oxazine)
thionine (thiazine)
azure A (thiazine)
azure B (thiazine)
toluidine blue (thiazine)

It will be noticed that the thiazines are pre-eminent in providing us with valuable metachromatic dyes.

Certain dyes show their metachromatic effect when used vitally. These will be considered in one of the chapters devoted to vital dyeing (p. 281).

It will be convenient to consider first an ideal dye that has a nearly symmetrical absorption curve with a peak in the middle of the visible spectrum. Basic fuchsine would serve as a fairly good example, though the curve is not very symmetrical. Such a dye will necessarily transmit light from the two ends of the spectrum (compare fig. 17, p. 161), and if the transmission curve is regular on the two sides, the colour will be purple. (Basic fuchsine transmits a high proportion of red and the colour is magenta.)

If now we somehow influence our dye in such a way that the peak of its absorption curve (or the trough of its transmission curve) is slowly shifted towards the right (that is, towards the longer wave-lengths), the transmitted colour will change gradually from purple through violet to blue and then to green. The colour is said to be 'lowered', and the mean wave-length of the transmitted light has indeed been lowered. The influence we have brought to bear is therefore said to be 'bathochrome'. If on the contrary we somehow shift the peak of the absorption curve of our purple dye towards the left, we 'heighten' the colour through magenta to red, orange, and then yellow. The influence has been 'hypsochrome', since the mean wave-length of the transmitted light is now higher (longer). It is obvious that if either a bathochrome or a hypsochrome influence were pushed far enough, so that the absorption curve were moved right outside the visible spectrum, the transmitted light would brighten to white, and that is indeed why the two ends of the scale (green and yellow) are rather similar.

We are now in a position to lay down a general law about metachromasy. *The metachromatic effect is hypsochrome.* This law is subject only to the exceptions mentioned on pp. 258 and 259. Colours may be arranged in an order of increasing hypsochrome effect, thus: green, blue, violet, purple, magenta, red, orange, yellow. The metachromatic colour given by a dye is to the right of the orthochromatic in this list. It follows that no metachromatic dye can be yellow, for this would render chromotropes colourless. All the most valuable metachromatic dyes are blue or violet, and the colour-shift is generally from blue to purple or magenta, or from violet to red. The normal human eye is very sensitive to these changes. If a red dye had its absorption-maximum shifted by the same amount in wavelength as one of these dyes, the change in colour would not appear to us so striking. We seldom choose a red dye, such as neutral red or safranine (azines), when we want metachromatic effects.

In passing from the orthochromatic to the metachromatic colour the peak of the absorption curve may move a particularly long distance; or, alternatively, the colour-shift may be one that is particularly evident to the human eye, even though the peak of the curve may not have moved very far. Thus a dye may be considered highly metachromatic for one of two alternative reasons, which may be called respectively objective and subjective.

The colour-shift could be pictured mentally as a bodily movement of the absorption curve across the spectrum, without any change in its form; but in fact the shift is more complex than this.[347, 346, 467, 468]

When a metachromatic dye is dissolved in absolute alcohol or other organic solvent, the absorption curve shows a single main peak, called the α band. With toluidine blue the peak is at 630 mμ or a little less, in the reddish orange. The transmission curve naturally shows a trough, which will here be called the α trough. A glance at the transmission curves (fig. 27, A) suggests rightly

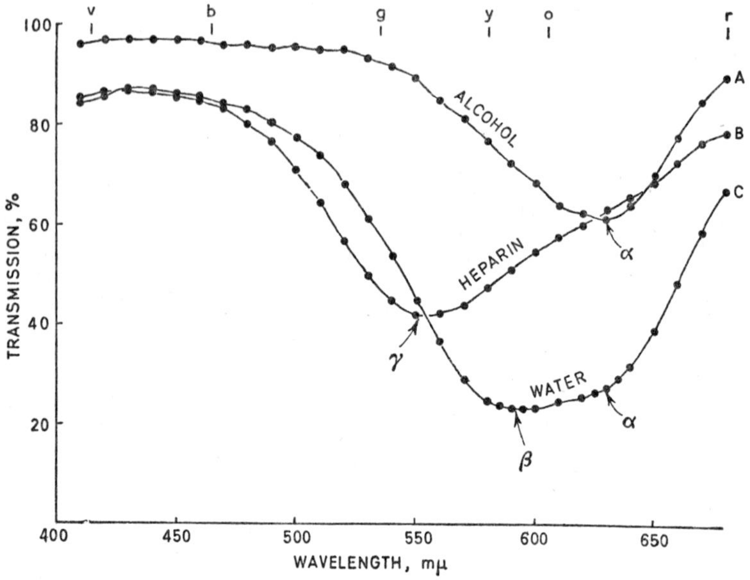

FIG. 27. Graph showing the transmission of light of various wave-lengths through a layer of toluidine blue solution 1 cm thick.[38]

A, the dye was dissolved at 0·0002% in 80% alcohol; B, at 0·001% in distilled water; C, the same as B, with the addition of 2 drops of heparin solution to 3·4 ml of solution. The heparin solution used contained 5000 international units per ml. The α, β, and γ troughs in the curves are marked by arrows.

that the colour of the solution is blue. When the dye is dissolved in aqueous solution a second or β trough appears, corresponding to a β band or hump in the absorption curve. It is claimed that this can just be detected even in non-aqueous solutions.[468] It is not ordinarily detectable in very dilute aqueous solutions, but becomes more and more marked as the concentration rises, while the α trough becomes shallower. The β trough is always situated on the side of the α trough towards the shorter wave-lengths. In fig. 27, C, the β trough is deeper than the α, and the two are almost smoothed out into one. The wave-length of the β trough of toluidine blue is about 590–600 mμ (yellowish orange).

When a strongly chromotropic substance is added to the solution, a new trough (γ trough) appears, still further in the direction of shorter wave-lengths. The addition of quite a small amount of an intensely chromotropic substance, such as the heparin of *Mastzellen*, suffices to hollow out a considerable γ trough at the expense of α and β (fig. 27, B). With toluidine blue the wave-length of the γ trough is about 550 mμ (yellowish green). The colour of the solution is purple, since light is freely transmitted from both ends of the spectrum.

Varying degrees of colour-shift will be produced by variations in the tendency of chromotropes to flatten out the α transmission trough and deepen the β and γ. Most chromotropes produce their main visible effect by lowering the γ trough, though they lower the β at the same time. RNA, however, a feebly chromotropic substance, makes a low β trough at certain concentrations, without affecting the γ region.[346]

Some metachromatic dyes show quite a low trough in the transmission curve in the region of the ultra-violet, but this is not affected by the presence of chromotropes.[347]

We naturally ask ourselves whether there are any features of chemical composition that separate metachromatic from orthochromatic dyes. Certain general remarks may be made under this head.

Dyes in which all the $-NH_2$ auxochrome groups have the hydrogen atoms replaced by $-CH_3$ or $-C_2H_5$ are not metachromatic. This applies, for instance, to crystal violet and methylene blue. These dyes are often regarded as somewhat metachromatic, but there is reason to believe that they would be quite orthochromatic if perfectly pure.[308] The former is generally contaminated with the highly metachromatic methyl violet, and methylene blue always gives rise by allochromasy to the azures, so that perfectly pure specimens have not been obtained.

Not all dyes that have an unsubstituted $-NH_2$ group are metachromatic. It is a remarkable fact that the dyes that are metachromatic are, in general, those that are capable of being transformed to imino-bases. The change from an $=\overset{+}{N}H_2$ group to an imino ($=NH$) group does not involve a loss of quinonoid structure in the ion. The imino-bases are therefore coloured, unlike the leucobases. The chemical structure of the imino-base of pararosaniline, for instance, may be compared with that of the leucobase (p. 163). The transformation from dye to imino-base involves a hypsochrome effect, for the dye is magenta, the base reddish orange. Pararosaniline is metachromatic, and it exemplifies the general rule that the transformation of a metachromatic dye to its imino-base involves a heightening of colour: that is to say, a change

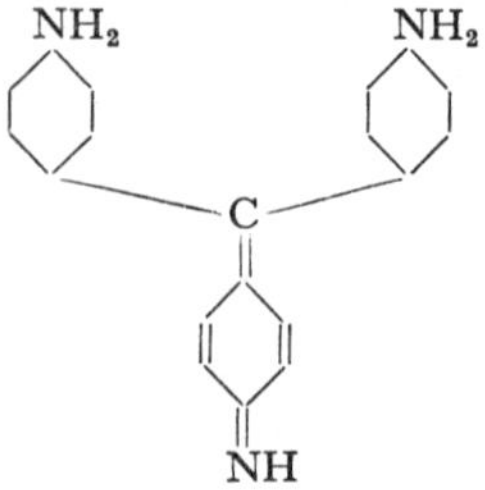

Imino-base of pararosaniline

in the same direction as the metachromatic shift. Because the colour changes when the imino-base is produced, and because a certain degree of alkalinity is necessary for its production, it follows that metachromatic dyes can be used as indicators of pH.

It is most tempting to assume, as Hansen [211] did nearly half a century ago, that chromotropes take up imino-bases from solutions of metachromatic dyes. He considered that these bases were present in ordinary solutions of the dyes and that they were somehow specifically stored up by the chromotropic tissue-constituents. He remarked that if one shakes up benzene, xylene, or chloroform with an aqueous solution of thionine, the little oily droplets give the impression under the microscope that they are composed of 'mucin', because they have the metachromatic colour.

Although there may be some roundabout connexion between the capacity of a dye to form an imino-base with heightened colour on one hand and its capacity to give a metachromatic effect on the other, yet Hansen was in fact mistaken, for these two kinds of colour change are not due to the same cause. As Lison [308] pointed out, metachromatic dyes give their colour-shift at a pH far below that at which an imino-base could exist. Brilliant cresyl blue, for instance, is pure blue at pH 10 and only becomes orange-red at higher pH than this, but the dye can act metachromatically down to pH 3 or even lower. There is no question of the chromotropes being local alkalinizing agents, since, as we have seen, they are acidic. At a low pH, at which the dye will still act metachromatically, no imino-base can be extracted by chloroform. If any imino-base is in fact present in a particular solution of a metachromatic dye, it can be extracted with chloroform without affecting the capacity of the solution to show its metachromatic effect. Even more conclusive than these arguments is the fact that although the metachromatic colour and that of the imino-base are similar, yet they are distinct spectroscopically.[308]

It was pointed out long ago [238] that concentrated solutions of metachromatic dyes show to some extent the metachromatic colour. It is, in fact, quite clear that although there may be no allochromasy whatever, yet the dye may appear in two forms—orthochromatic and metachromatic—in aqueous solution. When, however, the dye is dissolved in ethanol, only the orthochromatic colour is seen, however concentrated the solution may be. This is shown in fig. 27. It has already been remarked (p. 250) that when ethanol is the solvent, the wave-length of minimum transmission (or maximum absorption) by toluidine blue is 630 mμ, while in water it is about 590–600 mμ (at the dye-concentrations used in the observations).

The factors that promote and antagonize the presence of the metachromatic form of the dye in solution have been specially studied by Lison.[308] As an aqueous solution is made more and more concentrated, there is a progressive approximation towards the metachromatic colour, while at extreme dilutions the colour is purely orthochromatic.

The following experiments would demonstrate the spectroscopic difference between an orthochromatic and a metachromatic dye. In practice one would carry them out rather differently, but the methods used would not differ in principle.

Take two exactly similar glass tanks, flat-sided and rectangular, and place them in a spectrophotometer in such a way that the light must pass through both. Fill both with water. Now add a weighed quantity of an orthochromatic dye to the water in the tanks, allow it to dissolve fully, and draw a curve to show the absorption at different wave-lengths. Methyl green would be a particularly suitable dye, because it shows no trace of metachromasy. A series of such experiments will show that if the same weight of the dye is used each time, it makes no difference to the curve whether all the dye is put in one of the tanks, or some in each, in any proportion. If, however, a metachromatic dye be used, this is not so. The absorption maximum will be at one wave-length when the dye is equally distributed between the two tanks, and at another (shorter) wave-length when it is all put in one tank.

Another experiment might be performed thus. Take a flat-sided, wedge-shaped glass tank, and arrange it so that it may be pushed across in a spectrophotometer and the absorption of different thicknesses of the dye in the tank thus measured. Take two solutions of an orthochromatic dye at different concentrations and find the thickness of the dye solutions that give absorption-curves with the peaks rising to the same height. It will be found that the curves are exactly the same. Now measure the thickness of dye-solution through which the light passed in the two cases. Suppose they stand in the ratio x : 1. Then the dye measured at thickness 1 is x times as concentrated as that measured at thickness x. A substance of which this is true is said to obey Beer's law. Now it is characteristic of metachromatic dyes that they do not obey it. One can find the concentration of any solution of an orthochromatic dye by use of the spectrophotometer when a single absorption-curve has been obtained with a solution of known concentration, but this is not possible with a metachromatic dye in aqueous solution.

Increase of temperature acts on metachromatic dyes in the same way as dilution. Toluidine blue at about 0·6% is reddish violet at ordinary temperatures, but blue at boiling point. Certain neutral salts (sodium chloride, sodium sulphate, potassium chloride, and especially barium chloride) act like dilution or increase of temperature.[308] It is remarkable, however, that potassium acetate does not have much effect.[211] Increase of acidity acts like increase of temperature, though, as has already been remarked, metachromatic effects are sometimes seen even below pH 3.

Dehydrating agents, such as ethanol and glycerol, are antagonistic to metachromasy. Even quite strongly chromotropic substances, such as heparin and agar, show less and less capacity to produce metachromatic effects in increasing strengths of ethanol and lose it entirely when the concentration reaches about 50%; chondroitin sulphate loses it at about 30%.[494] The chromotropic property manifests itself again, however, when water replaces ethanol.

Chromotropes are not all alike in their reactions to anti-metachromatic substances. Thus DNA is extremely sensitive to salts but less so to ethanol.[494]

At any particular degree of dilution, at any temperature or pH, and at any concentration of salt or dehydrating agent, there is an equilibrium between the orthochromatic and metachromatic forms of the dye, and this equilibrium will be affected by a change in any of the factors mentioned. Every alteration in the equilibrium between the two forms of the dye is reversible. It is stressed by Lison [308] that chromotropes are substances that shift the equilibrium to an extreme extent in favour of the metachromatic form, without producing any irreversible effect.

The factors that influence the degree of metachromasy exhibited by solutions of a dye also influence the appearance of microscopical preparations coloured by that dye. The metachromatic effect is most strongly shown when the section is still in the dye solution, but is also well seen when this is replaced by distilled water. Unfortunately this is not a good mounting medium, because of the wide difference in refractive index from that of the fixed proteins. Hansen [211] recommended a saturated aqueous solution of potassium acetate, which is indeed usable, though there is some loss of metachromatic colour. If the dyed preparation be heated or acidified, or if the salts listed above or a dehydrating agent be added, the metachromatic effect will be reduced or abolished. Ordinary glycerine-jelly is acidic and contains a high proportion of a dehydrating agent: it therefore abolishes all except rather strong metachromatic reactions. Passage through absolute alcohol has an even more extreme effect, and preparations mounted in Canada balsam therefore show metachromasy only in the granules of *Mastzellen* and other particularly chromotropic objects, and even these would presumably not show it if dehydration had really been complete.

It has been suggested [310] that one should not speak of metachromasy unless the colour-shift is shown in preparations mounted in glycerine-jelly, balsam, or other commonly-used media. It must be observed, however, that this would be a very arbitrary decision. These mounting media have been chosen, not because they are adapted to studies of metachromasy, but solely because they give good optical results and preserve ordinary microscopical specimens permanently. It would be a strange chance if they happened to be the ideal media for quite another purpose. It would seem more rational to use various media that slightly oppose the metachromatic effect to varying degrees. One might bring dyed sections into 1% and 2% sodium chloride, for instance, or 25% and 45% ethanol, or into solutions at pH 4 and 3. In this way one would be able to distinguish grades of metachromasy, according to the ability of various tissue-constituents to resist the anti-metachromatic effect of these solutions. A start in this direction has been made by Sylvén.[494]

Certain substances give metachromatic colours some resistance to dehydrating agents. Potassium dichromate, potassium ferricyanide, ferrous sulphate, and uranyl nitrate have been especially recommended.[514, 246] No satisfactory explanation of the action of these salts has been provided.

The aluminium lake of coelestine blue is remarkable for the resistance of its metachromatic colour to extraction by ethanol during dehydration.[296]

There are several indications that polymerization may be concerned in the metachromatic colour-shift. Thus dilution would antagonize polymer-formation and it is also antagonistic to metachromasy. The same applies to increase of temperature. If polymer formation were the cause of metachromasy, we should have an explanation of the fact that the metachromatic constituent of the dye-solution cannot be isolated from the orthochromatic as a dry substance. Non-conformity with Beer's law implies polymerization or the formation of complexes of some kind.

The American chemists, Sheppard and Geddes,[467] suggested that the cause of non-conformity with Beer's law was the coupling of dye-ions in pairs. They considered that the conformist (orthochromatic) dyes were prevented from making dimeric associations of this kind by the particular shapes of their cations, for the ions of certain dyes would fit together less easily than others. Since resonance would occur differently in a monomer and dimer, colour would be affected.

The idea that metachromasy (whether of simple aqueous dye solutions or of dyed chromotropes) is due to polymerization has been especially advocated by Michaelis.[347, 346] His contention is that the α band in the absorption curve represents the monomeric form of the dye, the β band its dimeric form, and the wide γ band its various polymeric forms, with overlapping effects. Sheppard and Geddes [468] look at the problem in a slightly different light. They claim that the β band can just be detected even when the dye is dissolved in organic solvents, in which dimers (and polymers) were thought not to exist. In the dimeric form of the dye the resonance that exhibits itself as the β band is greatly augmented. These authors doubt whether the dye ever exists in a polymeric form, though they do not dispute the reality of the γ-band.

The necessity for the presence of water if metachromasy is to appear was recognized long ago, and Hansen [211] even gave instructions for making balsam preparations in such a way that some water would be retained in the final mount. It has been claimed [468] that a molecule of water is actually incorporated in the dimeric ion of the dye, and indeed lies between and links the two ions. If this could be substantiated, the effect of dehydrating agents would be readily understandable.

When a metachromatic dye is presented to the tissues, there is a selective uptake of the forms of the dye responsible for the α, β, and γ troughs. These forms will here be provisionally called respectively the monomeric, dimeric, and polymeric. The dye may be at such a low concentration that it is monomeric, yet certain tissue-constituents will take it up almost entirely in the polymeric form; or again, it may be so concentrated that it contains a high proportion of dimers, yet certain tissue-constituents will take it up as monomers.[346] It is evident that the various tissue-constituents are extremely diverse in their affinity for monomers, dimers, and polymers.

In general, small molecules have no affinity for the polymeric forms of dyes. It is when a substance becomes itself polymeric, and particularly when it exhibits itself as a gel or film, that it affords an opportunity for the attachment of polymeric dyes. Gels or films, however, have no tendency to take up metachromatic dyes unless they possess many negatively charged points on their surfaces, ready to attract the positive charges of the basic dye-ions.

Sylvén [494] lays special stress on the need for numerous negative charges on the surface to be dyed, suitably spaced out. If the conditions of the surface are appropriate, the arrangement of the dye in polymeric form may be facilitated. If the negative charges on the surface are too distant from one another, or too irregular, or too weak, the tendency to attract the dye-ions and hold them firmly in polymeric form will be small. We see here a hypothetical explanation for the variation between different chromotropes in the intensity of their metachromatic reactions. It is quite possible that when the dye is taken up in polymeric form, a water-molecule is intercalated between each dye-ion and the next; and if so, a substance would be particularly chromotropic if its negative charges happened to be so spaced as to conform with this requirement. If there were a regular alternation of water-molecules with dye-ions, the latter would be about 0·4 mμ apart.

As Sylvén points out, each dye-ion consists of a large hydrophobe and a small hydrophil (auxochrome) part, and this in itself will facilitate the orderly arrangement of the ions. It is certain that some objects can arrange the ions of dyes in an orderly way, not necessarily with the production of metachromasy. This was shown in the last century by Ambronn,[8] in the course of his studies with polarized light. If a needle-shaped crystal of methylene blue be placed above a Nicol, with its long axis parallel to the longer diameter of the lozenge-shaped top of the Nicol, it appears dark violet; if it be now rotated through a right angle, it nearly loses its colour and becomes a pale greenish blue. Ambronn showed that if cell-membranes were dyed with methylene blue and then rotated above a Nicol, the dichroic effect was again exhibited, for in one position they appeared strongly dyed, while on rotation through a right angle the intensity of the colour was reduced. We need a full investigation along these lines of the behaviour of dichroic metachromatic dyes taken up by chromotropic substrates.

Sylvén [494] has shown that when carboxyl groups are introduced into cellulose fibres, metachromasy starts as soon as every second glucose unit of the chain has become carboxylated. The average distance between the charged groups is now about 1 mμ. Much stronger metachromasy is exhibited by polysaccharides containing one sulphate and one uronic group per disaccharide unit: the distance between the charged groups now alternates along the chain between 0·6 mμ and 0·4 mμ.

Attention must now be directed to a most curious departure from the ordinary rules of metachromasy.

It has been remarked above that the nucleic acids, especially RNA, are rather feebly metachromatic, but chromatin does often give a violet reaction with toluidine blue in microscopical preparations (and a distinct purple with the azures; see p. 269). Occasionally, however, a most strange reaction is noted: chromatin is dyed *blue-green* by this blue dye, and the colour-change is therefore bathochrome.

When microscopical preparations are strongly dyed with toluidine blue, chromatin is violet; when feebly, it appears blue, blue-green, or even green, as though dyed by methyl green. Now toluidine blue itself, however dilute, is never green, nor is the alcoholic solution.

Lison and Mutsaars [313] took a very dilute solution of toluidine blue (about 0·0015%) and added varying amounts of DNA or RNA to it. When the nucleic acid was very dilute, a small hypsochrome change occurred, the colour going from blue towards red, and red itself was reached when the concentration of the nucleic acid was 0·0025%. When the concentration rose higher than this, the colour became violet-blue, and finally blue-green.

A study with the spectrophotometer gave results in conformity with what could be seen with the eye. As the concentration of the nucleic acid increased, the peak of the original absorption curve sank while a new one (β band) arose on the side of it towards the shorter wavelengths; with further increase in the concentration of nucleic acid the curve moved bodily in the opposite direction, its peak passing the wave-length of the original peak and going beyond it towards longer wave-lengths. At no concentration was the actual *shape* of the original curve restored, and it would therefore be wrong to refer to a reversion to the orthochromatic form of the dye, with subsequent movement beyond it.

When the blue-green colour had been obtained, heating abolished it and restored the truly orthochromatic form of the dye.

The concentration of DNA is particularly high in the heads of spermatozoa, and it is interesting to notice that these are rather easily dyed blue-green by toluidine blue.

No satisfactory explanation has been given of bathochrome or 'negative' metachromasy.

We now leave the metachromasy of basic dyes and turn to a subject that has received much less attention. A few acid dyes are metachromatic in peculiar ways. The rules that govern the metachromasy of basic dyes do not apply here. Thus azo dyes can be metachromatic, there is no correlation between metachromasy and the capacity to form imino-bases, and the effect is usually bathochrome.

Indigocarmine (indigo-dye) and orange G (mono-azo) are metachromatic towards solutions of clupeine, which is a protamine in combination with nucleic acid, extracted from the heads of the spermatozoa of certain fishes; it is strongly basic (acidophil). The same dyes show metachromasy in the presence of certain large crystalloid cations, especially those of quinine and strychnine.[43]

These alkaloids were tried as chlorides, in experiments carried out in solution. It would be of interest to make a full study of microscopical preparations of the parts of those plants and animals that contain these substances in suitable form. The alkaloids themselves, as they occur in the bark or seeds of plants, might be unreactive, but sections of the testes of the appropriate fishes would presumably be suitable research material. It might be possible to extend the list of acid dyes known to be metachromatic, and to find some indication of chemical similarity between them.

Haematein, used as an acid dye without lake-formation, can in certain circumstances show a metachromatic effect.[432, 433] In solution in 40% alcohol it colours the cytoplasm of certain nerve-cells red, but the nucleus homogeneously blue. The orthochromatic colour of this dye, used substantively, is reddish, and the metachromatic shift is thus bathochrome. It would appear that the nuclear sap, not the chromatin, is coloured. The facts have not been explained, but it is tempting to suppose that the histone of the nuclear sap may be the chromotrope.

Certain acid dyes, particularly those that are disulphonates, show a strange form of bathochrome metachromasy. The two best examples are Congo rubin (not Congo red) and Bordeaux red, both azo dyes. If one of these is injected into the abdominal cavity of a mouse, coloured particles are later found in the phagocytic cells of various parts of the body, that is, in the histiocytes and cells of the reticulo-endothelial system. The uptake of certain dyes by these cells will be considered in a general way in a later chapter (p. 276); here it is only necessary to say that the dyes do not colour pre-existing objects, but are segregated in the form of granules. The interesting fact about the particular dyes with which we are concerned here is that many of the granules show the metachromatic (blue) colour.

It might be thought that the colour-change was connected with pH, since Congo rubin, like Congo red, is an indicator; but this is not so, for the metachromatic change is not dependent on pH, and Bordeaux red and other metachromatic acid dyes are not indicators. It seems almost certain that this kind of metachromasy is due to polymerization of the dye.[443] In concentrated solutions the dyes concerned show the beginnings of a bathochrome colour-shift. There are marked differences, however, from the polymerization of basic metachromatic dyes. Not only is the shift in the opposite direction, but it is *aided* by the addition of neutral salts such as potassium chloride, which turns Congo rubin from red to violet and finally flocculates it as a blue substance. This change of colour accords with the general rule (disobeyed by the basic metachromatic dyes) that increase in particle-size has a bathochrome effect. Gold hydrosols, for instance, are light red, but if coagulated by electrolytes change their colour through violet to blue and finally form a blue precipitate. If a red hydrosol is injected into a mouse, the gold is seen in the phagocytic cells as blue-black particles. Thus the gold hydrosol is metachromatic.[443] This strongly suggests that the same process accounts for the metachromasy of Congo rubin and similar dyes.

CHAPTER 14

The Blood Dyes

Special dyes are used in medical practice for the differential colouring of blood-smears. They are adapted to the easy and rapid diagnosis of disease. They achieve this end by distinguishing clearly the different kinds of leucocytes and colouring brilliantly any protozoan parasites that may be present in the red corpuscles or plasma. These dyes are of considerable theoretical interest and deserve a chapter to themselves.

The blood-dyes evolved under the influence of Ehrlich's idea of 'neutral' dyes. Probably the first dye to which such a name could at all reasonably be given was Ranvier's [412] *carmine picrique*, which was made known in 1875, before Ehrlich had published his first paper. This dye, which is useful in general microtechnique, is made by adding a saturated aqueous solution of picric acid to a saturated aqueous solution of ammonium carminate and evaporating the mixture. A crystalline precipitate is formed, which is separated from the fluid and dissolved in distilled water. As we have seen (p. 193), carminic acid acts as a basic dye when acidified, and it is possible that the precipitate is really a picrate of the red dye. If so, it was formed by the combination of a coloured cation with a coloured anion. The chemistry of the precipitate has not, however, been worked out, and anyhow this curious substance cannot be regarded as a typical example of a neutral dye.

Ehrlich's plan was to allow a basic dye to react with an acid one, so as to produce a new substance with new properties—a dye in both halves of its molecule. His procedure was given in detail in his joint work with Lazarus on anaemia.[158] Ehrlich added a solution of an acid dye such as orange G drop by drop to a solution of a basic dye, such as methyl green. A coloured precipitate was formed. Orange G is a sodium salt and methyl green a chloride. Ehrlich thought that double decomposition occurred, with production of sodium chloride and the 'neutral' dye, methyl green/

orange G. He discovered that the addition of a small quantity of the acid dye in excess allowed the neutral dye to dissolve.

Such neutral dyes were used by Ehrlich for colouring blood-smears. They were found to be particularly good for bringing out the basiphil constituents in one colour and the acidophil in another, and thus making it easy to distinguish the different kinds of leucocytes. Ehrlich attributed this to the fact that the dyeing of the two kinds of constituents was simultaneous instead of successive. The dyeing cations and anions are present in almost exactly equal numbers, and this circumstance must give better indications of basiphilia and acidophilia than any arbitrary mixture or succession of dyes. In addition, the granules of polymorphonuclear leucocytes are coloured by both the components of the neutral dye, and that is why Ehrlich called them 'neutrophil'. He regarded the specific dyeing of these granules as an important property of the neutral dyes, not to be obtained without their use.

Methyl green is unusual among triarylmethane dyes in possessing two positively charged nitrogen atoms. Orange G possesses two negatively charged sulphonate groups. Thus one molecule of the basic dye could react with one molecule of orange G. Ehrlich, however, was under the mistaken impression that all three nitrogen atoms of methyl green combined with acid dye-radicles when a neutral dye was prepared. He therefore gave the misleading name 'Triacid' to the best-known of his neutral dyes, to emphasize his

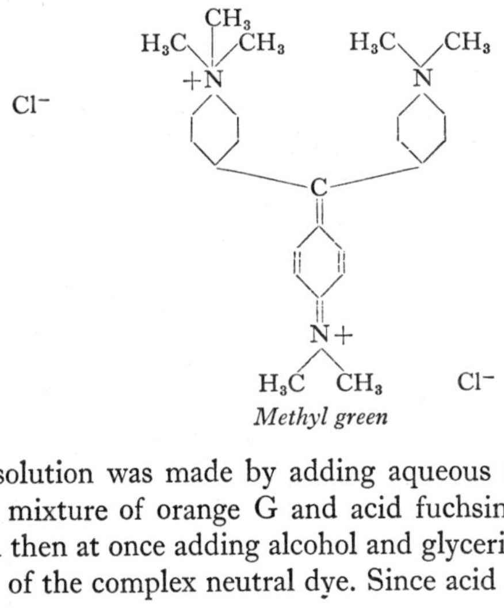

Methyl green

belief. The solution was made by adding aqueous methyl green solution to a mixture of orange G and acid fuchsine in aqueous solution, and then at once adding alcohol and glycerine to prevent precipitation of the complex neutral dye. Since acid fuchsine, like orange G, is a sodium salt, the mixture must contain sodium and chloride ions, methyl green cations, and the anions of orange G and acid fuchsine, with perhaps also some undissociated neutral dye. It is not possible to isolate methyl green/orange G and methyl green/acid fuchsine from the Triacid solution as separate, dry substances.[345]

When used for blood-smears fixed by heat (or less well by alcohol/ether), the Triacid dye colours chromatin greenish with methyl green, red blood-corpuscles orange with orange G, and eosinophil granules in copper-colour by the action of both the acid dyes; the neutrophil granules are dyed violet, which sharply distinguishes them from everything else in a blood-smear. (See fig. 28, A.)

Ehrlich considered methyl green, methylene blue, and amethyst violet (azine) the most suitable basic dyes for forming neutral compounds. Among the acid dyes he preferred those that had more than one sulphonic group, because they formed neutral dyes that could be dissolved without much difficulty.

Ehrlich's Triacid and similar fluids are useful in haematology, but it is doubtful whether great interest would have been aroused by neutral dyes had not methylene blue been tried in combination with eosin. No one could have foreseen the extraordinary value of this particular combination, or the complications that would result from the association of two apparently commonplace dyes.

Before 1891 the nucleus of the malarial parasite had never been seen. A Russian protozoologist, Romanowsky,[423] set out to try to dye it differentially. Aware of Ehrlich's work on the use of neutral dyes for the differential colouring of blood, he tried a new combination of the same sort. He added a 1% aqueous solution of eosin to a saturated aqueous solution of methylene blue, apparently until the acid dye was present in slight excess (in accordance with Ehrlich's practice). On dyeing malarial blood-smears with this, he found a remarkable range of colours in the blood-corpuscles. The dyeing of the parasites was successful beyond any reasonable expectation: for the cytoplasm of the trophozoite was Prussian blue, the nucleus 'carmine-violet'. Thus was the nucleus of the malarial parasite discovered.

Nearly half a century later the attempt was still being made to disclose the principles underlying Romanowsky's results, and indeed, as we shall see, there is plenty of room for research today.

A

B

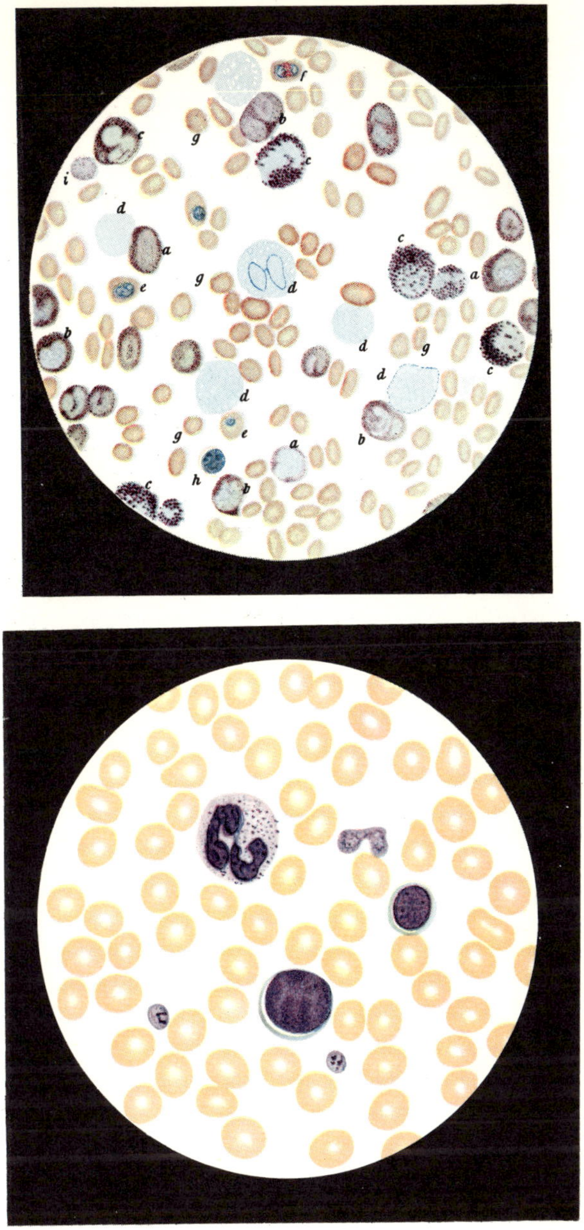

FIG 28 (A) human blood from a patient with myeloid leucaemia; coloured by Ehrlich's Triacid dye.

a, neutrophil myelocytes; *b*, polymorphs; *c*, eosinophil myelocytes; *d*, *Mastzellen*; *e*, normoblasts; *f*, normoblast with mitotic figure; *g*, normal erythrocytes; *h*, megaloblast; *i*, dwarf polymorph.

(B) normal human blood dyed by Leishman's method. Two lymphocytes, one polymorph, and three platelets are seen, as well as many red blood-corpuscles.

Meanwhile Romanowsky [424] went on to study the multiplication of the nucleus and the life-history of the parasites, and clinicians soon made use of his method as an invaluable aid to diagnosis.

The complications behind Romanowsky's dye can best be elucidated by the historical method. The accompanying evolutionary diagram will help to explain the course of progress. The reader may care to turn back to this diagram from time to time while following the development of the story. It leads from Romanowsky's brilliant but empirical and not always repeatable results to

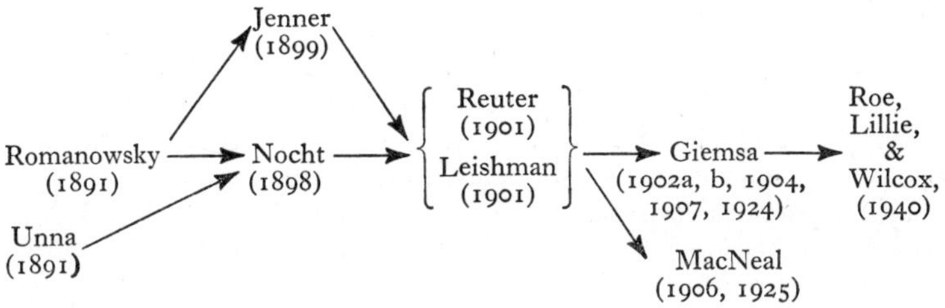

The evolution of blood dyes

the sure and scientific preparations of Giemsa, Roe and his colleagues, and MacNeal. Care has been taken to exclude from this account all those methods—some of them very well known—that did not lead towards the goal of understanding. Bernthsen [66] and Kehrmann,[263] who played particularly important parts, have been omitted from the diagram because their influence on the course of research cannot be indicated without making it too complicated.

A Londoner, Jenner,[251] discovered that the precipitate formed on mixing solutions of eosin and methylene blue was particularly soluble in methanol. Subsequent workers have mostly made use of this fact instead of relying on the solubility of the neutral compound in excess of the acid or basic dye (though some excess of the *basic* has usually been allowed). Jenner, however, did not obtain the wide variety of colours seen in a successful Romanowsky preparation. Similar results were obtained in Germany by May and Grünwald,[328] who sometimes used a fresh mixture of eosin and methylene blue before it had time to precipitate, sometimes allowed precipitation to occur and then dissolved it in methanol. Preparations made by these methods are useful enough for certain purposes. The red blood-corpuscles are bright red, eosinophil granules deep red, the nuclei of leucocytes and the basiphil cytoplasm of lymphocytes and of the malarial parasite blue. The blue of the parasite stands out against the red of the blood corpuscle that contains it, but its nucleus is not differentially dyed. One might choose this method in an investigation of eosinophils, which are brilliantly shown, but as a general rule the haematologist naturally prefers the much wider range of coloration seen in Romanowsky preparations.

In successful preparations made by the original method of the Russian author or by any of the numerous modifications that have been introduced subsequently, the general scheme of colouring is this:—

chromatin of leucocytes	purple
nucleus (or part of nucleus) of parasitic Protozoa	red or carmine
basiphil cytoplasm of lymphocytes, monocytes, and parasitic Protozoa	blue
eosiniphil granules	pink
neutrophil granules	purple
red blood-corpuscles	pink (sometimes bluish)

The expression 'Romanowsky dyes' will be used here to cover all mixtures or compounds of eosin with methylene blue and allied dyes, which give these or similar colours. The different techniques give somewhat different results, the red corpuscles in particular being rather variable; but unless the general scheme applies, it is not justifiable to speak of the Romanowsky effect.

It is natural enough that basiphil cytoplasm should be blue and eosinophil granules pink, but not obvious why the chromatin of leucocytes should be purple and far from obvious why the nucleus of parasitic Protozoa (or at any rate part of it) should be red.

Attempts to invent a dye that would give the Romanowsky effect every time led gradually to understanding. The first step was taken by Nocht, a port medical-officer of Hamburg, later Director of the *Institut für Schiffs- und Tropenkrankheiten* in that city. This institute played an important part in the development of modern Romanowsky techniques.

Nocht's efforts to make a reliable dye-solution failed so long as he worked on the assumption that a neutral dye, in Ehrlich's sense, was responsible for the unexpected colouring. He came to the conclusion that the cause must lie in the presence of impurities in the dyes used. He was influenced by Romanowsky's remark [423] that the best result was obtained when a mould was growing on the surface of the stock solution of methylene blue. He remembered that Unna [513] had investigated the changes undergone by methylene blue solutions on standing or treatment with dilute alkali. A new substance appeared, which Unna called methylene red; the 'ripened' solutions coloured *Mastzellen* red. Unna tried various ripening agents, and found certain carbonates, especially potassium carbonate, the best. From these researches resulted his polychrome methylene blue.

Nocht [369] tried a methylene blue/eosin solution in which the polychrome variant had been substituted for pure methylene blue. He found it necessary to neutralize the alkaline solution. When this was done, the nucleus of the malarial parasite was regularly dyed red. The basiphil cytoplasm, however, was violet, not blue, and this gave insufficient contrast with the red of the nucleus. Nocht overcame this difficulty by simply adding pure methylene blue to his polychrome neutral dye: the basiphil cytoplasm was now blue.

Nocht worked with a freshly-made solution containing a small excess of the basic dyes. The Romanowsky effect could be produced with certainty, but one practical defect remained: the proper balancing of the eosin with the basic dyes had to be carefully done every time one wanted to dye blood-smears. The obvious necessity was a stable stock solution of the dyes, requiring nothing but dilution.

The problem was solved independently in Germany by Reuter [418] and in England by Leishman.[293] Wright's [547] well-known mixture is a mere variant of Leishman's. All used a partly polychromed methylene blue and dissolved the precipitated neutral dye in alcohol. Reuter used absolute ethanol, while the others made use of Jenner's discovery that absolute methanol is a particularly good solvent for methylene blue/eosin. Leishman, an Assistant Professor in the British Army Medical School, made the important discovery that a single solution could act first as fixative for the blood-smear, because the solvent was methanol, and subsequently, on dilution on the slide with distilled water, as a quickly-acting dye. Leishman's and Wright's techniques are often used to the present day and generally give excellent results, though they are not quite so uniform as the dyes made up entirely from known ingredients, in accordance with the principles that will now be explained.

It is necessary to know what substances besides methylene blue itself are present in the polychromed dye. It will be convenient to represent the various thiazines by shortened formulae. The

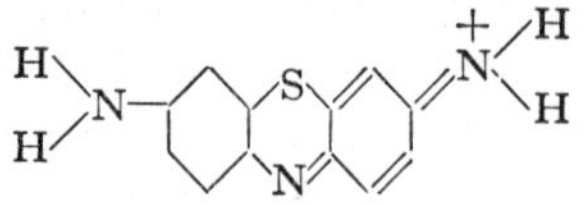

Thionine (cation)

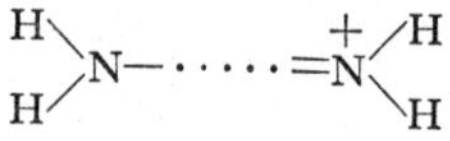

Shortened formula for thionine (cation)

simplest is thionine. The four hydrogens shown in the shortened formula can be replaced one by one by methyl, till the tetramethyl

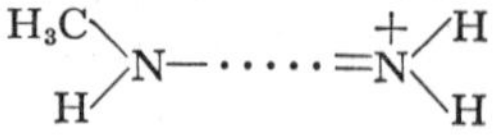

Monomethyl thionine (azure C)

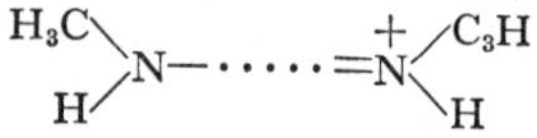

Symmetrical dimethyl thionine (azure IV)

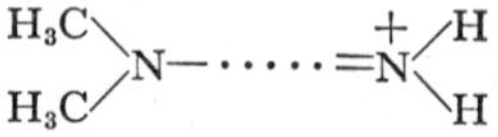

Asymmetrical dimethyl thionine **(azure A)**

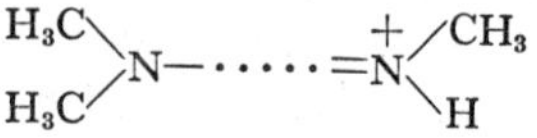

Trimethyl thionine **(azure B,** *main component of azure I)*

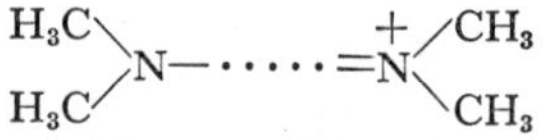

Tetramethyl thionine **(methylene blue)**

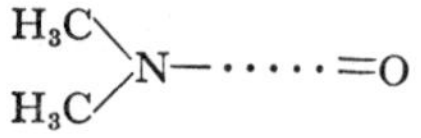

Dimethyl thionolin (*Bernthsen's* **methylene violet**)

compound, methylene blue, is reached. In methylene violet an oxygen atom replaces one of the substituted amino-groups of methylene blue.

Our understanding of the composition of polychrome methylene blue is due to the researches of Bernthsen [66] and Kehrmann.[263] The former isolated methylene violet from the oxidation products of methylene blue, and announced the presence of yet another dye, which he called *Methylenazur*; the latter proved that Bernthsen's *Methylenazur* was a mixture of what are now called azure B and azure A. The other thionines shown above are not present in polychrome methylene blue. Apart from thionine itself, they have no important uses in microtechnique.

More than to any other man we are indebted to Giemsa for the placing of Romanowsky dyeing on a scientific basis. It was his purpose to avoid methylene blue that had been polychromed at random, and to use instead known quantities of known dyes. Throughout the 22 years during which he published papers on this subject, he remained an assistant in Nocht's institute.

Giemsa made use of a dye that he called methylene azure or

azure I. This was evidently azure B, or else azure B somewhat contaminated with related dyes.[131] He mixed this with eosin and showed that the main features of the Romanowsky effect were obtained, in the absence of methylene blue.[186] Now azure B, unlike methylene blue, is a metachromatic dye. Itself blue, it has a strong tendency to impart a reddish colour to chromatin. Giemsa attributed the reddish or red colour of the nuclei of leucocytes and of the malarial parasite in Romanowsky preparations to the presence of azure I in polychrome methylene blue.

Giemsa subsequently came to the conclusion that his azure/eosin alone was imperfect, because the colour of basiphil cytoplasm was greyish rather than pure blue and therefore gave less striking contrasts. It was for this reason that he put methylene blue in his mixture, in addition to azure and eosin. This gave the desired result.[187] He mixed his azure with an equal weight of methylene blue, and from this mixture ('azure II') he made a neutral dye by precipitation with eosin. He added a small extra quantity of azure II so as to obtain a preponderance of the basic dyes.

It is fortunate that azure B, a blue dye, gives a strongly metachromatic colour to the chromatin of leucocytes, while methylene blue has a special affinity for the basiphil cytoplasm of lymphocytes and monocytes. Otherwise it would have been difficult to dye the chromatin of these cells in a different colour from the cytoplasm. The acidic substance in the cytoplasm that is coloured by these basic dyes is presumably RNA. In the form in which it occurs in the tissues, RNA is luckily less apt to act as a chromotrope than DNA.

Giemsa experimented also with methylene violet/eosin dissolved in alcohol and water, but did not obtain the full Romanowsky effect. On adding methylene violet to his azure mixture he found that dyeing was inhibited and the tendency to precipitation increased. For these reasons he never used methylene violet in his later mixtures.

Giemsa's subsequent papers [188, 189, 191] are mainly concerned with solvents and practical details of technique. He followed Leishman and Wright in using methanol as a solvent for his stock solution, but added glycerol to increase the capacity to dissolve the dyes and allow easier mixture with water.

To rationalize Giemsa's technique still further, it was necessary to use an azure of definitely known chemical composition. A comparison of azures C, IV, A, and B was undertaken in the U.S.A, by Roe, Lillie, and Wilcox.[421] Each of these was tried on blood-smears in conjunction with eosin; so also were toluidine blue and thionine. The Romanowsky effect was given by all the azures, but best of all by azure B; it was given faintly by toluidine blue but not at all by thionine. Azure B was also found to be superior to A in giving the Romanowsky effect in the presence of methylene blue and eosin; it should therefore be substituted for the azure I of Giemsa's formula. It will be remembered that azure I is in fact azure B, more or less adulterated. The American authors found no conclusive evidence that any benefit could be obtained by adding azure A as well as azure B to the mixture.

A different line from Giemsa's was followed by MacNeal, though both were animated by the desire to place Romanowsky dyeing on a scientific basis. MacNeal placed the emphasis on methylene violet. This substance is insoluble in water unless other thiazine dyes are present, but soluble in alcohols and to some extent in mixtures of alcohols and water. MacNeal[318] found that by itself or in the presence of eosin it would not dye chromatin at all, but in the presence of methylene blue it showed its metachromatic property by giving the usual Romanowsky purple to the chromatin of leucocytes. Strangely enough, it did not colour the chromatin of the malarial parasite.

MacNeal did not deny that the Romanowsky effect could be produced by the azures, but thought that the metachromatic colouring of the chromatin of leucocytes was best achieved by methylene violet in the presence of methylene blue. He thought that the ordinary method of polychroming methylene blue to produce the Romanowsky effect resulted in a preponderance of methylene violet over azure, and that the former was chiefly responsible for the metachromatic dyeing of chromatin.

MacNeal used azure A to give a reddish colour to the nucleus of parasitic Protozoa, for this is not given by methylene violet. His excellent tetrachrome mixture[319] therefore consists of methylene blue, methylene azure A, methylene violet, and eosin, dissolved in methanol. It is used exactly like Leishman's. One might suppose that it could be improved by the substitution of azure B for azure A, but for some unexplained reason this is not so; on the contrary, the change results in a markedly inferior dye.[38]

Methylene blue itself, if it could be obtained in a perfectly pure state, would probably be orthochromatic (p. 251). The ordinary product, freshly dissolved, is somewhat metachromatic towards strong chromotropes such as the granules of *Mastzellen*, but not at all towards DNA. Certain facts, then, stand out clearly from what has already been said in this chapter. On standing in solution, but particularly in the presence of alkaline carbonates, methylene blue gives rise to new, metachromatic substances: azure B, azure A, and methylene violet. It is these that are responsible for the purple or red colouring of the chromatin of leucocytes by the Romanowsky techniques, while methylene blue gives its own, contrasting colour to basiphil cytoplasm; eosin dyes the eosinophil granules and red blood-corpuscles. In order to have a solution of known and invariable composition, it is best to use weighed amounts of methylene blue, azure B or A, and eosin, with or without the addition of methylene violet. The basic components should slightly predominate so as to help the dyes to remain temporarily in solution in aqueous media. A stock solution should be made by dissolving them in methanol or in a mixture of this with glycerol. This solution should be strongly diluted with water when staining is to begin, so as to permit ionization of the dyes. No dye that does not arise spontaneously in the polychroming of methylene blue has any special virtue in Romanowsky dyeing.

Certain problems remain. The reader may have noticed that less emphasis has been placed on neutral dyes towards the end of this chapter than at the beginning. The Romanowsky dyes originated by the application of Ehrlich's idea, and neutral dyes do in fact exist in the dry state. It is doubtful, however, whether they play any important part in obtaining the Romanowsky effect. When a neutral dye is dissolved in water, ionization must occur, and the dyes presumably act mainly as ions. Some undissociated molecules may, indeed, remain, but these would be very unlikely to react with DNA. They might conceivably *dissolve* in lipids, and this could perhaps account for the colouring of the neutrophil granules; but the strongly acidic chromatin would attract only the basic dye-ions.

In an ordinary solution of methylene blue or the azures, chloride ions are present; in an ordinary solution of eosin, sodium ions. If an exactly balanced compound of methylene blue or azure with eosin be precipitated, the precipitate dissolved in methanol, and the methanol then diluted with water, there will be no chloride or sodium ions in the solution. This fact cannot account for the Romanowsky effect, for the dyes act in almost exactly the same way if dissolved together in water without elimination of the inorganic ions; and anyhow chloride ions are present in the usual mixtures, since the basic dyes are present in excess.

The evidence suggests, then, that the formation of a neutral dye does not in itself confer important special properties; yet it is a fact, established more than half a century ago by Giemsa,[186, 187] that azure chloride makes the nucleus of the malarial parasite only feebly violet, while azure/eosin makes it the desired brilliant red. Similarly, polychrome methylene blue alone does not give the Romanowsky effect with the nuclei of parasitic Protozoa; but if the smear be treated previously with eosin, the characteristic red colouring results.[518] Since eosin alone will not dye the nuclei of these organisms, there is obviously a problem to be solved.

The dyeing of the chromatin of leucocytes is not the same process as the dyeing of the nucleus (or part of the nucleus) of Protozoa. It is to be noticed first of all that the colour is different. The nuclei of Protozoa are dyed a much purer red than the chromatin of leucocytes, which is usually coloured purple. Further, the chromatin of leucocytes is similarly coloured whether eosin be present or not, while Romanowsky dyeing of the protozoan nucleus is dependent on the use of eosin. When the granules of *Mastzellen* are coloured purple by a Romanowsky dye, we are concerned simply with what Pappenheim [386] meant to call the substantive metachromatic basiphilia of acidic objects (though he accidentally said 'subjective' when he meant substantive). The same applies to the chromatin of leucocytes. Pappenheim regarded the red coloration of the nuclei of Protozoa as an example of adjective metachromatic neutrophilia: that is to say, he considered that the nuclei of Protozoa were 'neutrophil' in his sense and that the eosin played the part of a mordant in the process of dyeing.

The idea that eosin might act as an acidic mordant for a basic dye (unlike the usual basic mordants for acid dyes) seems to have suggested itself to Nocht [370] long before. Nocht had reason to suppose that resorcinol might replace eosin in Romanowsky dyeing. This idea was further developed by Unna,[518] who found that potassium tribromophenolate, a colourless substance related to one part of the eosin molecule, could replace eosin in Romanowsky dyeing. The nuclei of Protozoa were dyed in a colour closely resembling that given when eosin was used. (Both Unna [518] and Giemsa [190] considered that the colour of eosin itself slightly influenced the final result of the dyeing of the protozoan nucleus by Giemsa's method, even though eosin by itself does not colour it.)

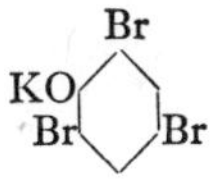

Potassium tribromophenolate

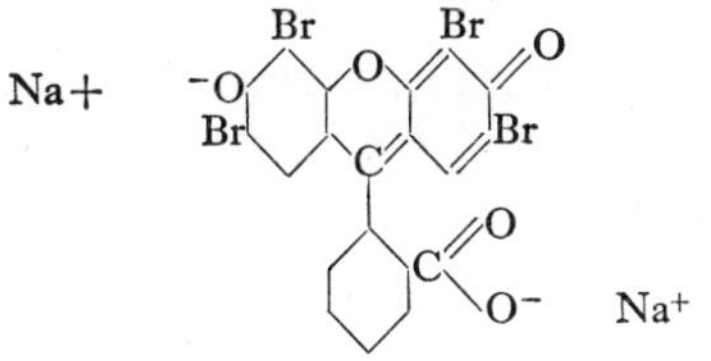

Eosin Y

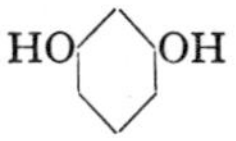

Resorcinol

Partly as a result of his researches on the nuclei of free-living amoebae,[519] Unna came to the conclusion that the substance in the nuclei of Protozoa that reacts with Romanowsky dyes to give a red colour is not acidic chromatin, but the highly basic protein, protamine. He considered that eosin (or a substitute) combined chemically with this and also with the basic dye. The latter he believed to be the imino-base of asymmetrical dimethylthionine (azure A). The imino-bases of the azures are red. If one of these is in fact responsible for the colouring of the protozoan nucleus, no question of metachromasy arises. The evidence suggests strongly that the desired colour is best given when azure B and eosin (or a substitute) are present; but whether the azure acts as a base or metachromatically as an ion, whether the substance dyed is really protamine, and how the eosin affects the result, are still unsolved problems.

A full explanation of Romanowsky dyeing has not yet been provided, but the main outline of the process is understood and sufficient knowledge has been gained to allow the preparation from pure dyes of solutions that can be relied upon to give the striking effect that is desired.

CHAPTER 15

Introduction to Vital Colouring

It is about 80 years since Ranvier, the well-known French histologist, stated unequivocally that cells could only be dyed when dead. Vital dyeing of a very special kind, by natural colouring agents, had actually been described by Trembley [506,29,35] more than a century and a quarter before; but Trembley's methods had not been followed by others, and Ranvier's remark was in accordance with the beliefs of his time. As a matter of fact the elaborately-named Polish pathologist, Chrzonszczewsky, had already in 1864 injected ammoniacal carmine into the blood-stream of mammals and shown, by the examination of sections, that the cytoplasm of the convoluted tubules and Henle's loop was coloured,[122] and indigo-carmine had been used successfully for the same purpose;[223] but no one had actually watched the imbibition of a synthetic dye by a living cell. This was first done in 1878, when Brandt, in Berlin, coloured the lipid droplets in the cytoplasm of living *Actinosphaerium* with Bismarck brown.[88]

There was sporadic use of this and other vital dyes in the following years,[115-117, 89] and the golden age of vital dyeing started about the middle of the eighties. Pfeffer [398] in Tübingen was the first to investigate the action of dyes on the living cells of plants. In a comprehensive study involving the use of many kinds of cells and many different dyes, he showed that the cytoplasm and various granules and vacuoles could be tinged during life, while the dye often accumulated in the cell-sap, sometimes in the form of crystals. Ehrlich followed Pfeffer and became the main motive force behind research into the action of vital dyes on the cells of animals (fig. 29). He showed that methylene blue (already used by Pfeffer) would colour living axons and nerve-endings and thus demonstrate their course.[156] He introduced [157] into this field of study the *colorant vital par excellence*, as neutral red has with some justification been called.[51] The work was ably developed by Arnold,[15-20] Michaelis,[343] Fischel,[167, 168] Himmel,[233] and de

FIG. 29. Ehrlich at about the time when his work on vital dyes was beginning to merge into chemotherapy.

(From Marquardt,[324] by kind permission of Messrs William Heinemann Medical Books, Ltd.)

Beauchamps.[51] It is a curious fact that while many others were playing unscientifically at non-vital dyeing, these men were making such a profound study of the living cell that their papers, published from 1886 to 1911, can be studied with profit at the present day. Ehrlich regarded his work on vital dyes as the basis of his later pharmacological researches. He was struck by the strange specificity of vital dyes, and looked for chemical agents that would be equally specific in attaching themselves to harmful parasites.[324] In introducing methylene blue as a vital dye for nerve-axons,[156] he indicated that he had been trying to find how poisonous substances might distribute themselves differentially among the tissues of the body, and used a dye in his experiments because its colour would announce its distribution.

Certain non-living parts of organisms may be coloured equally well and in the same way whether the animal be alive or dead, and without any necessity for the dye to enter living cells. The jelly of colonial Radiolaria, the tubes of various Protozoa, the intercellular matter of many Metazoa, and the peritrophic membrane of insects provide examples of what de Beauchamps [51] called 'pseudovital' colouring. This is a process that does not require special consideration here, because there is no important difference from the dyeing of any other kind of non-living matter.

The use of vital coloration has become less frequent since the introduction of phase-contrast and interference microscopy, because these methods enable one to study the living cell in as natural a state as possible. Extremely valuable though they are, phase-contrast and interference depend wholly on differences in refractive index between an object and its surroundings. They cannot, therefore, give us any direct information about chemical composition. The main virtue of vital dyeing is that it calls attention to the heterogeneity of cell-inclusions. Some of them colour with one dye, some with another. These are indications of chemical diversity, and vital dyes sometimes give useful pointers towards composition. Further, an inclusion that is of the same or nearly the same refractive index as the ground cytoplasm and therefore invisible or nearly so by phase-contrast or interference, may take a vital dye strongly.[464] Beyond this, as we shall see, the vital activity of the cell in responding to the presence of the dye may give us information of particular interest. The examination of the untreated cell by the newer optical methods should go hand-in-hand with studies made by direct microscopy with vital coloration,

It is necessary to distinguish between two quite different ways in which coloured substances can be used in the study of living cells. On one hand certain particular cells actively take up coloured particles in the course of their ordinary function of eliminating foreign bodies. On the other hand cells of all kinds allow certain dyes to infiltrate into them and to colour certain pre-existent cellular constituents. For the clear distinction between these two kinds of vital colouring we are largely indebted to von Möllendorff.[353, 354] It is obvious that the second kind of vital colouring has much more general interest than the other. It will be convenient to consider the more special kind first, in order to leave the field clear for a detailed study of the wider subject.

The cells that actively take up coloured particles are those of the excretory and phagocytic systems, especially the reticulo-endothelial system of vertebrates, and particular coloured substances may indeed be injected into the blood-stream of an animal with the deliberate intention of finding out which cells have the function of excreting or storing foreign particles. The coloured substances used for this purpose need not be dyes: it is only necessary that the particles should be sufficiently small for uptake by single cells. The carbon particles of Indian ink are suitable. Carmine may be used in the form of minute, insoluble particles. The only dyes that are suitable for use in solution are those that have a strong tendency to flocculate into particles of colloidal dimensions. Certain acid disazo dyes, especially trypan blue, are particularly suitable. A considerable number of such dyes has been used, and some have even been synthesized especially for the purpose,[160] but it would not appear that any of them are superior to trypan blue for general use.

This kind of vital colouring was invented by Baron von Gleichen [194] nearly two centuries ago, in the course of his investigation of the way in which ciliates nourish themselves. He knew that madder coloured the bone of animals fed on it, and sought to apply a similar process to microscopical organisms. Particles of carmine were eaten by the animals and seen in the food-vacuoles. The method was copied by other protozoologists, especially Ehrenberg,[151] but appears not to have been noticed by histologists until Ranvier [412] began injecting various insoluble colouring agents (aniline blue SS, vermilion, sepia, carmine, and others) into the dorsal lymph-sack of the frog and subsequently examining the leucocytes. He also saw the latter ingest coloured particles on a microscopical slide.

Nowadays it is usual to make one or more subcutaneous injections and then to kill the animal, generally after the lapse of several days, and fix the tissues with a suitable fixative (often formaldehyde or mercuric chloride or Zenker).[103] Sections are prepared and stained with a dye that contrasts in colour with the vital colouring agent. Particles of carmine and other insoluble pigments naturally remain in position, but it is a remarkable and convenient fact that the soluble disazo dyes used in this kind of work do not quickly dissolve out of the cells during the after-treatment, presumably because their escape is impeded by the coagulation of proteins round them. As a result, it is usual to make observations on sections mounted in Canada balsam, although the uptake of the coloured substance was an active, vital function of the cells.

It is not perfectly clear whether the disazo acid dyes sometimes flocculate outside the cells and are then engulfed, or whether they always penetrate in a finely dispersed state and are then aggregated into microscopically visible particles within the cells. It seems likely that the fate of such acid dyes as can penetrate cells depends on their ability to flocculate. Those that exist as small molecules or ions with little tendency to clump together are not used in vital studies of this kind. They diffuse out of cells as easily as they diffuse in. Those that exist as large molecules or ions—notably the disazo and trisazo [27] dyes—have a tendency to clump; these are captured if they enter certain cells.[443, 312] In some cases, especially in excretory as opposed to phagocytic cells, it seems certain that aggregation occurs after absorption. In the kidney tubules of vertebrates, droplets containing such dyes first appear in the region of what is commonly called the Golgi apparatus, and subsequently spread into the ground cytoplasm.[316]

Whether insoluble particles or soluble dyes are used, it is characteristic of this kind of colouring that pre-existent objects in cells do not take up the colour. Thus, even if the colouring agent used is a dye, it does not act as a dye in these circumstances, since a prerequisite for dyeing is something that can be dyed. Ranvier was perfectly logical in using vital colouring agents of this kind while claiming that cells could only be dyed when dead. The soluble colouring agents used in this kind of work may give a slight, diffuse colour to cellular or intercellular matter,[103] but the reaction for which they are used is active aggregation by the cells into new, microscopically-visible droplets or irregular particles.

The rest of this chapter and the whole of the next will be devoted to what may be called general vital colouring, that is to say the kind of vital colouring that is applicable to cells of all kinds.

The number of colouring agents that have been shown to be really useful in this kind of work is small. We scarcely need more than the following, though a few others have limited uses and will be mentioned below:—

dahlia (triarylmethane)
brilliant cresyl blue (oxazine)
Nile blue A (oxazine)
azures A and B (thiazine)
methylene blue (thiazine)
neutral red (azine)
Janus green (azine-azo)
Bismarck brown (azo)

Fluorescent dyes are sometimes used vitally,[399] but it would not appear that they are superior to ordinary vital dyes in most cytological studies, and ultra-violet light does not commend itself as an illuminant for living cells.

The dyes used may be presented in very different ways. Aquatic animals may simply be placed in a solution of the dye. This is indeed the only practical way of colouring most Protozoa, and it was employed by Brandt [88] when he introduced the use of dyes as vital colouring agents. Ehrlich [157] used the same method with tadpoles when he introduced neutral red as a vital dye. Low concentrations must necessarily be used, even with the least toxic dyes. Ehrlich used neutral red at concentrations from 0·01% down to one-tenth of this. If the organism is small, it may be examined intact under the microscope. If it is not small, loose cells or minute pieces of tissue must be removed, still living, for examination in a suitable saline medium.

An alternative method, applicable to all animals that are big enough, is to inject a solution of the dye either subcutaneously or into a blood-vessel or else into one of the cavities of the body. For this purpose it is best to use a freshly-made solution at high concentration (about 1% or ½%). Saline solutions of vital dyes should not be kept for a long time, since these dyes have a special tendency to be flocculated by salts (p. 290). It is usual to give several injections at intervals of hours or days and subsequently to remove fragments for study while still alive in a saline medium.

Another method is to mix the dry dye with the food. The fact that bones become coloured if an animal eats madder appears to have been known to the Chinese of antiquity,[51] but this should be regarded as an example of pseudovital colouring. Lipid-soluble colouring agents may be dissolved in edible fats. The method is simple but has the disadvantage that dosage cannot be easily controlled.

Yet another method is to remove loose cells or minute fragments of tissue from the body and place them in a solution of the dye in a saline fluid of the right osmotic pressure. A concentration of 0·01% or less is usually best. It is convenient to keep stock solutions at 1% or ½% in distilled water and to dilute these with the appropriate saline solution just before use. Michaelis [343] called this important method of vital colouring *Die postmortale Färbung*, but there is, of course, no reason why the animal from which the cells were taken should die. Arnold [15, 16] made a careful study of the dyeing of *überlebender* cells. He chose this very suitable term because the cells, having been removed from the body, *survived* during the period of coloration. This should properly be called survival or supervital dyeing (see p. 329).

If one prefers to leave cells in their natural medium during the process of supervital dyeing, it is possible to dissolve the dye directly in the natural medium and thus avoid the use of any artificial saline solution. To do this, the dye is dissolved in absolute ethanol, and the solution spread on microscopical glass slides and evaporated to dryness.[436] When a drop or two of blood or any other natural fluid that contains cells is placed on a slide that has been treated in this way, the dye dissolves in the fluid and colours the living cells. It is desirable to take precautions to ensure that the ethanol is not contaminated by anything that could damage the cells.

A routine microscopical preparation of dead tissues focuses the observer's attention on the distribution of chromatin. The appearance of cells after vital dyeing is entirely different. Usually the nucleus and ground cytoplasm are scarcely or not at all affected, and attention is focused on certain cytoplasmic inclusions, the existence of which is not even suggested in the routine slide. The principal inclusions that are coloured are certain vacuoles, lipid globules, and mitochondria. The vacuoles that are coloured are of various kinds. The food vacuoles of phagocytic cells have a particularly strong tendency to take up vital dyes. The lipid globules that colour easily are commonly those that do not consist simply of triglyceride.

Either the ground cytoplasm or the nucleus or both are occasionally tinged; the nucleolus is sometimes more strongly coloured. The nuclear colouring, apart from that of the nucleolus, is generally diffuse: a chromatic network, like that often seen in fixed preparations, does not appear. Strangely enough, the very first paper that recorded direct vital dyeing with a synthetic dye included also a description of vital nuclear dyeing: for in his work on *Actinosphaerium* Brandt [88] dyed not only the lipid globules of some specimens with Bismarck brown, but also the nuclei of others with haematein. Unfortunately he did not state the concentration of his 'Haematoxylin'. He must have used the dye without mordant. A few years later [89] he obtained nuclear colouring of various amoebae with the same dye. Most of the colour was in the *Nucleinkugeln* (presumably the peripheral bodies now known to be nucleoli). He also achieved the *tour de force* of double vital dyeing, by first colouring the lipid globules of a living Protozoon with Bismarck brown, and then its nucleus with haematein.

The nuclei of many different cells of plants were successfully dyed in life by the British botanist Campbell,[101] working under Pfeffer's supervision at Tübingen. He used mauveine, dahlia, and methyl violet. His studies were carried out chiefly with species of *Tradescantia*, the staminal hairs providing most of his material; he also used the spermatozooids of Characeae, and various other cells. Protoplasmic streaming provided evidence of the vitality of the stationary cells, while the spermatozooids with dyed nuclei showed that they were alive by moving about. In a few cases Campbell was able to dye the 'segments' (chromosomes) of dividing *Tradescantia* cells, and to watch them moving apart at anaphase; he also saw them begin to undergo their telophase transformations.

The curious but careful observations of Brandt and Campbell, dating from about 70 years ago, deserve to be repeated.

It has been suggested that Heidenhain and Neisser [223] dyed nuclei vitally in 1874. This is extremely improbable. These distinguished investigators injected indigocarmine into the bloodstream of rabbits, waited till the urine was blue, then fixed the kidneys in absolute ethanol and cut thin sections. A blue colour was certainly seen in the nuclei, but there is every reason to suppose that the chromatin was fixed by the ethanol and then dyed by indigocarmine taken up from the tissue by the fixative.

The fact that chromatin is usually not coloured may be due in part to the dye having difficulty in penetrating the nuclear membrane. It is probable, however, that another cause is more important. The DNA of chromatin is combined with protein during life, and its phosphoric groups may not be free to react with basic dyes unless a fixative has acted.[346]

There is not one set of dyes that colours one sort of cytoplasmic inclusion exclusively and another set that colours another. Nevertheless, the dyes do not act quite unpredictably. Neutral red has a strong tendency to colour vacuoles, and the phrase 'neutral red vacuoles' was at one time commonly used. There is, however, no reason to suppose that all vacuoles colouring with neutral red have important features in common, and it must be remembered that this dye also colours many lipid globules. Most vital dyes colour some kinds of lipid globules, while few of them colour mitochondria. Dahlia, however, colours both lipid globules and mitochondria, and Janus green B shows some degree of specificity for the latter (see p. 292). This specificity has, however, been exaggerated, for it dyes strongly the oxyflavones that colour the vacuoles of plant cells, and also certain plastids, though these less strongly than mitochondria [200]; it also dyes the external part of the paranuclear bodies ('Golgi apparatus') in the primary spermatocyte of the snail.[426] Brilliant cresyl blue has a tendency to dye the nucleolus, as the Lewises noted long ago in the course of their work on cell-culture.[301]

Certain cytoplasmic inclusions are coloured metachromatically by particular vital dyes. A striking instance of this was reported by Lauterborn [286, 287] well over 60 years ago. He found that when certain diatoms were stained vitally with methylene blue, the inclusions known as 'Bütschli's granules' were coloured reddish violet. These granules contain or consist of the substance later called 'volutin', that is to say, of macromolecular metaphosphates (p. 246). It is probable that the metachromatic colouring attributed by Lauterborn and other early workers to methylene blue was in fact due to the highly metachromatic azures A and B, which always form in old solutions of methylene blue (p. 268). At about the same time Bolles Lee [291] noted vital metachromatic staining of the *Nebenkern* of the primary spermatocytes of the snail with an old solution of methylene blue. The hyaline substance of the region became violet, while the external parts (so-called *bâtonnets*) were blue. A few years later Ancel [9] reported the red vital colouring by methylene blue of globules in this region. This has recently been confirmed by Roque,[426] who used toluidine blue.

These early records are mentioned here because some writers suppose vital metachromasy to be a modern discovery.[520] The dyes used in more modern work have been new methylene blue,[59] azure B,[283] and toluidine blue.[138, 139, 426]

The inclusions that are coloured metachromatically during life are by no means necessarily 'volutin' granules. In the higher animals they commonly contain or consist of mucopolysaccharides.[138, 139] They sometimes occur in the part of the cell that is blackened by the 'Golgi' techniques.[283]

There is one striking difference between orthochromatic and metachromatic vital colouring. An object that is chromotrope in life is commonly chromotrope also in fixed preparations. In other words, the chromotrope substances are not easily altered by the fixation of the cell, but remain in position and continue to show their characteristic effect upon dyes. Lipid globules, on the contrary, are either not retained in preparations of dead tissues, or else do not colour specifically with the same dyes that coloured them in life.

Since the cytoplasmic inclusions could not be coloured unless the dye passed through the ground cytoplasm, it is natural enough that the latter is often slightly tinged. This is usually not helpful, but in one particular case the colouring of the ground cytoplasm is the object desired. It was discovered by Ehrlich [156] in 1887 that certain nerve-fibres and nerve-endings, especially in the taste-papillae of the frog, are coloured during life by methylene blue. He says that thionine and its dimethyl derivative give metachromatic colouring of nerve-endings, and this is probably the earliest mention of vital metachromasy. This colouring of nerve-fibres is due to the uptake of the dye by the axon. The cytoplasm of the body of the nerve-cell also becomes coloured. Other kinds of cells may be dyed in the same way, but they begin to lose their colour while the axons are becoming darker; in the mammals, hair-follicles and sebaceous glands retain the colour longer than other non-nervous tissues.[545] Ehrlich's discovery was of great importance to neurology, for it simplified the tracking of nerve-fibres, while avoiding artifacts due to metallic impregnation.

When a living cell, stained supervitally, has been under examination for an hour or so, without any precautions having been taken to keep it alive, its appearance begins to change. Water often separates from close association with the proteins of the cytoplasm and appears in the form of vacuoles. These vacuoles have a tendency to colour with neutral red. It is important to distinguish between these new formations and the pre-existing bodies that have a particular affinity for this dye. At length, whether this vacuolation has taken place or not, certain irreversible changes occur, which are made very evident by the presence of the dye. Previously, as has been remarked, the cell bore little resemblance to what is seen in a permanent preparation. Now, suddenly, that is no longer so. A network appears in the nucleus and takes up the dye strongly, and cytoplasmic inclusions that were coloured in life become invisible.

In general, one should not expect a dye to give the same picture when used vitally as it gives in fixed preparations. This is not simply because fixation and embedding dissolve out certain colourable cell-inclusions and change the reactions of others. In fixed preparations the dye penetrates most of the tissues at the concentration at which we present it, and we determine the end-point of its reaction, either by removing it when a desired effect has been produced, or by differentiating. In vital colouring, on the contrary, as de Beauchamps [51] pointed out half a century ago, our solvent is not the solvent in which the dye is presented to the objects contained in cells, nor can we directly control the concentration at which the dye will act. A state of equilibrium is gradually built up between the dye as we present it and the fluid of the cell, and that equilibrium is not under our control.

CHAPTER 16

The Mode of Action of Vital Dyes

Three requirements must be satisfied if a dye is to be used in general vital work. It must be able (1) to enter cells of various kinds; (2) to diffuse through the protoplasm without killing the cells; and (3) *either* to colour certain pre-existent cell-inclusions distinctively, *or* to colour the whole of the cytoplasm of particular cells so strongly that the cells and their processes stand out from the surrounding intercellular material and from other cells. In short, the three requirements are penetration, harmlessness, and some degree of specificity.

The dyes that satisfy the three requirements are all basic, but by no means all basic dyes satisfy them. A glance at the list of particularly useful general vital dyes on p. 278 will show that they do not all belong to any particular chemical group. Indeed, nearly all the groups mentioned in the chapter on the classification of dyes (p. 169) provide examples that can be used vitally. Heidenhain [218] claimed that most vital dyes, and the best, were oxazines, thiazines, and azines, and that when, as in safranine, one of the central nitrogen atoms became quinquevalent (or as we should say today, positively charged), vital dyeing was no longer possible. There is some truth in these statements, but in fact it is not possible to generalize about the chemical composition of vital dyes in such simple terms. The special characters of vital dyes will here be considered under the separate headings of penetration, harmlessness, and specificity.

Penetration

A few kinds of cells are so large that dyes may be injected into them, but the danger of mechanical damage usually outweighs the advantage of certain and controlled entry, and anyhow the method is not applicable to most cells. In ordinary vital work one arranges that the cell shall be bathed in a solution of the dye, and the process of penetration is not under the direct control of the experimenter.

The ability of dyes to enter living cells was first studied systematically by Overton.[377] He was engaged on the researches that proved the existence of a cell-membrane having different properties from the ground cytoplasm. He found that basic dyes entered cells readily, but sulphonated acid dyes slowly or not at all. He noticed that basic dyes were soluble in melted cholesterol and in lecithin dissolved in warm benzene, and also that lecithin and other phospholipids suspended in water take up basic dyes strongly. Most acid dyes, on the contrary, are insoluble in these solvents, and scarcely or not at all taken up by suspended phospholipid. Overton concluded that the penetration of basic dyes into cells was made possible by the presence of lipid in the cell-membrane. He was confirmed in this view by the facts that methyl orange and tropaeolin, though acid dyes, have some capacity to enter cells, and differ from most acid dyes in being somewhat soluble in lecithin solution and capable of absorption by suspended lecithin.

Fischel [167] made an interesting generalization about the chemistry of vital dyes. He pointed out that in these dyes the hydrogens of the amino-groups were commonly (though not always) substituted by methyl or ethyl. Pararosaniline and rosanilin, for instance, are not usable vitally, but the related methyl violet (with methyl groups substituting some of the amino-hydrogens) and especially dahlia (with ethyl groups) are useful vital dyes. Fischel's generalization applies to seven of the eight especially valuable vital dyes listed on p. 278. The substitution by methyl and ethyl would tend towards solubility in lipids. Not all basic dyes are lipid-soluble. As Seki [460] showed, it is especially those that are soluble in solutions of lecithin that are able to enter living cells.

The capacity of dyes to pass through a layer of a liquid lipid or of a lipid-solvent such as chloroform may be studied by the use of quite simple apparatus.[248] A horizontal glass tube is turned vertically upwards at each end and a vertical tube is connected with it at its middle point. Enough lipid or lipid-solvent is put in the apparatus to fill the horizontal tube and rise some distance in the vertical ones. A solution intended to represent a cellular fluid may now be added to one of the end-tubes. In the original experiments actual sap taken from the very large vacuole of *Valonia* (Siphonocladiales) was used. The aqueous solution of a dye is put in the tube at the other end. Mechanical stirrers in the end-tubes keep the dye-solution and sap (or other fluid) moving, while the middle tube admits a stirrer to the lipid-solvent. After the lapse of an hour the concentration of the dye that has diffused through the lipid-solvent into the sap is determined colorimetrically.

There is a general correspondence between the rapidity with which a dye passes through a layer of chloroform and its ability to enter cells of various kinds, but the relation is by no means exact.[248, 92] In the circumstances of the experiment the chloroform contains a certain amount of water, and it has been pointed out [92] that passage through the layer of chloroform does not actually prove the necessity for lipid-solubility. It has been held [92] that vital dyes may enter the living cell by either an aqueous or a lipid path, and that the former is open to the cations of basic dyes. Seki,[460] however, found no correlation between the diffusibility of basic dyes in aqueous media and their ability to enter living cells.

In his paper of 1887, in which he introduced methylene blue as a vital colouring agent for nerve-fibres, Ehrlich [156] mentioned that the dye tends to be reduced to its leucobase in the tissues and in this state to diffuse easily *out of* cells. Today, the ready diffusibility of leucobases *into* cells is of greater interest. It is usual to reduce methylene blue to its leucobase in order to make it enter axons more easily. The reduction of dyes to leucobases for this purpose may be done by adding sodium thiosulphate and acidifying. The resultant sulphurous acid acts as reducer. Alternatively one may use a compound of sodium sulphoxylate ($NaHSO_2$) with formaldehyde. This substance (with two molecules of water) is sold as white lumps or powder under various trade names (Rongalit, etc.). It was used by Unna for the reduction of methylene blue long ago.[515, 517] Its power to reduce dyes derives from its capacity, when in solution, to take oxygen from water and thus convert itself to the compound of acid sodium sulphite with formaldehyde:—

$$NaHSO_2.CH_2O + H_2O = NaHSO_3.CH_2O + H_2$$

The hydrogen formed by this reaction combines with the dye. The commercial product contains some sodium sulphite (Na_2SO_3), which gives an alkaline reaction to the solution. Since the leucobase is formed more slowly in alkaline solution, and also tends to be precipitated, it is usual to acidify slightly with hydrochloric acid.

For an example of the practical use of sodium sulphoxylate/formaldehyde in the reduction of methylene blue for vital studies, see Smith.[479]

The penetration of leucobases has been investigated especially by Harris and Peters,[216] who used methylene blue. The leucobase is, of course, colourless (non-chromophoric) and non-ionic (p. 163). It is about 25 times as soluble in chloroform as in water. Ionized methylene blue scarcely penetrates, except where cell surfaces have

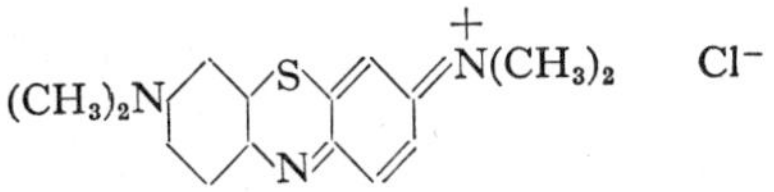

Methylene blue

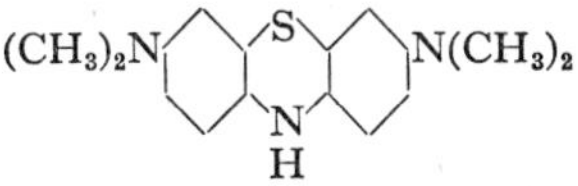

Leucobase of methylene blue

been damaged. When the dye is used in this form, it is probable that part of it is reduced in the vicinity of the cells and enters as the leucobase. For some reason that has not been explained, acidity makes the cell-membrane more permeable to the leucobase; but if strongly acid solutions are used, hydrogen ions are

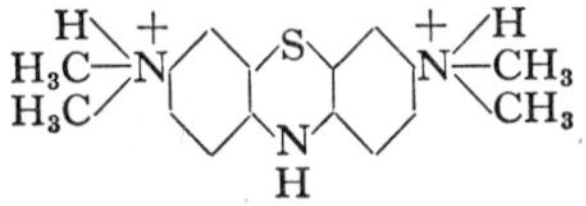

The cation derived from the leucobase of methylene blue, in strongly acid solution

added to one or both of the dimethylamino-groups, and new, ionized forms of methylene blue are thus produced, which cannot readily penetrate. As the pH is lowered from alkalinity, the increased permeability of the cell-membrane increases the uptake of methylene blue until pH5 is reached; from this degree of acidity onwards the uptake is less, because there is less leucobase.

If ciliates are put in solutions of the leucobases of vital dyes, a most surprising appearance is given, for the animals swim about actively with coloured macronuclei. It is as though a fixed and stained preparation had come alive. Thionine, brilliant cresyl blue, and several other dyes are suitable.[427] When a metachromatic dye such as thionine is used, the whole organism is at first blue, but the cytoplasm becomes violet and then raspberry-coloured, while the macronucleus remains blue. It is rather strange that the use of the leucobase makes it easier to colour the macronucleus, for lipid-solubility would not be thought likely to have this effect; but perhaps the cause is simply the abundance of the colouring agent in the cytoplasm, caused by the ready penetration of the cell-membrane by the leucobase.

The fact that the leucobase colours the living cell or parts of it shows that the cell is able to restore the dye by oxidation. As Ehrlich showed, the cell is also able to reduce the dye to its leucobase. The dehydrogenases of the cell take hydrogen from their substrates and use the dye as acceptor, unless something is done to keep the dye oxidized. Supervital preparations are usually exposed to atmospheric oxygen while they are being vitally coloured, so as to prevent this reduction. One really needs anaerobic conditions at the start, to help penetration, and then abundant oxygen to convert all leucobase into dye. A convenient technique has not been worked out for common use, though Harris and Peters [216] have achieved this end in a rather elaborate way.

The tendency of vital dyes to be reduced to their leucobases varies considerably. Thionine is particularly easily reduced, neutral red with difficulty. That is to say, thionine readily gives up its ionic form in the presence of reducers, by accepting an electron, while neutral red does not. Vital dyes arrange themselves in this order of oxidation-potentials, beginning with the most easily reduced (that is to say, the strongest oxidizer): thionine, brilliant cresyl blue, methylene blue, Janus green, neutral red.

It is rather surprising that neutral red should penetrate cells so particularly easily, in view of its strong tendency to retain the ionic form. Its resistance to anaerobic conditions are well seen in supervital work, for the colour is well retained even under a cover-slip sealed at the edges.

Despite the small value or uselessness of acid dyes in general vital dyeing, it is desirable to record shortly what is known about the capacity of some of them to enter living cells. A few acid dyes, such as orange G, that occur in aqueous solution in the form of particles not exceeding 0·64 mμ in radius, are able to enter certain cells that are impervious to those acid dyes that are dispersed in larger particles.[197] The indophenols as a group are also able to enter living cells. It has been pointed out [92] that these are acid dyes that scarcely ionize, and it seems that entry is denied, as a rule, not to acid dyes as such, but to acid dye-ions. The indophenols are not further considered in this book, since they have only limited applications in microtechnique. The ability of certain lipid-soluble acid dyes to enter cells has already been mentioned (p. 285).

Möllendorff [353] called attention to one of the rare exceptions to the rule that acid dyes do not colour pre-existent cell-inclusions. Eosin, an acid dye, will colour supervitally certain strongly refractile granules in connective tissue cells of the tadpole and frog.

Harmlessness

Dyes kill cells instantly if used at the concentrations that are customary in work with fixed tissues. It is easy to find a concentration (usually about 0·01%) at which a particular dye can be used supervitally without killing the cell; but one can neither control nor ascertain the concentration of the dye in the protoplasm, except in those rare cases in which the cell is so large that a measured quantity of the dye can be injected. It is probably for this reason that we have so little knowledge of the relation between the chemical structure and harmlessness of dyes.

We cannot account satisfactorily for the special tolerance of neutral red by cells. Protozoa may be cultured generation after generation in solutions of this dye at high enough concentrations to colour the food-vacuoles. Conjugation proceeds in *Paramecium* in the same circumstances.[271] Mitosis can occur normally in the roots of plants bathed in this dye at 0·02% or even higher concentrations.[52] Many aquatic animals can be kept for long periods in solutions strong enough to colour the tissues.

Fischel [167] made the generalization that in vital dyes the hydrogen of the amino-groups is not replaced by an aryl-group; among the azine dyes, similarly, those that have an extra aryl ring attached to one of the central nitrogen atoms (safranine, for instance, see p. 181) are not usable as vital dyes. Actually, Janus green B has an extra aryl ring; but this is rather a toxic vital dye, which would not be much used if it had not a special affinity for mitochondria (p. 292). Extra aryl rings may perhaps confer toxicity.

According to Seki,[460] those basic dyes that maintain their electric charge in alkaline solution tend to coagulate the cytoplasm and thus kill the cell.

Specificity

Diffuse colouring, such as can be obtained with certain acid dyes, is useless. The dye must either colour certain cell-inclusions strongly, while leaving the background unstained or nearly so, or else it must stain the cytoplasm of a particular kind of cell strongly, so that this cell and its processes show up clearly against intercellular material and other kinds of cells. The colouring of nerve-cells, axons, and dendrites is much the most familiar example of the latter sort of vital dyeing. The special suitability of methylene blue in this kind of work is unexplained.

Cell-inclusions are diverse chemically, and quite different processes are probably involved in the colouring of different inclusions. The subject is very complicated, and any attempt at facile presentation would be misleading.

When Nile blue colours a lipid inclusion pink or pinkish, there is every reason to suppose that the red oxazone present in solutions of this dye simply dissolves in the lipid, so that in this particular case there is no question of dyeing but only of colouring by solution (p. 302). Bismarck brown is exceptional among dyes in being fairly soluble in olive oil,[460] and it possibly colours certain lipid globules by dissolving in them. Neutral red is very slightly soluble in olive oil, with a pale yellow colour, but the other vital dyes are insoluble. However, since most vital dyes are somewhat soluble in mixtures of olive oil with lecithin or oleic acid, it might be thought that they colour lipid globules by solution. This is not so, for when microscopical droplets of such solutions are examined, there is not enough dye in them for the colour to be visible.[460]

Somehow or other most vital dyes are aggregated by certain lipid globules to a very high concentration. Seki relates this to their tendency towards flocculation. Many basic dyes are electropositive through a wide range of pH, but certain of them colour collodion (electronegative) *less* strongly in alkaline than in acid solution, which is the opposite of what one would expect. These exceptional basic dyes are dahlia, Nile blue, thionine, azure II (a mixture of azures with methylene blue), methylene blue, toluidine blue, neutral red, safranine, Janus green, and Bismarck brown.[460] It is a striking fact that every one of these except safranine is a vital dye. Seki noted that most of these dyes flocculate in alkaline solution, and he relates easy flocculation to the capacity to colour vitally. Vital dyes are often easily flocculated or indeed precipitated by salts, especially calcium chloride.[460] Four of the dyes mentioned above (Nile blue, neutral red, Janus green, and Bismarck brown) tend to become negatively charged in alkaline solution.[460]

When a vital dye colours a vacuole or other cytoplasmic inclusion, it is often seen first in the form of a speck on the surface of the inclusion.[353] Satellites of neutral red are sometimes formed round the food-vacuoles of Protozoa.[271] The dye may also be seen in the form of clumps within a vacuole. These facts suggest flocculation.

Although solution and flocculation no doubt play their parts in particular cases, yet it seems certain that dyeing in the strict sense also occurs. The facts are not in accordance with Fischer's [169] dogmatic statement, 'Without fixation there is no dyeing of histological preparations'. When brilliant cresyl blue colours a nucleolus there is no reason to doubt that the basic dye-ion is combining with RNA. Metachromatic colouring points in the same direction, for, as we have seen, the results are often the same in living and fixed cells. We can scarcely doubt that dyeing also occurs when a lipid globule colours strongly with a vital dye. The lipid globules that react in this way are not droplets of triglyceride or other lipid. On the contrary, they evidently contain a considerable amount of water and indeed are often spoken of as 'vacuoles'. In some cases the refractive index is quite low and it may even be rather hard to see the vacuole without using a vital dye. If the cell bursts, however, and the vacuoles escape from the cytoplasm, the lower refractive index of the saline solution in which the cell is suspended makes them easily visible.[464] That such vacuoles contain lipid may be proved by histochemical tests.[465] It is evident that the lipid is dispersed in water. Now phospholipids are easily dispersed in this way, and when so dispersed they present negative charges (on the phosphoric acid component), which could react with the dye-ion of a basic dye. It is probable that basic dyes often colour lipid globules and 'vacuoles' because these cytoplasmic inclusions contain acidic lipids dispersed in water.

The food-vacuoles of phagocytes are of quite a different nature from the vacuoles we have been considering. It has long been known that food-vacuoles are strongly coloured by neutral red. To prevent any possibility of misunderstanding, it must be remarked that this has no connexion with the uptake of insoluble pigments or the aggregation of acid azo dyes by phagocytic cells. We are concerned here with the colouring of pre-existent objects by a basic dye. It was at first thought that the object coloured was the protein matter of the foodstuff, in the course of digestion.[403] It was noticed, however, that when particles of talc were taken up by leucocytes, they appeared to become red, though talc itself has no affinity for the dye.[233] It was evident that the fluid of food-vacuoles had an affinity for the dye, though the ingested food often coloured as well. It was remarked by Marston [326] that all the azine dyes he tried were capable of precipitating trypsin from solution. The precipitate is coloured by the dye. This suggested that neutral red and other azine dyes might be used as indicators of the presence of proteolytic enzymes, an idea taken up with enthusiasm by Koehring,[271] who thought there was some necessary connexion between colouring with this dye and the synthesis or degradation of protein. There may indeed be such a connexion in particular cases, but neutral red can of course colour other cellular constituents than proteolytic enzymes.

The colouring of mitochondria by Janus green B calls for particular mention. This dye (called green from the colour it gives to certain textiles) generally imparts a blue colour to these particular organelles and nothing else in the cell, if used in sufficiently dilute solution; or if the cell as a whole takes it up, the mitochondria retain it when the rest of the cell has become colourless. This very special (though not complete) specificity of the dye was mentioned on p. 281, when the objects that are coloured by vital dyes were being enumerated.

Attempts have naturally been made to explain this specificity. It is held to be related to the differential reduction of the dye in various parts of the cell. Because Janus green is both an azine and an azo dye, its reduction is complicated. At the first stage the azine-azo structure is retained; but further reduction splits the azo linkage, and a red azine dye, rather closely related to safranine, results. This has a tendency to give a diffuse, very pale pink colour to the nucleus, and sometimes to the cytoplasm also. Still further reduction produces the leucobase of the azine. Molecular oxygen can oxidize the leucobase to the red azine, but cannot reassemble the broken azo-link.[289]

It has been suggested [200] that the dye is first taken up in its blue form by the mitochondria and subsequently reduced by them to the red compound, which escapes and colours the cell diffusely. Lazarow and Cooperstein,[289] who have made a special study of Janus green, have reached a very different conclusion. They consider that the dye is absorbed everywhere in the cell, but soon reduced except in the mitochondria, where the cytochrome C/cytochrome oxidase system maintains it in the oxidized (blue) form. They have shown that outside the cell Janus green B is taken up by diverse proteins and not by mitochondria only, and they consider the specific colouring to be due to the vital activities of the cell.

The vital activity of mitochondria can be exhibited in an even more striking way by the use of colouring agents of a very special kind, which develop colour or become more intensely coloured on reduction. These are the tetrazolium salts. Their reduction-pro-

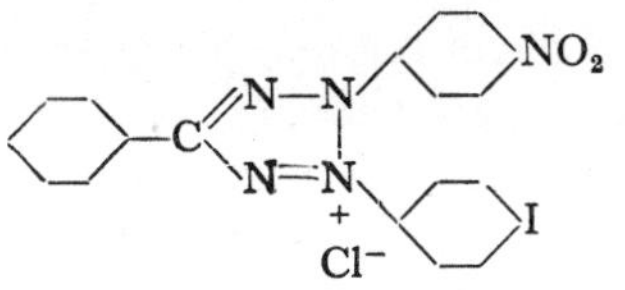

Iodo-nitro-tetrazolium

ducts, which are insoluble, are called formazans. Iodo-nitro-tetrazolium is a pale yellow substance, soluble in water. It is particularly suitable for the purpose because it reacts readily and

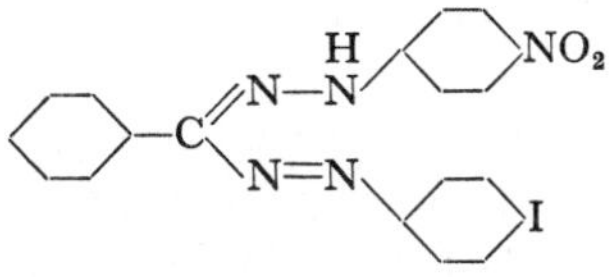

Formazan of iodo-nitro-tetrazolium

its reddish-purple formazan stays immobile at the place of its formation in the tissues.[422] The azo chromophore, to which it owes its intense colour, will be noticed; so will the absence of any auxochrome.

Succinic dehydrogenase is able to remove hydrogen from succinate and transfer it to tetrazolium compounds, which mark the site of this reaction by the deposition of formazan. It has been shown that if mitochondria, isolated by differential centrifuging, are treated with a tetrazolium salt in the presence of succinate, so much formazan is deposited on them that the matted clumps of mitochondria and coloured precipitate appear almost black. In the absence of succinate the reaction does not occur.[466]

Iodo-nitro-tetrazolium is luckily very innocuous, and indeed cells from chick-embryos can be cultured in solutions at 0·25%. In living nerve-fibres of the ray, Hughes [243] has been able to see particles of formazan, apparently deposited in rows along the mito-chondria. This very interesting observation requires confirmation. It seems to open the way towards a dynamic histochemistry of the living cell.

When a vital dye has been taken up by the cytoplasm of a cell or by a cell-inclusion, it is easily removed by alcohol and other solvents, and one cannot therefore transform a vital into a permanent preparation without taking special steps to immobilize it. In most cases observations are made and photomicrographs taken if necessary, and the specimen then discarded. It is sometimes convenient, however, to retain the dye permanently in the particular position that it took up during life. If nerves have been coloured by methylene blue, for instance, their relations with other tissues are often best seen in thick permanent preparations, rendered transparent by the use of suitable mounting media. Alternatively one may want to colour vitally and then make sections.

The way in which certain vital dyes may be fixed in the tissues appears to have been discovered by chance. It was customary in the eighties to make permanent preparations of small pieces of tissue by putting them in a solution of ammonium carminate neutralized by picric acid.[412] The chromatin was coloured red with carmine, the cytoplasm yellow by picrate. It was noticed that when nerve-preparations, dyed vitally by methylene blue, were treated with this picro-carmine, the colour of the axons was retained.[476] It was soon shown that the trapping of the colour was due to the ammonium picrate in picro-carmine, and that this salt could advantageously be used alone.[147] These discoveries were first reported in 1887.[147] The methylene blue picrate deposited in the tissues is unfortunately soluble in alcohol.[71]

The trapping of methylene blue in vital preparations was comprehensively studied by Bethe.[69–71] It was his object to find a salt of methylene blue that was insoluble in the various reagents commonly used in making permanent microscopical preparations. He precipitated methylene blue from its solutions by various anions and studied the solubility of the precipitates. He found that ferricyanide and molybdate gave precipitates with the necessary insolubility. When these were tried on tissues, molybdate gave a more finely granular precipitate. Bethe therefore used ammonium molybdate, $(NH_4)_6Mo_7O_{24}.4H_2O$, as a fixative in the ordinary sense and at the same time as a fixative or trapping agent for methylene blue. He sometimes added hydrogen peroxide to oxidize the dye from the leucobase, if necessary. The methylene blue molybdate formed in the axons was retained through paraffin or collodion embedding into permanent mounts in Canada balsam. This method of trapping methylene blue, introduced in 1895, is used to the present day.

Just as methylene blue can be fixed in nerve-axons, so this and other vital dyes can in certain cases be fixed in separate cell-inclusions. It is claimed that mitochondria can be coloured vitally by methylene blue and that the dye is retained if pieces of tissue are fixed in a solution of formaldehyde containing ammonium molybdate and picric acid.[508, 509] Certain granules in mammalian blastomeres colour metachromatically with toluidine blue during life.[138] These granules are tentatively identified as large mitochondria. The metachromatic colouring can be fixed by a solution of mercuric chloride containing ammonium molybdate and phosphotungstic acid; it is retained in balsam.[249] Neutral red is said to be fixed by Altmann's fluid.[509]

(For the use of trapping agents in non-vital work, see p. 224.)

CHAPTER 17

A Comparison between Dyeing and other Processes of Colouring

The words 'stain' and 'staining' have been avoided in this book, lest they should convey the impression of a particular, definable process. These words are often used in biology with such latitude that they no longer have recognizable meanings. It is said that basic fuchsine 'stains' chromatin, silver compounds 'stain' nerve-fibres, Sudan black 'stains' lipid. The three processes are fundamentally different from the chemical and physical points of view. It is a curious fact that the layman is more careful in his use of the word. He would not say that he 'stains' a fluid by dissolving a soluble coloured substance in it, yet that is what we do when we colour lipids with Sudan black. He would not say he 'stains' a piece of wood when he applies an insoluble pigment to it, yet the expression 'vital staining' is often used in biology in reference to the uptake of insoluble pigments by cells.

We need a phrase to cover all materials that we use to confer colour (or blackness) on the parts of an organism. The simplest and most self-explanatory general expression seems to be 'colouring agent' or 'colorant', and the corresponding verb is to 'colour'. These words suggest nothing but the demonstrable facts.

Anyone may define words as he chooses, yet his usage will not be logical unless the objects or processes called by his names possess genuine similarities that are not shared by objects or processes to which the names are not applied. Now the words 'dye' and 'dyeing' do have meanings, even if the limits of what is meant may be rather difficult to define. We shall here pass in review a series of processes departing in varying degrees from the process of dyeing as described in the preceding chapters. We shall start with colorations that obviously do not involve dyeing, and pass on step by step through various intermediates until we reach once again the substances and reactions with which we have familiarized ourselves.

We shall then be in a position to formulate more exactly what we mean by dyes and dyeing.

It must be noted at the outset that noun and verb are not inseparable, for we may use dyes without dyeing. The most obvious example is the use of a dye such as carmine as a component of an injection medium. Since carmine can be dispersed in water, though insoluble, we may suspend it in a suitable vehicle (such as a gelatine sol) and force it into the vessels or other cavities of an organism, through the walls of which it cannot escape. When microscopical preparations are subsequently made, the colour of the dye defines the form of the cavities. There is here no question of a process of dyeing, or of any reaction with tissue-constituents. Any strongly coloured substance whatever, provided it were insoluble but capable of dispersion in water, could be used instead.

The uptake of carmine and certain other dyes by phagocytic cells (p. 276) is another example of a process that has nothing to do with dyeing. The particles may equally well be of carbon in the form of lamp-black: the cells make no distinction, but pile up the substance in their cytoplasm. Neither the cytoplasm nor any pre-existent object in it is coloured. The process shows some resemblance to the sweeping up into a dustpan of a powdered dye that has fallen on the floor, but none at all to the dyeing of one's clothes.

The simplest way in which a colouring agent can act is by solution in a fluid contained within the tissues. Water itself is not a suitable solvent for this purpose, since it is distributed almost throughout the bodies of organisms, and very diffuse results would therefore be obtained; beyond this, many coloured substances that are soluble in water would be taken up also by proteins, so that the distribution of the water itself in the tissues would not be displayed. Separate liquid droplets lend themselves particularly to colouring by solution. The chief liquids that occur in the tissues of organisms in this form are lipids and the essential oils of plants. These can be coloured by solution. Solid (crystalline) lipids will not dissolve colouring agents.[99, 100] (The essential oils will not be mentioned again. It is to be remembered that most colouring agents that are soluble in lipids dissolve also in essential oils and their thickened products, the resins.)

A substance that is used in microtechnique to colour tissue-constituents by dissolving in them will here be called a lysochrome (Greek, *lúsis*, solution). Since these colorants find scarcely any application except in the study of lipids, the word will be used by itself in what follows to mean lipid-soluble lysochromes, unless the contrary is indicated.

To present a lysochrome to a section or other microscopical preparation, it is necessary to use a suitable solvent. Lysochromes are freely soluble in benzene, chloroform, ether, etc., but these solvents are unusable, since they would dissolve out the tissue-constituents that it is desired to show. A solvent that will not dissolve lipids is required. It is important also that the colouring agent should be less soluble in the solvent than in lipids. An example of a fluid fulfilling these requirements is 70% ethanol. If the lysochrome is presented to the tissues as a solution in this solvent, a partition will take place, some of it remaining in the solvent and the rest dissolving in the lipid. The amount of colour shown by the lipid will be governed by the partition-coefficient applying to the particular colouring agent in the presence of the two solvents concerned. Ideally the coefficient should strongly favour the lipid.

If a preparation that has been coloured in this way is placed in 70% ethanol (or other such solvent), the minute amount of lysochrome contained in the lipid tissue-constituents has to undergo partition between this lipid on one hand and the whole of the solvent in the jar on the other. This partition greatly favours the solvent, on account of its immense superiority in volume. As a result, no detectable amount of lysochrome remains in the lipid. There are no electrostatic or other bonds to hold the lysochrome in position. No more than a dip in the solvent is permissible (to prevent precipitation of the lysochrome, which is usually dissolved in the ethanol near saturation); the preparation must then be brought quickly into a fluid in which the colour is insoluble.

It follows from what has been said that a lysochrome, in the widest sense, must fulfil the following requirements. It must

(1) be strongly coloured;
(2) be very soluble in the substances it is intended to show;
(3) have no capacity to attach itself to any tissue-constituent except by solution;
(4) be capable of presentation to the tissues in a solvent having the following characters. (*a*) The solvent will not dissolve the substances the lysochrome is intended to

show. (*b*) The lysochrome is much less soluble in the solvent than in these substances.

Lipid-soluble lysochromes that fulfil all these conditions are necessarily insoluble in water.

A single Japanese worker [360] claims to have tested about 3,000 substances to find which of them could be used as lysochromes. Less than a dozen in all, however, have found favour with histochemists in general, and none of these is among the 3,000. The chosen few are diverse in chemical composition. Some of them are not related to any group of dyes. Coloured lipids are themselves suitable, and it may thus be said that a lipid is set to catch a lipid. The carotene of the fruit of red pepper (*Capsicum* sp., Solanaceae), for instance, is easily extractable and gives good results.[374] The related substance, annatto, from the fruit of a small tree (*Bixa orellana*, Bixaceae), could also be used; it is a straight-chain compound, one of the very few wholly aliphatic compounds that are strongly coloured. It is used for tinting butter. Alkannin, another plant-product (from the root of *Alkanna tinctoria*, Boraginaceae), was formerly much used as a lysochrome, especially in botanical studies. The molecule somewhat resembles alizarine, but one of the lateral rings has been broken and spread out as a straight chain.

Nowadays most workers prefer the synthetic lysochromes, which rely on the same chromophores that give colour to dyes, but lack the typical auxochromes. The latter, if present, would cause ionization, and as a consequence the substance would be unusable as a lysochrome; for it would be soluble in aqueous media, insoluble in lipids, and capable of dyeing proteins and other tissue constituents. It is characteristic of lysochromes that they are non-ionic compounds.

The first reasonably satisfactory synthetic lysochrome was Sudan III, introduced in Italy towards the end of the last cen-

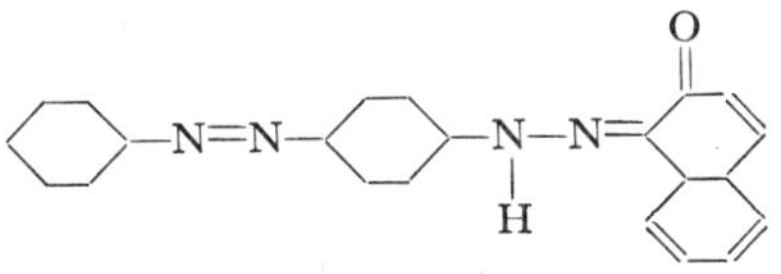

Sudan III

tury.[137] It is a red compound, related to the disazo dyes. The absence of typical auxochromes will be noticed. Sudan IV [344] differs only in possessing methyl groups attached to two of the aryl rings. It has the advantage of being rather darker and thus showing small lipid globules more clearly, and is one of the most

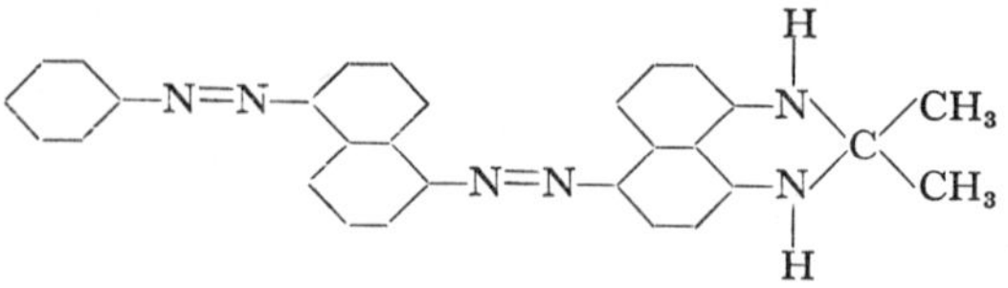

Sudan black B[305]

frequently used lysochromes. Sudan black B is a more complex, almost black lysochrome, introduced by Lison.[307,311] It colours lipids with particular intensity. This appears to be due not only to the darkness of its colour, but also in part to particularly high solubility in lipids.

Although colouring by lysochromes is so simple in principle, and sometimes in fact as well, yet some of the most commonly-used synthetic lysochromes do not act in exact accordance with expectation. This subject deserves rather careful study, because it throws an interesting sidelight on the nature of colouring processes in general.

Whereas the carotenoid lysochromes leave the cytoplasm untouched, Sudan III and IV show some tendency to colour it, as though they were capable of acting to a very small extent as dyes. This has not been explained. It is conceivable that in the presence of particular substances the $-\underset{\underset{\displaystyle H}{|}}{N}-N=$ group might be changed to a second azo group (–N=N–) and the =O to the auxochrome –OH. On the possibility of a transformation of this sort, see an old but very interesting paper by Michaelis.[344]

Sudan black also sometimes behaves unexpectedly, for it tends to be retained in certain tissue-constituents when the section is placed in a solvent such as 70% alcohol after colouring. The two $-\underset{\underset{\displaystyle H}{|}}{N}-$ groups suggest themselves as potential auxochromes. It has been shown[305, 110] that if these are blocked by acetylation with acetic anhydride, the colorant now dissolves out from the tissues like any other lysochrome.

It is a matter of particular interest that Sudan black, when not acetylated, has a special tendency to be taken up by phospholipids and retained in these against solvents.[110] It would appear that Sudan black is able to permeate phospholipids on account of its lipid-solubility, and to combine with phosphoric radicles on account of its possession of $-\underset{\mathrm{H}}{\underset{|}{\mathrm{N}}}-$ groups. Thus it has some of the characters of both lysochromes and dyes. Acidic substances other than lipids may sometimes be coloured by untreated Sudan black. Tests with the acetylated colorant should always be carried out if the colour is not readily removed by solvents. It must be remembered, though, that resistance to solution does not necessarily indicate that the substance coloured is not a lipid.

Nile red is a lysochrome of altogether special interest, on account of the complications that surround it in ordinary use. It is an oxazone; that is to say, a non-ionic substance formed by the

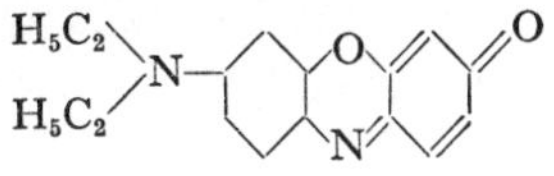

Nile red

oxidation of an oxazine dye (see p. 180). It can be dissolved in 70% ethanol and used in exactly the same way as Sudan III or IV. It has the advantage of possessing no potential auxochrome.

Nile red appears by spontaneous oxidation in aqueous solutions of Nile blue A. The anion of this dye is sulphate, because the chloride is rather insoluble. (The dye is often called Nile blue sulphate, but there is no more reason for doing this than for speaking of basic fuchsine chloride or sodium eosin.) Although Nile red is insoluble in distilled water, it is soluble in aqueous solutions of

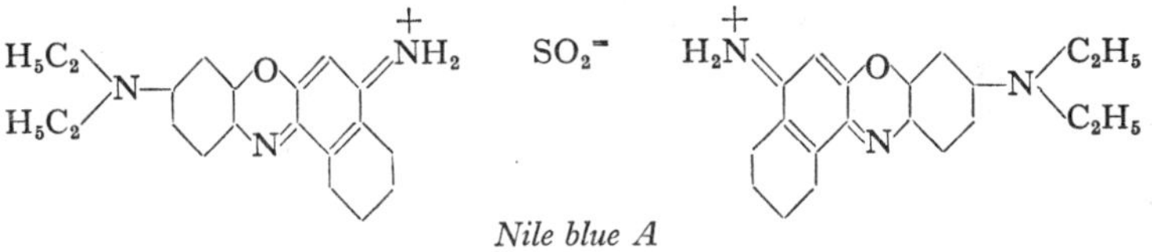

Nile blue A

Nile blue. Solutions of this dye have therefore very interesting properties, to which attention was called nearly half a century ago by the English pathologist, Smith.[480, 482] There has been much controversy about the action on tissues of solutions of Nile blue, but the main facts have been exposed by the careful studies of Lison [309] and Cain.[99]

An aqueous solution of Nile blue contains these substances:—

cation of the dye (blue);
anion of the dye (sulphate);
Nile red (red or rose, maximum absorption at about 500 mμ, fluorescent);
imino-base of Nile blue (orange-yellow, maximum absorption at about 482 mμ, not fluorescent).

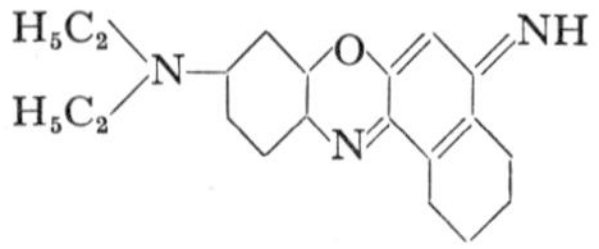

Imino-base of Nile blue [357, 503]

The proportions in which these substances occur vary according to the concentration at which the Nile blue was made up. Nile red is abundant in strong solutions. The amount can be increased by boiling the solution with a little sulphuric acid.[480] The imino-base of Nile blue is relatively more abundant in dilute than in strong solutions. Like Nile red, it is soluble in toluene and in aqueous solutions of Nile blue.

When a section is exposed to a solution of Nile blue, the Nile red present in it dissolves in all the liquid lipid components without distinction. Two substances are present in the solution, however, that are capable of reacting with acidic tissue-constituents; namely, the cation of Nile blue and the orange-yellow base. Either or both of these can react with acidic lipids (fatty acids) or lipids capable of acting as acids (phospholipids). Since the cation dyes blue and the salts of the base are of the same colour, we do not know, in any particular case, which of the two has been responsible for colouring these particular lipids blue. (The colour is sufficiently intense to mask the Nile red dissolved in them.) On one hand the lipid-soluble base would permeate lipids much more readily than the cation; on the other, there is plenty of cation but very little base in strong solutions of the dye. It seems certain, however, that cations can colour fatty acids, for basic fuchsine (rosaniline chloride) reacts with oleic acid to form rosaniline oleate and hydrochloric acid.[482]

Beyond these reactions, Nile blue naturally acts like any other basic dye in colouring all acidic tissue-constituents, such as chromatin and RNA. The orthochromatic colour is of course blue, but it may be remembered, as an added complication, that Nile blue is metachromatic (p. 278).

It follows that if any non-chromotropic tissue-constituent is coloured reddish by a solution of Nile blue, it contains or consists of a neutral (non-acidic) lipid. If any tissue-constituent *that has been proved to consist wholly of lipid* is coloured blue, it consists of fatty acid or phospholipid.

It is not necessary to rely on the amount of Nile red that happens to arise by spontaneous oxidation of Nile blue, for Nile red crystals may be prepared and added to a solution of Nile blue.[417]

Since nothing is easier than to colour the tissues differentially by soaking them in a solution of a dye or lysochrome, it might be thought that one could do the same with a solution of some coloured substance other than a dye or lysochrome. One would use a solution of a particular colour and would find that certain tissue-constituents had taken up a considerable amount of it and had assumed its colour, while others had taken up little or none.

It is not easy to specify a typical instance of the process just described. The difficulty forces on our attention the very special properties of dyes. The colouring of proteins by iodine probably provides as good an example of the process as one can find. It is far from satisfactory, however, for the reactions involved are not understood.

The facts are these. When iodine is dissolved in potassium iodide solution, a brown fluid is produced. If a section is placed in the solution, the proteins are coloured brown. Certain other tissue-constituents, such as cellulose cell-walls, also take up the colour, but to a much smaller extent.

The solution is brown owing to the presence of the I_3^- ion of potassium tri-iodide. This reacts with proteins by adding iodine to the phenyl ring of tyrosine, with the production chiefly of iodo-

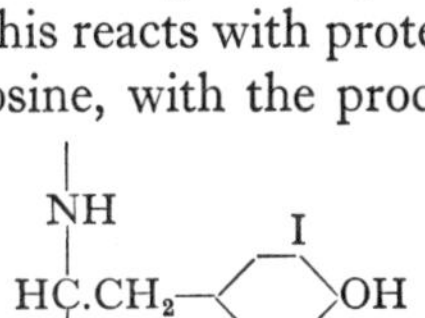

Iodogorgonic acid as part of a protein chain

gorgonic acid (di-iodotyrosine).[536, 376] This amino-acid was called *Jodgorgosäure* by its discoverer,[149] because he isolated it from the protein that forms the axial skeleton of the octactiniarian, *Gorgonia.*

The crystals of the pure substance are colourless [149, 536] or at most just tinged with yellow.[227] There is not much information about the colour of pure proteins in which iodine has been incorporated by reaction with the phenyl rings of the tyrosine residues, but casein that has been modified in this way and then washed free of loose iodine is white.[376] It is probable, therefore, that when proteins are browned by iodine/iodide solutions, the tri-iodide ion (or iodine in some other non-molecular form) is loosely adsorbed; but this rather evasive verb only veils our ignorance of the actual process involved.

In microtechnique we very frequently cause the *local formation* of a coloured substance that is not a dye. We soak the section in a colourless solution, and certain tissue-constituents become coloured (or black); or alternatively we soak it in a coloured solution, and certain tissue-constituents take on a different colour. Very many histochemical tests belong to this category of colouring processes.

It is an essential part of this method that the coloured (or black) reaction-product should not diffuse away from its site of production; otherwise vague or misleading results would be given. The substance produced must be either insoluble or else capable of adsorption to the object in or on which it originated.

There seems to be no limit to the variety of different processes that are used in microtechnique to achieve the local formation of coloured substances (other than dyes) in particular tissue-constituents. The reactions vary in complexity from the simple reduction of a metal from its salt to the trapping of the reaction-product of a cellular enzyme with a provided substrate, and its subsequent visualization.

One of the most familiar and striking examples of this kind of colouring is the blueing of starch by the same potassium tri-iodide solution that colours protein brown; but unfortunately, despite a great deal of study, the reactions concerned in this colour-change are not yet established with certainty.[411]

Colouring of this sort is particularly satisfactory when we can actually write the equation for the local reaction. Perls's test [392] for ferric iron is a good example. A section is soaked in a yellow solution of potassium ferrocyanide, and sites of ferric iron stand out sharply by the deposition in them of insoluble Prussian blue:—

$$4Fe^{+++} + 3Fe(CN)_6^{\equiv} \longrightarrow Fe_4[Fe(CN)_6]_3\downarrow$$

The reduction of a metal in a salt or other compound to the elementary state, with consequent formation of a black deposit, is a familiar process in microtechnique. One must choose a metal the ions of which take up electrons readily. This ensures easy reduction and stability after deposition. Silver and gold are suitable. Tissue-constituents that can provide electrons will reduce their salts:—

$$Ag^+ + e^- \longrightarrow Ag\downarrow$$

Complex silver compounds are often used, especially argentammine $[Ag(NH_3)_2]OH$.

The chief substances in the tissues capable of producing microscopically-visible deposits of metal are phenolic compounds, such as those present in the granules of the Kultschitzky cells of the intestinal crypts.

Metallic reduction is very largely used in tracing the course of the axons and dendrites of nerves, but a complex procedure is here necessary, for reducers are not present in sufficient amount to give microscopically visible deposits. It is uncertain what the reducers are, but there is evidence that sulphydryl groups are concerned.[542] The reducers, whatever they are, produce submicroscopic 'nuclei' of silver, similar to those present in an exposed but undeveloped photographic plate. In order to make a visible deposit, an enormously greater amount of silver must be deposited round these nuclei. Luckily the nerve-fibres take up a lot of unreduced silver, the ions of which associate themselves with histidine and other amino-acid constituents of the proteins of the fibre.[394] The double process of nucleus-formation and storage of unreduced silver is called 'impregnation'. The impregnated tissue remains transparent. When the tissue is transferred to a photographic developer, the stored silver is reduced and deposited round the nuclei. Thus the deposit is particulate. The particles can be resolved by the electron-microscope.[395] Together they form a black mass that appears continuous under the light-microscope. The term 'impregnation' should not be used unless the process of silvering (or the deposit of some other metal) involves two separate steps; the word is then applicable to the first step only, during which no visible deposit is made.

The fundamental dissimilarity between silvering and dyeing does not need to be stressed.

Impregnation methods have been much used in the study of

cytoplasmic inclusions. There is here no question of merely filling up a fibre with silver. The intention is to display the actual form of objects lying in the cytoplasm. Silver is piled on silver at the investigator's discretion. The method is useful for calling attention to particular parts of a cell, but may give a misleading idea of the shape of the objects on which the deposition occurs. A deposit on the surface of a spherical object may take the form of a ring or a cup (appearing as a crescent in optical section).

This technique is also apt to be misleading in another way. The preparations may give the false impression that all the blackened areas have a similar chemical composition. As Lison[310] has pointed out, the impregnation/development methods have no histochemical value. Various unrelated chemical substances can reduce silver if helped by a developer.[37]

Osmium tetroxide can also be used to give a black or blackish deposit on certain cytoplasmic inclusions, but this is an easily reducible, non-ionic compound, not requiring separate processes of impregnation and development. The nature of the reduction-product has already been mentioned (p. 119). It is usual to fix, wash out the fixative, and then leave the tissue for several days in a warm solution of osmium tetroxide; sections are cut subsequently. This 'postosmication' (p. 126), which is applicable to certain cytoplasmic inclusions, but not to axons or dendrites, is open to the same objections as silvering.[37] Both methods are applicable to certain problems of cytoplasmic cytology, but only if used with discretion.

Some of the processes that we have so far considered in this chapter have involved the use of a dye, but it is clear that none of them can be considered as a process of dyeing. We turn now to a realm of more subtle distinctions—to the borderland between not-dyeing and dyeing.

It is hard to decide whether we may properly speak of dyeing when we use substances, not themselves dyes, that react to form dyes in the tissues. This process is more familiar in the textile industry than in microtechnique. The oldest example is provided by indigo. The substance that attaches itself to the textile fibre is the colourless indigo-white (leuco-indigo). The precise method of attachment does not appear to have been worked out. When this process has been achieved, atmospheric oxygen is allowed access to the fibre, and the indigo-white becomes oxidized to the intensely blue substance, indigo. This possesses a chromophore and also side-groups corresponding to auxochromes, through which, presumably, attachment to the fibre is maintained. The inability

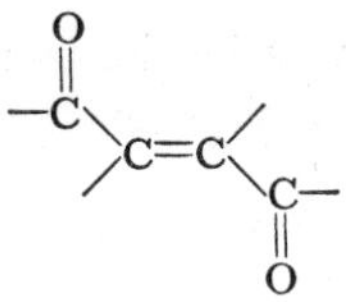

The chromophore of indigo [523]

of indigo to ionize directly in water or aqueous solutions prevents us from regarding it unequivocally as a dye, yet the resemblance is close. The chromophore is the same as in the microtechnical dye, indigo-carmine; but the latter, derived from indigo by sulphonation, is a typical, water-soluble acid dye of the slowly-diffusing kind (p. 236).

The 'azoic' dyes of the textile industry bear a resemblance to indigo, for here again the dye is synthesized as an insoluble substance in the fibre; but the reaction is not one of oxidation. These dyes received their name because of their resemblance to azo dyes. Indeed, they owe their colour to the same chromophore, but insolubility prevents their being used in the same way. The fibre is impregnated with a naphthol (β-naphthol, for instance), which attaches itself through its –OH group, and usually, in modern practice, through another group as well (often –CONH–, *ortho* to the –OH).[523] These attaching-groups resemble auxochromes, though the substance is colourless. The fibre is then treated with an aromatic diazo compound, and this snaps on to the naphthol by a –N=N– link. The resulting coloured substance is essentially a non-sulphonated azo dye. It can also be made in the absence of the fibre; but it cannot be used in this form, because it is insoluble.

A somewhat similar process is used in a histochemical technique for the recognition of tyrosine. Here, however, there is no preliminary impregnation of the tissue with a naphthol or related compound; on the contrary, one relies on the presence in the tissues of a phenol that will take the place of the naphthol used in azoic dyeing. It is because tyrosine possesses a phenyl ring that it can be histochemically demonstrated by the production of what is essentially an azoic dye. The reagent used for this purpose is tetra-azotized benzidine. In the presence of tyrosine this substance links itself to the phenyl ring of the amino-acid,[140] with the form-

Tetra-azotized benzidine

ation of what is essentially an insoluble, non-sulphonated azo dye. Since the phenyl ring of the tyrosine is part of the dye, no question

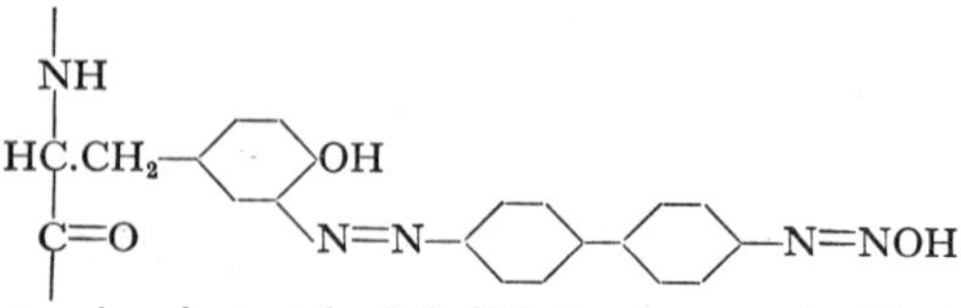

The tyrosine of a protein chain linked to tetra-azotized benzidine

of attachment through an auxochrome presents itself. The colour is brown. A bright red can be produced by coupling β-naphthol at the other end of the benzidine through another azo-link.[140] Histidine and tryptophane behave similarly to tyrosine. The three amino-acids can be distinguished by the use of 'blocking' agents; that is, substances that prevent one or other of the amino-acids from forming a link with tetra-azotized benzidine.[389]

An important histochemical test for aldehydes provides another

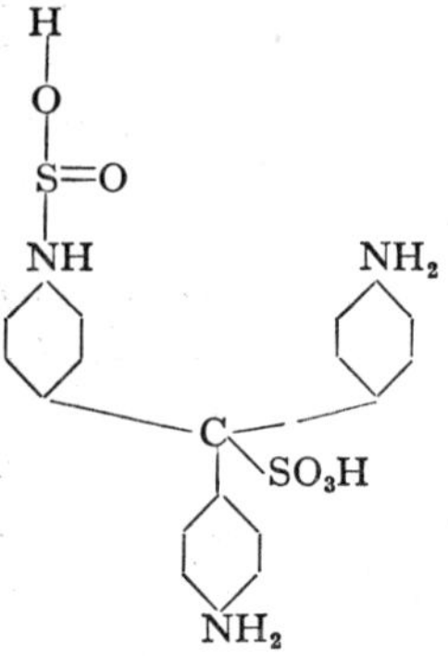

Schiff's reagent [539]

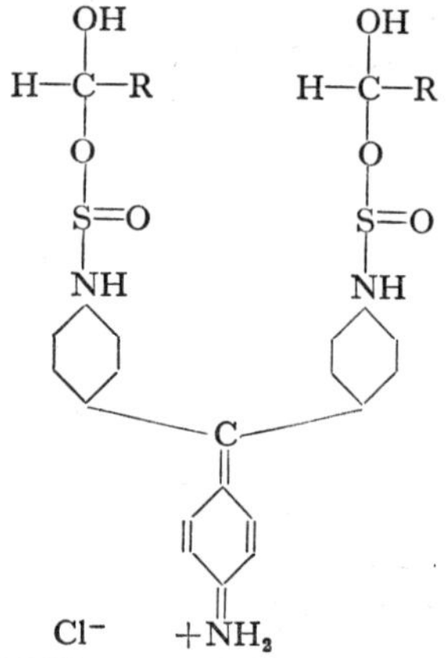

Schiff's reagent, in the presence of hydrochloric acid, after reaction with an aldehyde.[539, 310] (R *varies according to the aldehyde used.*)

example of the localized synthesis of a dye in certain particular tissue-constituents. Sections are treated with Schiff's reagent, which is made by the action of sulphurous acid on basic fuchsine.

(For simplicity, the reagent made from pararosaniline instead of basic fuchsine is shown here; see p. 160). Schiff's reagent is not itself a dye, for it lacks a chromophore and is therefore colourless. It is often regarded as a leucobase, but this is an error; for a leucobase becomes coloured on oxidation and could not possibly serve in Feulgen's reaction, which is not a test for oxidizing agents. When Schiff's reagent comes into contact with an aldehyde, the chromophore of the triarylmethane dyes is reconstituted. There seems to be no reason to deny the name of dye to the additive compound of aldehyde with Schiff's reagent. Since one treats the tissues with a colourless substance and a dye is formed in them, there is some resemblance to the use of indigo and the azoic dyes in the textile industry; but there are also important differences. With the textile dyes it is necessary to provide from outside the fibre *both* the constituents required for the making of the dye (indigo-white and oxygen, or naphthol and diazo-compound), while the tissues provide one of the reactants (aldehyde) when Schiff's reagent is used.

Another important difference is that the Schiff/aldehyde compound need not be insoluble. If one adds soluble aldehyde to Schiff's reagent in a glass vessel, one may then use the soluble, coloured reaction-product as a dye: it acts like other basic dyes in colouring acidic objects.[487, 488] Used in this way, the dye has naturally no affinity for aldehydes. The possibility exists that Schiff's reagent may form a coloured dye in the tissues by combination with aldehydes, and the dye may then wander off and colour any acidic objects.[487] Feulgen and Rossenbeck's [164] reaction for DNA relies on the visualization by Schiff's reagent of an aldehyde produced from deoxypentose. The exact position of DNA can only be revealed by this test if the reaction-product remains at the site of its formation.

When we use dyes in ordinary microtechnique, we soak the tissue in a coloured fluid and then observe the distribution of the colour among the tissue-constituents. To do this we shine visible light through the object and rely on the chromophores of the attached ions to lower the amplitude of rays of certain wavelengths. Our colour-sense is thus stimulated. That is generally the purpose of dyeing.

Sometimes, however, we use what are unquestionably dyes and allow them to become attached to the tissues by a process that must unquestionably be called dyeing, yet we do not follow this up in the way described in the last paragraph. On the contrary, we do not allow the chromophore to act in this way, for we deliberately prevent visible light from reaching the dyed object. A brief mention must be made of this technique, in which dyes are used but their most obvious character is not.

Quite a number of dyes are fluorescent, including some very common ones, such as eosin. The typical 'fluorochromes', however, have been chosen on account of their powerful fluorescence,

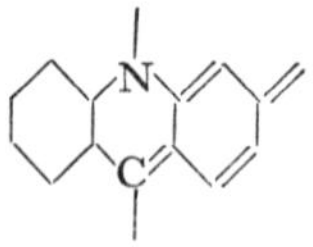

Skeleton formula of a typical acridine dye

and are not particularly adapted for use with visible light. Several of these, such as coriphosphine, are acridine dyes, but others belong to various other groups (the anthraquinonoid, for instance). The fluorochromes, in fact, possess ordinary chromophores and auxochromes, and attach themselves as basic or acid dyes.

When a fluorochrome has attached itself to the tissue, the preparation is illuminated by ultraviolet light. New rays of visible light originate wherever the dye is present. Different fluorochromes give light of different colours. The object has now become self-luminous. It is almost as though we were examining the abdomen of a glow-worm in the dark. The microscope uses the new rays originating in the fluorochrome. The image is produced in an unusual way, for there is no interference of direct and diffracted rays such as is always concerned in image-formation when an object is illuminated from below with visible light.

Since the intrinsic colour of a fluorochrome is irrelevant, some very feebly coloured ones are usable. Fluorescein itself is an example. This is an acid xanthene compound, too pale to be used as a dye. One could go further still and use a fluorescent substance that lacked a chromophore, provided that it could attach itself to the tissues. Quinine sulphate could be used in this way. This is a colourless, salt-like substance, the cation of which attaches itself like a basic dye and therefore interferes with subsequent dyeing by basic dyes.[346] It is fluorescent, but since it is not a coloured substance it cannot properly be called a fluorochrome. It might almost be said of substances such as this, whether fluorescent or not, that they lack chromophores but possess auxochromes.

Fluorochromes have proved of practical value in the rapid recognition of certain bacteria,[206] because the latter can be made to shine brilliantly in the dark and therefore can be seen easily with only moderate powers of the microscope. It would not appear, however, that the use of fluorochromes has resulted in many important discoveries in biology. This is not surprising, since they are not in general more selective than non-fluorescent dyes.

Fluorochromes are said to exhibit 'secondary' fluorescence. 'Primary' fluorescence is the direct origin of visible rays from tissue-constituents when irradiated with ultraviolet light. This can give valuable histochemical information. It does not fall within the scope of this book, since no colouring agent is used.

The foregoing analysis of the various processes of colouring used in microtechnique is summarized in table 11.

TABLE 11

Analysis of the processes of colouring in biological microtechnique

	See pages	*Examples*	
		Colouring of	*by*
A. Injection of suspended coloured particles into closed spaces	297	the cavities of blood-vessels	carmine/gelatine
B. Uptake of suspended coloured particles by phagocytic cells	297	uptake of injected carmine particles by histiocytes (no colouring of pre-existent space or object)	
C. Solution of a lysochrome	298	fat-droplets of adipose cells	Sudan IV
D. Adsorption from solution of a coloured substance, not a dye	303	protein	potassium tri-iodide
E. Local formation of a coloured substance, not a dye	304	ferric iron	potassium ferrocyanide
F. Local formation of a dye	307	protein	tetra-azotized benzidine
G. Dyeing	—	chromosomes	basic fuchsine

Group D is not very satisfactory, since few processes can be referred to it with certainty and the reactions involved are not understood. The other six groups are valid and clearly distinguishable, and many examples of all of them except F could be quoted.

The process of dyeing (G) is distinguished from all the others by the fact that the tissues are exposed to the action of a *solution* of a *dye*. A dye may be defined, for the purposes of microtechnique, as an aromatic, salt-like compound having these characters:—

(1) it ionizes in the presence of water;
(2) either the cations or the anions are coloured (sometimes both);
(3) the coloured ions are able to make chemical linkages with the proteins (and generally also with other constituents) of the fixed tissues of organisms (and in some cases with constituents of living cells as well);
(4) when the coloured ions make their linkages with the tissues, they do not lose colour, and generally they do not change it.

It is not necessary to mention mordants in the definition of dyes, since all dyes that work with mordants are able to colour the fixed tissues of organisms in the absence of any mordant, though the results are then very different.

The definition includes no substance that is not obviously a dye, and excludes scarcely any that deserve the name. The number of substances left in doubt is very small. Methylene violet (p. 268) is an example. This certainly acts as a dye, in the presence of related substances that are dyes, but when isolated is a non-ionic compound. It might fall into place under the definition if we knew its chemical structure in the presence of other dyes.

APPENDIX I

THE COMPOSITION OF SOLUTIONS EXPRESSED AS PERCENTAGES: CONVENTIONS ADOPTED IN THIS BOOK

It would be a convenience if the composition of solutions were generally expressed in terms of gram formula-weight per litre, or of normality in the case of acids. At present, however, it is usual in biological laboratories to express concentrations as percentages.

When percentages are expressed by the symbols w/w, w/v, v/w, and v/v, the meaning of the numerator is the weight (w) in g, or volume (v) in ml, of the solute. The denominator denotes the weight (100 g) or volume (100 ml) of the *solution* (not of the solvent). Thus, for instance, the expression 'formaldehyde, 4% w/v' means that in 100 ml of the solution there are 4 g of formaldehyde, and the expression 'acetic acid 5% v/v' means that in 100 ml of the solution there are 5 ml of glacial acetic acid.

As a general rule, solids should be dissolved in such a way that the percentage composition can at once be given as w/v, and liquids so that it can be given as v/v. In other words, a particular weight of a solid or a particular volume of a liquid should be made *up to* a particular volume by the addition of solvent. In biological laboratories the effect of the solute on the volume of the solution is often overlooked, a certain weight of substance being dissolved in a particular volume of water or other solvent. The solution, which contains x grams of solute to 100 ml of solvent, differs from an x% w/v solution, but many of the solutions commonly used in microtechnique are not concentrated enough to cause the difference to be of practical importance.

For certain particular purposes (p. 144) it is *desirable* to express the composition of a solution in terms of the number of g or ml of solute in 100 g of water (or other solvent). When this is done, the denominator may be written with a capital letter, to draw attention to the unusual way of expressing the concentration. Thus, a solution of 5 g of mercuric chloride in 100 g of water may be called a 5% w/W solution. It is convenient to remember that W here means 100 g of water (or other solvent).

APPENDIX 2

EXPERIMENTS ON FIXATION

CONTENTS

Preparation of gelatine/albumin gel

Dilute white-of-egg with twice its volume of distilled water. Centrifuge the solution until the supernatant is clear. To 100 ml of the supernatant add 15 g of powdered gelatine. Leave for 1 hour; then place in an incubator at 37° C and stir from time to time until the gelatine has dissolved. Filter through a cloth.

Pour the gel into metal pessary-moulds. The 30-grain size is convenient for most purposes. Cool in a refrigerator and remove the separate gels. Each has a volume of about 1·87 ml. Keep the gels in the refrigerator until required. They may be kept for several days, but moulds eventually grow on them.

The refractive index of this gel is about 1·365. The protein content is about 19·4%.

Preparation of nucleoprotein

Separate the fat from the fresh thymus glands of four or five calves. Pass the glands through a domestic mincing-machine. The instructions given below refer to 100 g of minced tissue.

Grind up the mince for a quarter of an hour in a mortar with a roughly equal volume of washed sand. Put the ground material in a flask and add 1000 ml of distilled water and 1 ml of chloroform (the latter to check decay). Shake vigorously from time to time over a period of about 20 hours.

Allow the gross sediment to fall. Decant the fluid and centrifuge it to get rid of the rest of the sediment.

Put the supernatant fluid in a flask and add 25 ml of 10% acetic acid to precipitate the nucleoprotein. Warm gently to 37° and leave in an incubator at this temperature for 15 minutes to complete the precipitation.

Shake the flask and pour the contents into a measuring-cylinder or other tall, narrow vessel. Leave for 1½ hour. Pour off the supernatant fluid and discard it.

Centrifuge the wet sediment and discard the supernatant fluid.

Wash the sediment by adding 150 ml of 0·25% acetic acid, stirring, and re-centrifuging.

Repeat the washing twice (three washes altogether).

The sediment is wet nucleoprotein. It may be dried in a desiccator over calcium chloride, but the fully dried substance is rather troublesome to re-dissolve and it is best to proceed as follows. Weigh a gram or two of the wet sediment accurately and desiccate to constant weight. The loss in weight will show how much of the wet substance must be taken to make a nucleoprotein solution of any particular concentration. The wet substance may lose about ⅞ of its weight on drying.

Preparation of gelatine/nucleoprotein gel

Put 15 g of powdered gelatine in a beaker and add 50 ml of potassium hydroxide, 0·25% aqueous solution. Place in an incubator at 37° C and stir occasionally until the gelatine dissolves. This is solution A.

Put some wet nucleoprotein containing 4 g of dry nucleoprotein in a measuring cylinder. Add 17 ml of potassium hydroxide, 10% aqueous solution. Stir. Make up to 50 ml with distilled water. Place in an incubator at 37° C and leave till warm, with occasional stirring. This is solution B.

Mix A and B, stirring well. Leave in the incubator for ½ hour, stirring from time to time. The solution may then be sucked into pipettes while still warm, for experiments on the rate of penetration of acetic acid (p. 39). It hardens to a gel on cooling. The gel may be kept temporarily in a refrigerator.

The coagulation of egg-albumin by various fixatives

Add to egg-white 4 times its volume of distilled water; mix; clear the solution by centrifuging. The albumin will now be present at about 2·6%.

Make up the following fixatives in aqueous solutions at twice the concentrations given on p. 24:—hydrochloric acid, nitric acid, trichloracetic acid, chloroplatinic acid, chromium trioxide, formaldehyde, osmium tetroxide, potassium dichromate, acetic acid.

Put 5 ml of the clear albumin solution in a test-tube and add an equal volume of the fixative solution to be tested. (The experiment may be done with smaller quantities when osmium tetroxide or chloroplatinic acid is being tested, on account of the high cost.) Close the tube and turn it upside down once. The fixative will now be at the concentration shown on p. 24.

Note the appearance of the contents of the tube immediately, at 5 min., and at 1 hour from mixing, and also on the following day. Record whether there is opalescence, flocculation, or clot-formation, and whether flocculi or clots fall to the bottom of the tube.

A different technique is necessary with fixatives that are used in practice at or near saturation or unmixed with water; namely, picric acid, mercuric chloride, methanol, ethanol, and acetone. To test these, mix egg-white with twice its volume of water and clear the solution by centrifuging. Put 3 ml of this stronger albumin solution in a test-tube and add 7 ml of the fixative at the concentration shown on p. 24. It will be noticed that the amount of albumin in the tube is the same as in the experiment described in the preceding paragraphs. Close the tube and turn it upside down once. Record the result as before.

The results of these experiments are briefly recorded on p. 32.

The coagulation of egg-albumin by ferric sulphate

Add to egg-white 4 times its volume of distilled water. Centrifuge to clear.

Put 5 ml of the albumin solution in each of two test-tubes marked A and B.

Make these solutions:—

A.	Ferric sulphate, anhydrous, 2% w/v aq. . .	2·5 ml
	Distilled water.	2·5 ml
B.	Ferric sulphate, anhydrous, 2% w/v aq. . .	2·5 ml
	Ammonium sulphate, 5·28% w/v aq. . . .	2·5 ml

Add solution A to test-tube A; close; turn upside down once. There is only a momentary formation of a few specks of coagulum: they dissolve at once.

Add solution B to test-tube B; close; turn upside down once.

A flocculent coagulum forms instantly throughout the fluid and remains there.

Note that the final concentration of ferric sulphate in each tube is ½%.

The coagulation of egg-albumin by mercuric chloride

The following experiments illustrate the effect of acidity on coagulation, and of sodium chloride and potassium iodide on the coagulates.

Add to egg-white twice its volume of distilled water. Centrifuge to clear.

Put 3 ml of the albumin solution in each of two test-tubes marked A and B.

Make these solutions:—

A.	Mercuric chloride, sat. aq. . . .	6·5 ml
	Distilled water	0·5 ml
B.	Mercuric chloride, sat. aq. . . .	6·5 ml
	Acetic acid, glacial	0·5 ml

Add solution A to test-tube A; close; turn upside down once. A thick white clot forms at once throughout the fluid; the next day it has fallen somewhat.

Add solution B to test-tube B; close; turn upside down once. Fine white flocculi form at once throughout the fluid; the next day the appearance is unaltered.

The experiment shows that acetic acid reduces the coagulating power of mercuric chloride.

To each tube add 2 ml of a saturated solution of sodium chloride; close the tube and turn upside down once. The coagulum in A dissolves; that in B remains unaltered.

The experiment may be repeated with the substitution of potassium iodide for sodium chloride. The result is essentially the same, but the coagulate in A turns bright orange momentarily before dissolving, while that in B turns bright orange momentarily and then reverts to its former state: it does not dissolve.

(Whether sodium chloride or potassium iodide be used, the fluid in A will very gradually become cloudy on prolonged standing.)

The coagulation of nucleoprotein by various fixatives

Make a 4% aqueous solution of nucleoprotein, prepared from the thymuses of calves by the method given on p. 314.

To the 4% nucleoprotein solution add two-thirds of its volume of distilled water.

Make up the following fixatives at twice the concentrations given on p. 24:—hydrochloric acid, nitric acid, trichloracetic acid, chloroplatinic acid, chromium trioxide, formaldehyde, osmium tetroxide, potassium dichromate, acetic acid.

Put 5 ml of the diluted nucleoprotein solution in a test-tube and add an equal volume of the fixative solution to be tested. (The experiment may be done with smaller quantities when osmium tetroxide or chloroplatinic acid is being tested, on account of the high cost.) Close the tube and turn it upside down once. The fixative will now be at the concentration shown on p. 24.

Note the appearance of the contents of the tube immediately, and again the next day.

A different technique is necessary with fixatives that are used in practice at or near saturation or unmixed with water; namely, picric acid, mercuric chloride, methanol, ethanol, and acetone. To test these, put 3 ml of the 4% nucleoprotein solution (undiluted) in a test-tube and add 7 ml of the fixative at the concentration shown on p. 24. It will be noticed that the amount of dry nucleoprotein in the tube is the same as in the experiment described in the preceding paragraphs (0·12 g.) Close the tube and turn it upside down once. Record the result as before.

The results of these experiments are recorded in table 2 (p. 35).

The penetration of fixatives into gelatine/albumin gel

Cut glass tubing, about 0·7 mm in internal diameter, into short lengths (about 45 mm long is suitable). Make a transverse mark with a diamond near one end of each tube, and scratch on a distinguishing number. Put a rubber pipette-bulb on the marked end. (Each tube will be called a pipette, though it is not narrow at the tip.) See fig. 2, p. 38.

Melt the gelatine-albumin gel (p. 314) at 37° C and draw it into the pipettes. Support them vertically until the fluid has solidified. (The exact position of the top of the gel is of no consequence, but the tube must be filled nearly or quite to the bottom. There should be no gel within the rubber bulb.)

Measure and record the distance from the transverse diamond-mark to the bottom of the gel. Measurements should be made to the nearest quarter of a millimetre.

Put 150 ml of the fixative to be tested in a large specimen-tube.

(A smaller volume should be used when osmium tetroxide is the fixative.) Drop a prepared pipette into the specimen-tube. It will float, bulb uppermost. Note the time. The fixative will begin to penetrate into the gel from the bottom of the tube.

Make observations at intervals on the distance penetrated by the fixative. It is best (though rather inconvenient) to make them at 4, 9, 16, and 25 hours after the fixative has begun to act.

The method of measuring the rate of penetration of the fixative depends on whether the latter is coagulant or not. If it is, simply note where the coagulation has reached, measure the distance from the diamond-mark to the limit of coagulation, and subtract this distance from the measurement made at the beginning of the experiment. The limit of coagulation by most fixatives can easily be seen, on account of the resulting opacity. (It is rather difficult to see the limit when the fixative is hydrochloric acid or mercuric chloride, unless one is careful to arrange suitable lighting and background.)

A more complicated procedure is necessary with non-coagulant fixatives, such as formaldehyde, because one cannot see the limit between the fixed and unfixed parts of the gel. To determine the position of the limit, remove the rubber bulb from the tube, put it on the other end, and then float the pipette in water at 37° C. In a few minutes the unfixed gel will have run out of the tube, and it will then be obvious how far the fixative has penetrated. It is necessary to use a separate pipette for each observation with each fixative.

The method described in the preceding paragraph is applicable also to osmium tetroxide. The fixative produces a black gel at the bottom of the tube, rather sharply marked off from a brownish yellow region above, which fades into the unaffected gel at the top. It is easy to show that the part of the gel that has been blackened is the only part that has been rendered insoluble in water at 37° C, and the limit of the blackened region may therefore be used as an index of the distance penetrated by osmium tetroxide at fixative strength.

It is to be noticed that what is measured is *the thickness of the layer of the original gel that has been penetrated by the fixative at its effective concentration.* This is by no means necessarily the same as the thickness of the fixed gel, for many fixatives shrink or swell it.

The results of the experiment are shown in fig. 3 (p. 39).

Acetic acid and potassium dichromate do not coagulate the

proteins of the gel nor render them insoluble in water at 37° C, and the method is therefore not applicable to them; but the rate of penetration of acetic acid may be tested by substituting gelatine/nucleoprotein gel (p. 315) for gelatine/albumin.

The action of formaldehyde in rendering albumin not coagulable by ethanol

Mix egg-white with twice its volume of distilled water. Clear by centrifuging. Put 1·5 ml of the solution in each of two test-tubes, A and B. To A add 1·5 ml of distilled water, to B 1·5 ml of pure ('reagent-grade') formaldehyde, 8% aqueous solution. Close the tubes and turn them upside down once. Leave for 15 min. Add 7 ml of absolute ethanol to each. Turn each tube upside down once. The fluid in tube A immediately becomes cloudy throughout, while that in tube B remains clear. The next day there are visible flocculi throughout the fluid in tube A, while there is only slight cloudiness in tube B.

APPENDIX 3

EXPERIMENTS ON DYEING

CONTENTS

Preparation of pararosaniline

Put about 1·5 ml of colourless (freshly distilled) aniline and a large crystal of paratoluidine in a test-tube. Scatter some powdered mercuric chloride down the side of the test-tube.

Heat strongly over a Bunsen burner, shaking the aniline on to the mercuric chloride. A magenta colour will suddenly develop. Stop heating while there is still some fluid in the tube.

Electrophoresis of dyes

The apparatus is briefly described on p. 189 and illustrated in fig. 23.

Prepare buffer solution from 'reagent-grade' chemicals as follows:—

$\frac{M}{7 \cdot 5}$ Na_2HPO_4. Make up 9·466 g of the anhydrous salt to 500 ml with distilled water.

$\frac{M}{7 \cdot 5}$ KH_2PO_4. Make up 9·078 g of the salt to 500 ml with distilled water.

For an experiment at pH 7·1, mix 66 ml of the Na_2HPO_4 solution, 34 ml of the KH_2PO_4 solution, and 100 ml of distilled water. This is 'buffer solution A'.

Prepare agar gels as follows:—

For the large U-tube. Put 1·2 g of agar in 120 ml of distilled water. Leave overnight at room-temperature (or for several hours at 37° C); then add 120 ml of buffer solution A. Heat over double asbestos gauze, with occasional shaking, until the fluid boils. Fix the large U-tube in a perfectly vertical position. Allow the fluid to cool to about 48° C and then pour some of it into the U-tube. Allow the agar to set. Mark the height of the agar with labels.

For the two small U-tubes (agar bridges). Put 0·5 g of agar in 100 ml of distilled water. Leave overnight at room temperature (or for several hours at 37° C). Heat over double asbestos gauze, with occasional shaking, until the fluid boils; add 10 g of sodium chloride; shake. Fix the small U-tubes in a vertical position. Allow the fluid to cool to about 48° C and then pour some of it into the U-tubes so as to fill them to the brim. Allow the agar to set.

Put copper sulphate solution, 10% aqueous, in the small beakers.

Dissolve the dye at 0·2% in distilled water. Before the experiment starts mix this solution with an equal volume of buffer solution A (p. 321). Methylene green is particularly suitable because its cataphoretic movement is rapid.

The only difficulty in the experiment is that the agar in the large U-tube sometimes comes unstuck from the glass, and the dye solution then slips down in between. The glass must be perfectly clean and the dye must be added very gently with a pipette above the agar on both sides of the large U-tube. Add a little alternately to each side.

The water in the agar gel in the large U-tube moves slowly towards the negative pole by electro-endosmosis. The columns of dye solution above the agar on the two sides must be kept level by occasional transference from one side to the other.

Diffusion of dyes

Prepare 'buffer solution A' according to the instructions given under the heading 'Electrophoresis of dyes' (p. 321). The buffer is at pH 7·1.

Put 1 g of agar in 100 ml of distilled water. Leave overnight at room-temperature (or for several hours at 37° C); then add 100 ml of buffer solution A. Heat over double asbestos gauze, with occasional shaking, until the fluid boils. Arrange several fairly long, narrow test-tubes in a rack. Allow the agar solution to cool to about 48° C and then pour some of it into the test-tubes so as to fill them to within about 4 cm from the top. Allow the agar to set. Mark the height of the agar with labels.

Dissolve the dyes that are to be tested at 0·5% in buffer solution A diluted with an equal volume of distilled water. Orange G, methyl blue, and aniline blue WS are particularly suitable.

Add one of the dye solutions to each of the tubes. This is best done with a pipette. If the solution slips down between the agar gel and the glass, discard the tube. Note the time.

Note the distance to which the dyes have penetrated after the lapse of suitable periods (e.g. 12, 24, and 48 hours).

The results of this experiment are recorded on p. 237.

The effect of alkalinity and acidity on dyeing with basic and acid dyes

The following are required:—

methylene blue, 0·1% aqueous
eosin Y, 0·1% aqueous
boric acid/NaOH buffer at pH 8

boric acid, 0·2 M	100 ml
NaOH, 0·2 N	8 ml
distilled water	up to 1 l.

acetic acid at pH 3
Make up 100 ml of 33% w/v acetic acid to 1000 ml with distilled water.

4 30μ *collodion sections* not containing any tissue, washed in distilled water. Any kind of nitrocellulose used for embedding in microtechnique is suitable.

4 40μ *gelatine sections* not containing any tissue; washed in distilled water.

Experiments with collodion sections. The collodion sections are used to represent the acidic tissue-components (DNA, etc.).

Put the following in four glass capsules labelled A to D:—

Capsule marked	*Methylene blue, ml*	*Eosin, ml*	*Boric acid/* NaOH *buffer at* pH 8, *ml*	*Acetic acid at* pH 3, *ml*
A	4	—	12	—
B	4	—	—	12
C	—	4	12	—
D	—	4	—	12

Put a collodion section in each of 4 porcelain pots with perforated bottoms (e.g. Gooch crucibles). Put one of the porcelain pots in each of the capsules, A to D. Leave for 15 minutes. Transfer the porcelain pots to distilled water and thus give the sections a preliminary rinse. Remove the sections from the pots and wash them in two lots of distilled water for about 5 minutes altogether.

Record the intensity of dyeing on an arbitrary scale (maximum +++++). The results will be as follows:—

A, methylene blue at pH 8	+++
B, ,, ,, ,, 3	+++
C, eosin Y ,, ,, 8	+
D, ,, ,, ,, 3	+++

Experiments with gelatine sections. The gelatine sections are used as models of amphoteric tissue-constituents (ground cytoplasm, etc.).

Repeat the experiment exactly as before, apart from the use of different sections.

The results will be as follows:—

A, methylene blue at pH 8	+++++
B, ,, ,, ,, 3	o
C, eosin Y ,, ,, 8	o
D, ,, ,, ,, 3	+++++

The effect of acid alcohol on tissues coloured with basic and acid dyes

Fix short lengths of the small intestine of the mouse in Zenker. Embed in paraffin. Cut sections at 8 μ.

Solutions required:—

Basic fuchsine, 0·5% aq.
Acid fuchsine, 0·5% aq.
Hydrochloric acid, conc., 1% in 70% ethanol.

Bring 4 slides to water (through iodine and sodium thiosulphate solutions, as usual after fixatives containing mercuric chloride).

Put one slide in basic fuchsine solution and another in acid fuchsine solution. Leave for 5 minutes. Rinse with distilled water. Pass through 70% alcohol (5 seconds), 90% (5 seconds), 1st absolute alcohol (dip), 2nd absolute alcohol (2 minutes), into xylene and then Canada balsam.

Treat another pair of slides in exactly the same way, with this exception: after rinsing with distilled water, leave the slides for 1 minute in the solution of hydrochloric acid in 70% alcohol.

The results will be as follows:—

Basic fuchsine. Chromatin and the mucosubstance of the goblet-cells are strongly coloured; the somewhat basiphil cytoplasm of certain cells takes the dye feebly; nothing else coloured.

Basic fuchsine followed by acid alcohol. There is no dye in the section.

Acid fuchsine. Chromatin and cytoplasm are both strongly coloured, especially the cytoplasm at the bases of the epithelial cells. The mucosubstance of the goblet-cells is not coloured. Collagen is coloured.

Acid fuchsine followed by acid alcohol. Same as acid fuchsine alone, but even more strongly coloured.

The effects of fixatives on the subsequent action of dyes

Fix testes of the mouse (quartered) and small pieces of the kidney-cortex of the same animal in these fluids:—

(1) ethanol, absolute
(2) picric acid, sat. aq., 100 ml
  sodium chloride, 0·7 g
(3) mercuric chloride, sat. aq.
(4) chromium trioxide, 0·5% in 0·7% sodium chloride
(5) formaldehyde, 4% in 0·7% sodium chloride
(6) osmium tetroxide, 1% in 0·7% sodium chloride
(7) potassium dichromate, 1·5% in 0·7% sodium chloride
(8) acetic acid, 5% aq.

For a small experiment it is best to choose nos. (4) and (5) only, since these will give the extremes of cytoplasmic acidophilia and basiphilia respectively. No. (2) should be chosen if one wants to show strong colouring of the meiotic chromosomes by acid dyes.

Wash out in the appropriate ways; embed in paraffin; cut sections at 8 μ. Each slide may carry sections of material fixed in each of the fixatives under test, so that subsequent treatment will be identical. Sections of tissues fixed in osmium tetroxide should, however, be attached to separate slides, bleached overnight in sodium iodate (1% aqueous solution), and then washed in distilled water before dyeing.

Colour the sections on each slide with one dye only, not two in succession or simultaneously. Any of the following dyes may be used for the times stated:—

basic dyes

basic fuchsine, 0·5% aq., 1 min.
toluidine blue, 0·5% aq., 5 min.
methylene blue, 1% aq., 15 min.

basic dye-lakes

Ehrlich's haematoxylin, 20 min. (Alkalinize, but do not differentiate.)
Mayer's carmalum, 1 hr.

acid dyes

acid fuchsine, 0·5% aq., 5 min.
eosin, 0·1% aq., 1 min.

For a small experiment choose toluidine blue and acid fuchsine only.

Dehydrate and pass through xylene into Canada balsam or similar mountant.

Study the depth of colour in the cytoplasm and chromatin of the spermatogonia and spermatocytes, and in the convoluted tubules of the kidney-cortex.

APPENDIX 4

USE OF THE WORD 'CHROMATIN'

In general, it is best to avoid words that are supposed to denote chemical composition but stand outside the system of nomenclature adopted by chemists. It seems unlikely that anything is gained by using such words as 'linin' and 'plastin'. A case can, however, be made for the retention of the word 'chromatin'.

This word was introduced by Flemming in 1880. It is commonly stated to have been introduced by him in 1879, presumably because E. B. Wilson [540a] said so; but it does not occur in the papers by Flemming quoted by Wilson in support of his contention. Flemming introduced it as follows.

> 'For further study of the phenomena of division, there is the question of a shorter word for what I have hitherto called the "colourable substance of the nucleus". Since the expression "nuclear substance" is obviously exposed to many misunderstandings, I shall for the time being coin the word chromatin for it. From this name no preconception ought to arise that this substance must be a definitely constituted chemical substance, remaining unchanged in all nuclei. Although this is indeed possible, we do not yet know enough about the nuclear substances to assume it. This only should be denoted by the word chromatin: *that substance in the cell nucleus which takes up the dye in the treatments with dyes known as nuclear colouring*.' [171a]

Two years later he added these remarks:

'The [nuclear] network owes its refractivity, the nature of its reactions, and its remarkable affinity for dyes, to a substance which, in consideration of the latter character, I have provisionally named chromatin.' He goes on to say that it may be the same substance as nuclein. 'I retain the name chromatin', he continues, 'until decision on this shall be given by chemistry, and I denote by it, *wholly empirically*, "the substance in the cell nucleus that takes up the colour in nuclear dyeing".' He remarks, 'As soon as anyone is able to say exactly what the colourable substance in the nucleus is, in terms of *chemistry*, such a name as chromatin will perhaps become useless, unless even then it still commends itself on account of its brevity.' [172]

In this book the word is used precisely in Flemming's sense. Despite all that has been done since his time to enlarge our knowledge of nucleoproteins and DNA, we still do not know exactly to what substance or substances in nuclei and chromosomes the special affinity for particular dyes, to which Flemming refers, is due. It may perhaps be nucleoprotein, but DNA split off from protein by the action of a fixative seems more likely; possibly, in some cases, the protein may itself hold basic dyes after detachment of DNA. It seems safest for the present to retain Flemming's word when describing what we see in microscopical preparations coloured with the usual dyes.

APPENDIX 5

NOTES ON SPELLING

Acheta domesticus (the house-cricket). It appears from the Copenhagen decisions of the International Commission on Zoological Nomenclature that the generic name *Acheta* is to be regarded (provisionally at least) as masculine. The specific name must therefore end in *-us*.

Acidophil. See *Basiphil.*

Aniline, etc. As a general rule the names of basic organic substances such as aniline and quinine end in *-ine*, while those of neutral ones such as dextrin end in *-in*. The first aniline dye to be produced on a commercial scale was named by its inventor *mauveine*. The ending of the word suggests the relationship with the parent substance. Other dyes introduced soon afterwards received names with the same ending (e.g. fuchsine). For the sake of conformity the ending was applied to acid as well as basic dyes. The final *-e* is often omitted from the names of dyes, but it has been retained in this book for the names of nearly all the synthetic ones, because the derivation supports this spelling.

Artifact. It was presumably a mere slip of the pen that caused the poet Coleridge to write *artefact* when coining this term, for English words do not include Latin ablatives such as *arte*. *Ars* being a Latin noun of the third declension, its stem in compound words is *arti-*.

Basiphil. Ehrlich was interested in the classics as a young man and frequently used Latin tags in conversation throughout his life.[324] It is strange that he should not have recognized the error in 'basophil'. The Latin noun *basis* is of the third declension and its stem in compound words is therefore *basi-*. 'Basophil' is as wrong as 'basosphenoid', 'matroarchal', and 'regocide' would be. There is no reason for adding *-ic* at the end of the word: it should have the same termination as the English adjective 'Francophil'. Similarly one should write 'acidophil', not 'acidophilic'.

Supervital. Arnold [15, 16] made a careful study of the dyeing of *überlebender* cells. He chose this very suitable term because the cells, having been removed from the body, *survived* while being dyed. Instead of leaving well alone, he later [18] coined the expression *Die supravitale Methode*, and the word 'supravital' is generally used. The Latin preposition *super* would have been the proper one to include in the compound word, since it is contained in *supervivo* (I survive). This method of staining should be called the 'survival' or 'supervital' method. The English word 'survive' is derived through French from the Latin *supervivo*.

LIST OF REFERENCES

Note. Some of the references to the *Zeitschrift für wissenschaftliche Mikroskopie* relate to abstracts published in that journal.

1. Ackroyd, W., 1876*a*. *Chem. News*, **33**, 60.
2. —— 1876*b*. Ibid., **34**, 75.
3. Allen, E., 1916. *Anat. Rec.*, **10**, 565.
4. Altmann, R., 1889. *Arch. Anat. Physiol., Anat. Abt.*, Supplement-Band, 86.
5. —— 1894. *Die Elementarorganismen und ihre Beziehungen zu den Zellen.* 2nd ed. Leipzig (Veit).
6. Amann, J., 1896. *Z. wiss. Mikr.*, **13**, 18.
7. —— 1899. Ibid., **16**, 38.
8. Ambronn, H., 1888. *Ber. deut. bot. Ges.*, **6**, 226.
9. Ancel, P., 1903. *Arch. de Biol.*, **19**, 389.
10. Anderson, L. H., & Yost, D. M., 1938. *J. Amer. chem. Soc.*, **60**, 1822.
11. Anon., 1908. Article on 'Embalming' in *The Illustrated Chambers's Encyclopaedia*, Vol. 4. London (Chambers).
12. Anson, M. L., 1938. Chapter on 'The coagulation of proteins' in *The chemistry of the amino-acids and proteins*, ed. by C. L. A. Schmidt. Springfield, Ill. (Thomas).
13. —— 1945. *Adv. prot. Chem.*, **2**, 361.
14. Aoyama, F., 1929. *Zeit. wiss. Mikr.*, **46**, 489.
15. Arnold, J., 1899. *Virchows Arch.*, **157**, 424.
16. —— 1900*a*. Ibid., **159**, 101.
17. —— 1900*b*. *Arch. mikr. Anat.*, **55**, 479.
18. —— 1903. *Anat. Anz.*, **24**, 1.
19. —— 1908. Ibid., **32**, 361.
20. —— 1911. *Sitz. Heidelb. Akad. Wiss.*, **2B**, 1.
21. Arnstein, C., 1887. *Anat. Anz.*, **2**, 551.
22. Auerbach, L., 1874. *Organologische Studien*, Vol. 1. Breslau (Morgenstern).
23. Bahr, G. F., 1954. *Exp. cell Res.*, **7**, 457.
24. Baker, H., 1744. *Phil. Trans.*, **42**, 616.
25. Baker, J. R., 1932. *Nature*, **130**, 741.
26. —— 1933. *Cytological Technique.* 1st ed. London (Methuen).
27. —— 1941. *Nature*, **147**, 744.
28. —— 1944. *Quart. J. micr. Sci.*, **85**, 1.
29. —— 1945*a*. *Monog. Quek. micr. Club*, No. 1.
30. —— 1945*b*. *Cytological Technique.* 2nd ed. London (Methuen).
31. —— 1946. *Quart. J. Micr. Sci.*, **87**, 441.
32. —— 1947. Ibid., **88**, 463.
33. —— 1949. Ibid., **90**, 293.
34. —— 1950. *Cytological Technique.* 3rd ed. London (Methuen).

35. Baker, J. R., 1952. *Abraham Trembley of Geneva, scientist and philosopher.* London (Arnold).
36. —— 1956. *Quart. J. micr. Sci.*, **97**, 621.
37. —— 1957. Symp. Soc. exp. Biol., **10**, 1.
37a. —— 1957. *Quart. J. micr. Sci.*, **98**, 425.
38. —— Results not previously published.
39. Baker, J. R., & Jordan, B. M., 1953. *Quart. J. micr. Sci.*, **94**, 237.
40. Bancroft, E., 1813. *Experimental researches concerning the philosophy of permanent colours.* 2 vols. London (Cadell & Davies).
41. Bancroft, W. D., 1922. *J. physical Chem.*, **26**, 447.
42. Bang, J., & Sjövall, E., 1916. *Beitr. path. Anat.*, **62**, 1.
43. Bank, O., & Jong, H. G. B. de, 1939. *Protoplasma*, **32**, 489.
44. Barer, R., & Joseph, S., 1955*a*. *Quart. J. micr. Sci.*, **96**, 1.
45. —— —— 1955*b*. Ibid., **96**, 423.
46. Bartholomew, J. W., & Mittwer, T., 1950. *Stain Tech.*, **25**, 103.
47. —— —— 1951. Ibid., **26**, 231.
48. Bartholomew, J. W., & Umbreit, W. W., 1944. *J. Bact.*, **48**, 567.
49. Bauer, H., 1933. *Jahrb. Morph. mikr. Anat.*, 2 Abt., **33**, 143.
50. Beale, L. S., 1880. *How to work with the microscope.* 5th ed. London (Harrison).
51. Beauchamps, P. de, 1906. *Année biol.*, **11**, p. XVI.
52. Becker, W. A., 1929. *Zeit. wiss Mikr.*, **47**, 411.
53. Bělar, K., 1928. Article on 'Die Technik der descriptiven Cytologie' in Peterfi.[393]
54. Benda, C., 1886. *Arch. Anat. Physiol.* (no vol. number), 562.
55. —— 1901. *Verh. anat. Ges. Bonn*, **15**, 155.
56. —— 1903. *Anat. Hefte*, 2 Abt., **12**, 743.
57. Benians, T. H. C., 1920. *J. Path. Bact.*, **23**, 401.
58. Bensley, R. R., 1911. *Amer. J. Anat.*, **12**, 297.
59. Bensley, R. R., & Bensley, S. H., 1938. *Handbook of histological and cytological technique.* Chicago (University Press).
60. Bentley, R., & Trimen, H., 1880. *Medicinal plants*, Vol. 2. London (Churchill).
61. Berg, W., 1903. *Arch. mikr. Anat.*, **62**, 367.
62. —— 1905. Ibid., **65**, 298.
63. —— 1908. Quoted by Tellyesniczky.[500]
64. —— 1926. Article on 'Chromsäure' in R. Krause's *Enzyklopädie der mikroskopischen Technik*, Vol. 1. Berlin (Urban & Schwarzenberg).
65. —— 1927. Article on 'Osmiumsäure' in R. Krause's *Enzyklopädie der mikroskopischen Technik*, Vol. 3. Berlin (Urban & Schwarzenberg).
66. Bernthsen, A., 1885. *Liebigs Ann.*, **230**, 169.
67. Berthold, G., 1882. *Jahrb. wiss. Bot.* (Pringsheim), **13**, 569.
68. Berthold, G., 1886. *Studien über Protoplasmamechanik.* Leipzig (Felix).
69. Bethe, A., 1895. *Arch. mikr. Anat.*, **44**, 579.
70. —— 1896*a*. *Anat. Anz.*, **11**, 358.
71. —— 1896*b*. Ibid., **12**, 438.
72. —— 1905. *Beitr. chem. Physiol. Path.*, **6**, 399.

73. Bignardi, C., 1940. *Boll. Soc. ital. Biol. sperim.*, **15**, 594.
74. Bignardi, C., & Casella, C., 1941. *Boll. di Zool.*, **12**, 83.
75. Bird, C. L., 1951. *The theory and practice of wool dyeing.* Bradford (Society of Dyers and Colourists).
76. Bird, C. L., & Newsome, O., 1950. *J. Soc. Dyers and Col.*, **66**, 423.
77. Bjorksten, J., 1951. *Adv. protein Chem.*, **6**, 343.
78. Black, C. E., 1938. *J. lab. clin. Med.*, **23**, 1027.
79. Blanchard, E., 1847. *Ann. Sci. nat. Zool.*, **8**, 271.
80. Blanchard, R., 1883. *Bull. Soc. zool. France*, **8**, 217.
81. Blank, H., & McCarthy, P. L., 1950. *J. lab. clin. Med.*, **36**, 776.
82. Blum, F., 1893. *Zeit. wiss. Mikr.*, **10**, 314.
83. Böeseken, J., & Giffen, J. van, 1920. *Rec. Trav. chim. Pays-Bas*, **39**, 183.
84. Bouin, P., 1897. *Arch. d'Anat. micr.*, **1**, 225.
85. Bowes, J. H., & Kenten, R. H., 1949. *Biochem. J.*, **44**, 142.
86. Brachet, J., 1953. *Quart. J. micr. Sci.*, **94**, 1.
87. Bradbury, S., 1956. Ibid., **97**, 323.
88. Brandt, K., 1878. *Arch. f. Physiol.* (no vol. number), 563.
89. —— 1881. *Biol. Centralbl.*, **1**, 202.
90. Brante, G., 1949. *Acta physiol. Scand.*, **18**, Suppl. **63**, 1.
91. Brasil, L., 1905. *Arch. de Zool. exp.*, **4**, 69.
92. Brooks, S. C., & M. M., 1932. *J. cell comp. Physiol.*, **2**, 53.
93. Brunn, A. v., 1884. *Arch. mikr. Anat.*, **23**, 108.
94. Buchsbaum, R., 1948. *Anat. Rec.*, **102**, 19.
95. Bull, H. B., 1937. *The biochemistry of the lipids.* New York (Wiley).
96. —— 1949. Chapter on 'Protein Denaturation' in *Outlines of biochemistry*, ed. by R. A. & W. A. Gortner. London (Chapman & Hall).
97. Burchardt, E., 1897. *Cellule*, **12**, 337.
98. Burke, F. V., 1933. *Amer. J. Path.*, **9**, 915.
99. Cain, A. J., 1947. *Quart. J. micr. Sci.*, **88**, 383.
100. —— 1950. *Biol. Rev.*, **25**, 73.
101. Campbell, D. H., 1888. *Unters. bot. Inst. Tübingen*, **2**, 569.
102. Cannon, H. G., 1950. *Endeavour*, **9**, 188.
103. Capell, D. F., 1929. *J. Path. Bact.*, **32**, 595.
104. Carleton, H. M., 1922. *Quart. J. micr. Sci.*, **66**, 501.
105. —— 1926. *Histological technique.* London (Oxford University Press).
106. Carleton, H. M., & Short, R. H. D., 1954. *Schafer's essentials of histology.* London (Longmans, Green).
107. Carnoy, J. B., 1884. *La biologie cellulaire.* Paris (Doin).
108. —— 1886. *Cellule*, **3**, 1.
109. —— 1887. Ibid., **3**, 227.
110. Casselman, W. G. B., 1954. *Quart. J. micr. Sci.*, **95**, 321.
111. —— 1955*a*. Ibid., **96**, 203.
112. —— 1955*b*. Ibid., **96**, 223.
113. —— Results not previously published.
114. Casselman, W. G. B., & Jordan, B. M., 1954. *Nature*, **173**, 1095.
115. Certes, A., 1881*a*. *C. r. Acad. Sci. Paris*, **92**, 424.

116. Certes, A., 1881*b*. *Zool. Anz.*, **4**, 208.
117. —— 1881*c*. Ibid., **4**, 287.
118. Champy, C., 1911. *Arch. d'Anat. micr.*, **13**, 55.
119. Chapman, L. M., Greenberg, D. M., & Schmidt, C. L. A., 1927. *J. biol. Chem.*, **72**, 707.
120. Chesterman, W., & Leach, E. H., 1956. *Quart. J. micr. Sci.*, **97**, 593.
121. Chou, J. T. Y. Results not previously published.
122. Chrzonszczewsky, N., 1864. *Virchows Arch.*, **31**, 153.
123. Ciaccio, C., 1910. *Anat. Anz.*, **35**, 17.
124. —— 1926. *Boll. Soc. ital. Biol. sperim.*, **1**, 47.
125. Clara, M., 1940. *Zeit. mikr.-anat. Forsch.*, **47**, 183.
126. Clark, J. B., & Webb, R. B., 1955. *Stain Tech.*, **30**, 89.
127. Clarke, J. L., 1851. *Phil. Trans.*, **141**, 607.
128. Collie, N., & Myers, W. S., 1893. *Journ. chem. Soc.* (Trans.), **63**, 122.
129. Collin, R., 1923. *C. r. Soc. biol.*, **89**, 562.
130. —— 1924. Ibid., **91**, 793.
131. Conn, H. J., 1953. *Biological stains*. Geneva, U.S.A. (Biotech).
132. Cornil, V., 1875. *C. r. Acad. Sci.*, **80**, 1288.
133. Corti, A., 1851. *Zeit. wiss Zool.*, **3**, 109.
134. Crawford, G. N. C., & Barer, R., 1951. *Quart. J. micr. Sci.*, **92**, 403.
135. Criegee, R., 1936. *Liebigs Ann.*, **522**, 75.
136. Criegee, R., Marchand, B., & Wannowius, H., 1942. Ibid., **550**, 99.
137. Daddi, L., 1896. *Arch. ital. Biol.*, **26**, 143.
138. Dalcq, A., 1952*a*. *C. r. Soc. Biol.*, **146**, 1408.
139. —— 1952*b*. *C. r. Assoc. Anat.*, **39**, 513.
140. Danielli, J. F., 1950. *Cold Spring Harb. Symp. quant. Biol.*, **14**, 32.
141. Davies, H. G., 1954. *Quart. J. micr. Sci.*, **95**, 433.
142. Dick, D. A. T., 1955. Ibid., **96**, 363.
143. Dietrich, A., 1910. *Verh. deut. path. Ges.*, **14**, 263.
144. Dimmock, G., 1884. *Amer. Nat.*, **18**, 324.
145. Dimroth, O., 1913. *Ann. der Chem.*, **399**, 1.
146. Dimroth, O., & Kammerer, H., 1920. *Ber. deut. chem. Ges.*, **53**, 471.
147. Dogiel, A. S., 1890. *Arch. mikr. Anat.*, **35**, 305.
148. Doutreligne, J., 1933. *Cellule*, **42**, 29.
149. Drechsel, E., 1896. *Zeit. f. Biol.*, **33**, 85.
150. Ehrenberg, D. C. G., 1838. *Die Infusionsthierchen als volkommene Organismen*. Leipzig (Voss).
151. Ehrlich, P., 1877. *Arch. mikr. Anat.*, **13**, 263.
152. —— 1879*a*. *Arch. Anat. Physiol., physiol. Abt.* (no vol. number), 166.
153. —— 1879*b*. Ibid., 571.
154. —— 1880. *Z. klin. Med.*, **1**, 553.
155. —— 1886. *Z. wiss. Mikr.*, **3**, 150.
156. —— 1887. *Biol. Centralbl.*, **6**, 214.
157. —— 1894. *Zeit. wiss. Mikr.*, **11**, 250.
158. Ehrlich, P., & Lazarus, A., 1898. *Die Anaemie*, 1 Abt. Wien (Hölder).
159. —— —— 1901. *Leukaemie. Pseudoleukaemie. Haemoglobinaemie.* Wien (Hölder).

160. Evans, H. McL., & Scott, K. J., 1921. *Carneg. Inst. Contrib. Embryol.*, **10**, 1.
161. Evans, J. G., 1949. Chapter XV in Preston.[406]
162. Fearon, W. R., & Foster, D. L., 1922. *Biochem. J.*, **16**, 564.
163. Feulgen, R., & Bersin, R., 1939. *Zeit. physiol. Chem.*, **260**, 217.
164. Feulgen, R., & Rossenbeck, H., 1924. *Hoppe-Seyl. Zeit.*, **135**, 203.
165. Feulgen, R., & Voit, K., 1924. *Pflüg. Arch. ges. Physiol.*, **206**, 389.
166. Firminger, H. I., 1950. *Stain Tech.*, **25**, 121.
167. Fischel, A., 1901. *Anat. Hefte*, 1 Abt., **16**, 417.
168. —— 1908. *Internat. Rev. Hydrobiol.*, **1**, 73.
169. Fischer, A., 1899. *Fixierung, Färbung und Bau des Protoplasmas*. Jena (Fischer).
170. Flemming, W., 1879*a*. *Virchows Arch.*, **77**, 1.
171. —— 1879*b*. *Arch. mikr. Anat.*, **16**, 302.
171a. —— 1880. Ibid., **18**, 151.
172. —— 1882. *Zellsubstanz, Kern und Zelltheilung*. Leipzig (Vogel).
173. —— 1884. *Zeit. wiss. Mikr.*, **1**, 349.
174. Fraenkel-Conrat, H., Cooper, M., & Olcott, H., 1945. *J. Amer. chem. Soc.*, **67**, 950.
175. Freeborn, G. C., 1888. *Amer. mon. micr. J.*, **9**, 231.
176. Freeman, B. L., Moyer, E. K., & Lassek, A. M., 1955. *Anat. Rec.*, **121**, 593.
177. French, D., & Edsall, J. T., 1945. *Adv. protein Chem.*, **2**, 277.
178. Frey, H., 1863. *Das Mikroskop und die mikroskopische Technik*. 1st ed. Leipzig (Engelmann).
179. —— 1881. Ibid., 7th ed. Leipzig (Engelmann).
180. Frey-Wyssling, A., 1948. *Sub-microscopic morphology of protoplasm and its derivatives*. London (Elsevier).
181. Fürnrohr, –, 1850. *Flora*, **8**, 641.
182. Gaunt, J. F., 1949. *J. Soc. Dyers & Col.*, **65**, 429.
183. Gehuchten, A. van, & Nelis, C., 1898. *Cellule*, **14**, 371.
184. Gerola, F. M., & Vannini, E., 1948. *Atti Accad. naz. Linciei*, **5**, 77.
185. —— —— 1949. *Boll. Soc. ital. Biol. sper.*, **25**, 644.
186. Giemsa, G., 1902*a*. *Centralbl. f. Bakt.*, 1 Abt., **31**, 429.
187. —— 1902*b*. Ibid., 1 Abt., **32**, 307.
188. —— 1904. Ibid., 1 Abt., **37**, 308.
189. —— 1907. *Deut. med. Woch.*, **33**, 676.
190. —— 1922. Personal communication to Unna.[518]
191. —— 1924. *Centralbl. f. Bakt.*, 1 Abt., **91**, 343.
192. Giles, C. H., 1944. *J. Soc. Dyers & Col.*, **60**, 303.
193. Gilson, G., 1898. Information supplied to Gehuchten & Nelis.[183]
194. Gleichen, W. F. von, 1778. *Abhandlung über die Saamen- und Infusionsthierchen, und über die Erzeugung*. Nürnberg (Winterschmidt).
195. Goeppert, H. R., & Cohn, F., 1849. *Bot. Z.*, **7**, col. 681.
196. Gomori, G., 1956. *Quart. J. micr. Sci.*, **97**, 1.
197. Gordon, H. K., & Chambers, R., 1941. *J. cell. comp. Physiol.*, **17**, 97.
198. Gram, C., 1884. *Forschr. Med.*, **2**, 185.
199. Grenacher, H., 1879. *Arch. mikr. Anat.*, **16**, 463.

200. Guillermond, A., & Gautheret, R., 1939. *C. r. Acad. Sci.*, **208**, 1061.
201. Gunther, R. T., 1925. *Early science in Oxford.* Vol. 3. Oxford (published privately).
202. Gustavson, K. H., 1943. *Kolloid Z.*, **103**, 43.
203. —— 1949. *Adv. protein Chem.*, **5**, 353.
204. Gutstein, M., 1926. *Virchows Arch.*, **261**, 846.
205. Haarmann, W., 1943. *Biochem. Zeit.*, **314**, 1.
206. Hagemann, P., 1937. *Deut. med. Woch.*, **63**, 514.
207. Häggqvist, G., 1933. *Z. wiss. Mikr.*, **50**, 77.
208. Hammar, J. A., 1924. *Z. mikr. Anat. Forsch.*, **1**, 85.
209. Hannover, A., 1840. *Ann. Anat. Physiol. wiss. Med.* (no vol. number), 549.
209a. Hansen, F. C. C., 1899. *Anat. Anz.*, **15**, 151.
210. —— 1905. *Zeit wiss Mikr.*, **22**, 45.
211. —— 1908. Ibid., **25**, 145.
212. Hantzsch, A., & Desch, C. D., 1902. *Liebigs Ann.*, **323**, 1.
213. Hardy, W. B., 1899. *J. Physiol.*, **24**, 158.
214. Hardy, W. B., & Gardiner, S., 1910. Ibid., **40**, p. LXVIII.
215. Harris, J. E., 1939. *J. exp. Biol.*, **16**, 258.
216. Harris, J. E., & Peters, A., 1953. *Quart. J. micr. Sci.*, **94**, 113.
217. Hartwig, G., 1863. *The tropical world.* London (Longman, Green, Longman, Roberts, Green).
218. Heidenhain, M., 1907. *Plasma und Zelle.* 2 vols. Jena (Fischer).
219. —— 1909. *Zeit. wiss. Mikr.*, **25**, 397.
220. —— 1915. Ibid., **32**, 361.
221. —— 1916. Ibid., **33**, 232.
222. Heidenhain, R., 1888. *Pflügers Arch. ges. Physiol.*, 33, Suppl., 1.
223. Heidenhain, R., & Neisser, A., 1874. *Arch. ges. Physiol.*, **9**, 1.
224. Helly, K., 1903. *Zeit. wiss. Mikr.*, **20**, 413.
225. Henry, H., & Stacey, M., 1943. *Nature*, **151**, 671.
226. —— —— 1946. *Proc. roy. Soc. B*, **133**, 391.
227. Henze, M., 1903. *Z. physiol. Chem.*, **38**, 60.
228. Hermann, F., 1889. *Arch. mikr. Anat.*, **34**, 58.
229. Hershberger, L. R., & Lillie, R. D., 1947. *Bull. internat. Assoc. Med. Mus.*, **27**, 162.
230. Hertwig, G., 1931. *Zeit. mikr.-anat. Forsch.*, **23**, 484.
231. Heschl, R., 1875. *Wien. med. Woch.*, **25**, 714.
232. Hill, J., 1774. *The construction of timber from its early growth.* 2nd ed. London (White & others). The first edition was published in 1770.
233. Himmel, I., 1902. *Ann. Inst. Pasteur*, **16**, 663.
234. Hintzelmann, U., 1922. *Z. wiss. Mikr.*, **39**, 216.
235. Hirsch, G. C., & Jacobs, W., 1926. *Zeit. Zellforsch. mikr. Anat.*, **3**, 198.
236. Hofmann, K. A., 1912. *Ber. deut. chem. Ges.*, **45**, 3329.
237. Hofmann, K. A., Ehrhart, O., & Schneider, O., 1913. Ibid., **46**, 1657.
238. Holmes, W. C., 1926. *Stain Tech.*, **1**, 116.
239. Hsu, T. C., 1952. *J. Hered.*, **43**, 167.
240. Hsu, T. C., & Pomerat, C. M., 1953. Ibid., **44**, 23.

241. Hughes, A., 1952. *Quart. J. micr. Sci.*, **93**, 207.
242. —— 1954. *Journ. Anat.*, **88**, 192.
243. —— 1956. *Quart. J. micr. Sci.*, **97**, 165.
244. Hughes, W. L., 1947. *Journ. Amer. chem. Soc.*, **69** (1), 1836.
245. —— 1950. *Cold Spring Harb. Symp. quant. Biol.*, **14**, 79.
246. Hughesdon, P. E., 1949. *Journ. roy. micr. Soc.*, **69**, 1.
247. Huskins, C. L., 1927. *J. Genet.*, **18**, 315.
248. Irwin, M., 1928. *Proc. Soc. exp. Biol. Med.*, **26**, 125.
249. Izquierdo, L., 1955. *Arch. de Biol.*, **66**, 403.
250. Jelinek, O., 1894. *Zeit. wiss. Mikr.*, **11**, 242.
251. Jenner, L., 1899. *Lancet*, **1**, 370.
252. Jones, W., 1920. *Nucleic acids.* London (Longmans).
253. Jordan, B. M., & Baker, J. R., 1955. *Quart. J. micr. Sci.*, **96**, 177.
254. —— —— 1956. *Entomologist*, **89**, 126.
255. Jürgens, R., 1875. *Virchows Arch.*, **65**, 189.
256. Kaiserling, C., & Germer, R., 1893. Ibid., **133**, 79.
257. Kaplan, M. L., & Kaplan, L., 1933. *J. Bact.*, **25**, 309.
258. Karpechenko, G. D., 1924. *J. Genet.*, **14**, 375.
259. Kaufmann, B. P., Gay, H., & McDonald, M. R., 1950. *Cold Spr. Harb. Symp. quant. Biol.*, **14**, 85.
260. Kaufmann, C., & Lehmann, E., 1926*a*. *Centralbl. allg. Path.*, **37**, 145.
261. —— —— 1926*b*. *Virchows Arch.*, **261**, 623.
262. —— —— 1928. Ibid., **270**, 360.
263. Kehrmann, F., 1906. *Ber. deut. chem. Ges.*, **39**, 1403.
264. Kelley, E. G., 1939. *J. biol. Chem.*, **127**, 55.
265. Kimmelstiel, P., 1929. *Z. physiol. Chem.*, **184**, 143.
266. Kingsbury, B. F., 1912. *Anat. Rec.*, **6**, 39.
267. Kirk, R. E., & Othmer, D. F., 1947–1955. *Encyclopedia of chemical technology.* 14 vols. New York (Interscience).
268. Kisskalt, C., 1901. *Centralbl. f. Bakt.*, 1 Abt., **30**, 281.
269. Kleinenberg, N., 1879. *Quart. J. micr. Sci.*, **19**, 206.
270. Knecht, E., Rawson, C., & Loewenthal, R., 1916. *A manual of dyeing.* 2 vols. London (Griffin).
271. Koehring, V., 1930. J. Morph., **49**, 45.
272. Kohl, E. J., & James, C. M., 1931. *Science*, **74**, 247.
273. Kramer, H., & Hill, R. G., 1956. *Quart. J. micr. Sci.*, **97**, 313.
274. Krönig, B., & Paul, T., 1897. *Zeit. f. Hyg.*, **25**, 1.
275. Kull, H., 1914, *Anat. Anz.*, **45**, 153.
276. Kurnick, N. B., 1947. *Symp. quant. Biol.*, **12**, 141.
277. —— 1952. *Stain Tech.*, **27**, 233.
278. La Cour, L., 1931. *J. roy. micr. Soc.*, **51**, 119.
279. —— 1941, *Stain Tech.*, **16**, 169.
280. Lang, A., 1878. *Zool. Anz.*, **1**, 14.
281. —— 1879. Ibid., **2**, 45.
282. Langeron, M., 1949. *Précis de microscopie. Paris* (Masson).
283. Lasfargues, E., & Di Fine, J., 1950. *Bull. d'Hist. appl.*, **27**, 25.
284. Lassek, A. M., 1949. *Anat. Rec.*, **105**, 429.
285. —— 1950. Ibid., **107**, 409.

286. Lauterborn, R., 1893. *Verh. naturhist.-med. Vereins Heidelberg*, **5**, 179.
287. —— 1896. *Untersuchungen über Bau, Kernteilung und Bewegung der Diatomen*. Leipzig (Engelmann).
288. Laveran, A., 1899. *C. r. Soc. Biol.*, **51**, 249.
289. Lazarow, A., & Cooperstein, S. J., 1953. *J. Histochem. Cytochem.*, **1**, 234.
290. Leathes, J. B., 1925. *Lancet*, **1**, 957.
291. Lee, A. B., 1896. *La Cellule*, **11**, 223.
292. Leeuwenhoek, A., 1719. *Epistolæ physiologicæ super compluribus naturæ arcanis. Delphis* (Beman).
293. Leishmann, W. B., 1901. *Brit. med. J.*, **2**, 757.
294. Lendrum, A. C., 1935. *J. Path. Bact.*, **40**, 414.
295. —— 1944. *Brit. med. J.*, **2**, 44.
296. —— 1951. Chapter on 'The routine diagnostic technique of the general morbid anatomist' in *Recent advances in clinical pathology*, edited by S. C. Dyke. London (Churchill).
297. Lesher, S., 1951. *Exp. cell. Res.*, **2**, 577.
298. Levene, P. A., 1901. *Zeit. physiol. Chem.*, **32**, 541.
299. Levy, M., 1934. *J. biol. Chem.*, **105**, 157.
300. Lewis, F. T., 1942. *Anat. Rec.*, **83**, 229.
301. Lewis, M. R. & W. H., 1915. *Amer. J. Anat.*, **17**, 339.
302. Lewitsky, G., 1911. *Ber. deut. bot. Ges.*, **29**, 685.
303. Lewkowitsch, J., 1901. *The laboratory companion to fats and oils industries*. London (Macmillan).
304. Liebermann, C., & Kostanecki, St. V., 1887. *J. Soc. chem. Ind.*, **6**, 724.
305. Lillie, R. D., & Burtner, H. J., 1953. *Journ. Histochem. Cytochem.*, **1**, 8.
306. Lison, L., 1932. *Bull. d'Hist. appl.*, **9**, 177.
307. —— 1934. *C. r. Soc. biol.*, **115**, 202.
308. —— 1935*a*. *Arch. Biol.*, **46**, 599.
309. —— 1935*b*. *Bull. d'Hist. appl.*, **12**, 279.
310. —— 1953. *Histochimie et cytochimie animales: principes et méthodes*. Paris (Gauthier-Villars).
311. Lison, L., & Dagnelie, J., 1935. *Bull. d'Hist. appl.*, **12**, 85.
312. Lison, L., & Fautrez, J., 1939. *Protoplasma*, **33**, 116.
313. Lison, L., & Mutsaars, W., 1950. *Quart. J. micr. Sci.*, **91**, 309.
314. Lloyd, D. J., Marriott, R. H., & Pleass, W. B., 1933. *Trans. Farad. Soc.*, **29**, 554.
315. Lubs, H. A., 1955. *The chemistry of synthetic dyes and pigments*. New York (Reinhold).
316. Ludford, R. J., 1928. *Proc. Roy. Soc. B.*, **103**, 288.
317. Luther, R., 1904. *Chem. Centralbl.*, **8** (1), 571.
318. MacNeal, W. J., 1906. *J. infect. Dis.*, **3**, 412.
319. —— 1925. Ibid., **36**, 538.
320. Mallory, F. B., 1900. *J. exp. Med.*, **5**, 15.
321. Mann, G., 1894. *Z. wiss. Mikr.*, **11**, 479.
322. —— 1902. *Physiological histology*. Oxford (Clarendon Press).

323. Markley, K. S., 1947. *Fatty acids, their chemistry and physical properties*. New York (Interscience).
324. Marquardt, M., 1949. *Paul Ehrlich*. London (Heinemann).
325. Marquez, J., 1934. *C. r. Soc. Biol.*, **97**, 986.
326. Marston, H. R., 1923. *Biochem. J.*, **17**, 851.
327. Maximov, A., 1909. *Zeit. wiss. Mikr.*, **26**, 177.
328. May, R., & Grünwald, L., 1902. *Centralbl. inn. Med.*, **23**, 265.
329. Mayer, A., Schaefer, G., & Rathery, F., 1913. *C. r. Soc. Biol.*, **74**, 241.
330. Mayer, F., & Cook, A. H., 1943. *The chemistry of natural colouring matters*. New York (Reinhold).
331. Mayer, P., 1887. *Internat. Monatsschr. Anat.*, **4**, 37.
332. —— 1892*a*. *Mitt. zool. Stat. Neapel.*, **10**, 480.
333. —— 1892*b*. Ibid., **10**, 505.
334. —— 1918. *Zeit. wiss. Mikr.*, **35**, 161.
335. Mayzel, W., 1878. Quoted by Flemming.[170]
336. McClung, C. E., 1937. *Handbook of microscopical technique*. Oxford (University Press).
337. Medawar, P. B., 1941. *J. roy. micr. Soc.*, **61**, 46.
338. —— 1942. Lancet, 21 March, 350.
339. Merry, E. W., 1936. *The chrome tanning process*. London (Harvey).
340. Metzner, R., see Meves.[342]
341. Metzner, R., & Krause, R., 1928. Abderhalden's *Handbuch der biologischen Arbeitsmethoden*, Abt. V, Teil **2**, Halfte **1**, 325.
342. Meves, F., 1911. *Arch. mikr. Anat.*, **76**, 683.
343. Michaelis, L., 1900. *Arch. mikr. Anat.*, **55**, 558.
344. —— 1901. *Virchows Arch.*, **164**, 263.
345. —— 1902. *Einführung in die Farbstoffchemie für Histologen*. Berlin (Karger).
346. —— 1947. *Cold Spring Harb. Symp. quant. Biol.*, **12**, 131.
347. Michaelis, L., & Granick, S., 1945. *J. Amer. chem. Soc.*, **67**, 1212.
348. Millon, E., 1849. *C. r. Acad. Sci.*, **28**, 40.
349. Millot, J., & Giberton, A., 1927. *C. r. Soc. Biol.*, **97**, 1674.
350. Mittwer, T., Bartholomew, J. W., & Kallman, B. J., 1950. *Stain Tech.*, **25**, 169.
351. Miyagawa, I., 1937. *Chem. Zentr.*, **108** (1), 885.
352. Mladenovic, M., & Lieb, H., 1929. *Z. physiol. Chem.*, **181**, 221.
353. Möllendorff, W. von, 1918*a*. *Arch. mikr. Anat.*, **90**, 463.
354. —— 1918*b*. Ibid., **90**, 503.
355. Möllendorff, W. & M. v., 1924. *Ergebn. Anat. Entw.*, **25**, 1.
356. Möllendorff, W. von, & Tomita, T., 1926. *Zeit. Zellforsch. mikr. Anat.*, **3**, 1.
357. Möhlau, R., & Uhlmann, K., 1896. *Liebigs Ann.*, **289**, 90.
358. Mojsvar, A. M. E. von, 1879. *Leitfaden bei zoologisch-zootomischen Präparirübungen für Studirende*. Leipsig (Engelmann).
359. Morgan, G. T., & Smith, J. D. M., 1925. *J. Soc. Dyers & Col.*, **41**, 233.
360. Mori, S., 1921. *Trans. Jap. path. Soc.*, **11**, 57.
361. Moses, M. J., 1952. *Exp. cell Res., Suppl.* **2**, 75.

362. Müller, H., 1860*a*. *Verh. phys.-med. Ges. Würzburg*, **10**, 138.
363. —— 1860*b*. Ibid., **10**, 179.
364. Nassonov, D., 1923. *Arch. mikr. Anat.*, **97**, 136.
365. Neale, S. M., 1947. *J. Soc. Dyers & Col.*, **63**, 368.
366. Neurath, H., Greenstein, J. P., Putnam, F. W., & Erickson, J. O., 1944. *Chem. Rev.*, **34**, 157.
367. Newton, W. C. F., 1926. *J. Linn. Soc. Bot.*, **47**, 339.
368. Nicholson, W. C., 1916. *Amer. J. Anat.*, **20**, 329.
369. Nocht, –, 1898. *Centralbl. f. Bakt.*, 1 Abt, **24**, 839.
370. —— 1902. Personal communication to Giemsa.[187]
371. Noël, R., & Mangenot, G., 1922. *C. r. Soc. Biol.*, **87**, 1130.
372. Ockenden, D. W., & Schofield, K., 1951. *Nature*, **168**, 603.
373. Oguma, K., 1934. *Arch. de Biol.*, **45**, 27.
374. Okajima, K., 1912. *Z. wiss. Mikr.*, **29**, 67.
375. Orth, J., 1896. *Berl. klin. Woch.*, **33**, 273.
376. Ostwald, A., 1911. *Z. physiol. Chem.*, **74**, 290.
377. Overton, E., 1890. *Zeit. wiss. Mikr.*, **7**, 9.
378. —— 1900. *Jahrb. wiss. Bot.*, **34**, 669.
379. Owen, G., & Steedman, H. F., 1956. *Quart. J. micr. Sci.*, **97**, 319.
380. Palade, G. E., 1952. *J. exp. Med.*, **95**, 285.
381. Palmer, R., 1930. *J. roy. micr. Soc.*, **50**, 221.
382. Panijel, J., 1951. *Les problèmes de l'histochimie et la biologie cellulaire.* Paris (Hermann).
383. Pantin, C. F. A., 1946. *Notes on microscopical technique for zoologists.* Cambridge (University Press).
384. Pappenheim, A., 1899. *Virchows Arch.*, **157**, 19.
385. —— 1901. Ibid., **166**, 42.
386. —— 1911. *Folia haematol.*, **11**, 194.
387. Partington, J. R., & Huntingford, D. B., 1921. *J. roy. micr. Soc.* (no vol. number), 15.
388. Patten, B. M., & Philpott, R., 1922. *Anat. Rec.*, **21**, 393.
389. Pearse, A. G. E., 1954. *Histochemistry, theoretical and applied.* London (Churchill).
390. Perényi, J., 1882. *Zool. Anz.*, **5**, 459.
391. Perkin, A. G., & Everest, A. E., 1918. *The natural organic colouring matters.* London (Longmans, Green).
392. Perls, M., 1867. *Virchows Arch.*, **39**, 42.
393. Peterfi, T. (ed. by), 1928. *Methodik der wissenschaftlichen Biologie.* Berlin (Springer).
394. Peters, A., 1955*a*. *Quart. J. micr. Sci.*, **96**, 84.
395. —— 1955*b*. Ibid., **96**, 317.
396. Peters, R. H., 1945. *J. Soc. Dyers & Col.*, **61**, 95.
397. Petrunkevitch, A., & Pickford, G. E., 1936. *Anat. Rec.*, **65**, 461.
398. Pfeffer, W., 1886. *Unters. bot. Inst. Tübingen*, **2**, 179.
399. Pick, J., 1934. *Z. wiss. Mikr.*, **51**, 338.
400. Pictet, A., & Spengler, T., 1911. *Ber. deut. chem. Ges.*, **44**, 2030.
401. Pischinger, A., 1926. *Zeit. Zellforsch. mikr. Anat.*, **3**, 169.
402. —— 1937. Ibid., **26**, 249.
403. Plato, J., 1900. *Arch. mikr. Anat.*, **56**, 868.

404. Policard, A., Bessis, M., & Bricka, M., 1952. *Bull. Micr. appl.*, **2**, 29.
405. Porter, K. R., & Kallman, F., 1953. *Exp. cell Res.*, **4**, 127.
406. Preston, J. M., 1949. *Fibre science.* Manchester (Textile Institute).
407. Proescher, F., & Arkush, A. S., 1928. *Stain Tech.*, **3**, 28.
408. Putnam, F. W., 1953. Chapter on 'Protein denaturation' in *The proteins: chemistry, biological activity, and methods*, Vol. 1, Part B. New York (Academic Press).
409. Quekett, [J.], 1848. *Practical treatise on the use of the microscope.* London (Baillière).
410. Race, E., Rowe, F. M., & Speakman, J. B., 1946. *J. Soc. Dyers & Col.*, **62**, 372.
411. Radley, J. A., 1940. *Starch and its derivatives.* London (Chapman & Hall).
412. Ranvier, L., 1875. *Traité technique d'histologie.* Paris (Savy).
413. Réaumur, – de, 1742. *Memoire pour servir à l'histoire des insectes.* Vol. 6. Paris (Imprimerie Royale).
414. —— 1750. *Phil. Trans.*, **45**, 304.
415. Regaud, C., 1910. *Arch. d'Anat. micr.*, **11**, 291.
416. Remak, R., 1854. *Arch. Anat. Path. wiss. Med.* (no vol. number), 99.
417. Rettie, T., 1931. *J. Path. Bact.*, **34**, 595.
418. Reuter, K., 1901. *Centralbl. f. Bakt.*, 1 Abt., **30**, 248.
419. Richards, A. G., Anderson, T. F., & Hance, R. T., 1942. *Proc. Soc. exp. Biol. Med.*, **51**, 148.
420. Rodd, E. H. (ed. by), 1954. *Chemistry of carbon compounds.* Vol. 3, A. London (Elsevier).
421. Roe, M. A., Lillie, R. D., & Wilcox, A., 1940. *U.S. Public Health Rep.*, **55**, 1272.
422. Rogers, G. E., 1953. *Quart. J. micr. Sci.*, **94**, 253.
423. Romanowsky, D., 1891*a*. *St. Petersburger med. Woch.*, **16**, 297.
424. —— 1891*b*. Ibid., **16**, 307.
425. Romeis, B., 1928. *Taschenbuch der mikroskopischen Technik.* München (Oldenbourg).
426. Roque, A. L., 1954. *J. roy. micr. Soc.*, **74**, 188.
427. Roskin, G., & Semenoff, W., 1933. *Zeit. Zellforsch. mikr. Anat.*, **19**, 150.
428. Ross, K. F. A., 1953. *Quart. J. micr. Sci.*, **94**, 125.
429. —— 1954*a*. *Nature*, **174**, 836.
430. —— 1954*b*. *Quart. J. micr. Sci.*, **95**, 425.
431. —— personal communication.
432. Rothig, P., 1906. *Z. wiss. Mikr.*, **23**, 316.
433. —— 1907. Ibid., **24**, 109.
434. Rowe, F. M., 1924. *Colour index.* Bradford (Society of Dyers and Colourists).
435. Russell, S. M. McG., 1955. *Nature*, **175**, 301.
436. Sabin, F. R., 1937. A chapter on *Chemical agents: supravital stains* in McClung.[336]
437. Sandritter, W., 1955. *Z. wiss. Mikr.*, **62**, 283.
438. Sanfelice, F., 1918. *Ann. Inst. Pasteur*, **32**, 363.
439. Schaffer, J., 1918. *Anat. Anz.*, **51**, 353.

440. Schmidt, C. L. A., 1938. *The chemistry of the amino-acids and proteins*. Springfield, Ill. (Thomas).
441. Schmidt, W. J., 1939. *Nova Acta Leop.*, **7**, 1.
442. Schneider, A., 1880. *Anat. Anz.*, **3**, 252.
443. Schulemann, W., 1917. *Biochem. Zeitschr.*, **80**, 1.
444. Schultze, M., 1864. *Sitz. niederrhein. Ges. Natur- u. Heilk.* Bonn (no vol. number), 61.
445. Schultze, M., 1865. *Arch. mikr. Anat.*, **1**, 124.
446. —— 1866. Ibid., **2**, 163.
447. Schultze, M., & Rudneff, M., 1865. Ibid., **1**, 299.
448. Schulze, F., 1850. Quoted by Fürnrohr [181].
449. Schumb, W. C., Satterfield, C. N., & Wentworth, R. L., 1955. *Hydrogen peroxide*. London (Chapman & Hall).
450. Schunck, E., & Marchlewski, L., 1894. *Ber. deut. chem. Ges.*, **27**, 2979.
451. Schwarz, F., 1887. *Beit. Biol. Pflanzen* (Cohn), **5**, 1.
452. Seki, M., 1932*a*. *Fol. anat. Jap.*, **10**, 621.
453. —— 1932*b*. Ibid., **10**, 635.
454. —— 1933*a*. Ibid., **11**, 1.
455. —— 1933*b*. Ibid., **11**, 15.
456. —— 1933*c*. Ibid., **11**, 377.
457. —— 1933*d*. Ibid., **11**, 391.
458. —— 1933*e*. *Zeit. Zellforsch. mikr. Anat.*, **18**, 1.
459. —— 1933*f*. Ibid., **18**, 21.
460. —— 1933*g*. Ibid., **19**, 289.
461. —— 1936. Ibid., **24**, 186.
462. —— 1937. Ibid., **26**, 305.
463. Seligmann, A. M., Chauncey, H. H., & Nachlas, M. M., 1951. *Stain Tech.*, **26**, 19.
464. Shafiq, S. A., 1953. *Quart. J. micr. Sci.*, **94**, 319.
465. Shafiq, S. A., & Casselman, W. G. B., 1954. Ibid., **95**, 315.
466. Shelton, E., & Schneider, W. C., 1952. *Anat. Rec.*, **112**, 61.
467. Sheppard, S. E., & Geddes, A. L., 1944*a*. *J. Amer. chem. Soc.*, **66**, 1995.
468. —— —— 1944*b*. *J. Amer. chem. Soc.*, **66**, 2003.
469. Sidgwick, N. V., 1942. *The electronic theory of valency*. Oxford (University Press).
470. —— 1950. *The chemical elements and their compounds*. 2 vols. Oxford (Clarendon Press).
471. Simpson, W. L., 1941. *Anat. Rec.*, **80**, 173.
472. Singer, M., 1952. *Internat. Rev. Cytol.*, **1**, 211.
473. Singer, M., & Morrison, P. R., 1948. *J. biol. Chem.*, **175**, 133.
474. Sjöbring, N., 1900. *Anat. Anz.*, **17**, 273.
475. Sjövall, E., 1906. *Anat. Hefte*, 1 Abt., **30**, 259.
476. Smirnow, A., 1888. *Anat. Anz.*, **3**, 258.
477. Smith, A. L., 1921. *Lichens*. Cambridge (University Press).
478. Smith, G. M., 1915. *Trans. Amer. micr. Soc.*, **34**, 71.
479. Smith, J. E., 1947. *Phil. Trans. B*, **232**, 279.
480. Smith, J. L., 1908. *J. Path. Bact.*, **12**, 1.
481. Smith, J. L., & Mair, W., 1908. Ibid., **12**, 134.
482. —— —— 1911. *Skand. Arch. Physiol.*, **25**, 247.

483. Smith, J. L., & Thorpe, J. F., 1909. *J. Path. Bact.*, **13**, 14.
484. Smyth, J. D., 1954. *Quart. J. micr. Sci.*, **95**, 139.
485. Starke, J., 1895. *Arch. Anat. Physiol., physiol. Abt.* (no vol. number), 70.
486. Stearn, A. E. & E. W., 1928. *Stain Tech.*, **3**, 81.
487. Stedman, E. & E., 1943. *Nature*, **152**, 267.
488. —— 1947. *Cold Spring Harb. Symp. quant. Biol.*, **12**, 224.
489. Steedman, H. F., 1947. *Quart. J. micr. Sci.*, **88**, 123.
490. Stoeltzner, H., 1906. *Zeit. wiss. Mikr.*, **23**, 14.
491. Strangeways, T. S. P., & Canti, R. G., 1927. *Quart. J. micr. Sci.*, **71**, 1.
492. Strasburger, E., 1879. *Sitz. Jena Ges. Med. Naturw.* (no vol. number), 93.
493. Sylvén, B., 1951. *Unio internat. contra cancrum, Acta*, **7**, 708.
494. —— 1954. *Quart. J. micr. Sci.*, **95**, 327.
495. Taft, E. B., 1951. *Exp. cell Res.*, **2**, 312.
496. Tarkhan, A. A., 1931. *J. roy. micr. Soc.*, **51**, 387.
497. Tellyesniczky, K., 1898. *Arch. mikr. Anat.*, **52**, 202.
498. —— 1902. Ibid., **60**, 681.
499. —— 1905. Ibid., **66**, 367.
500. —— 1926. Article on 'Fixation' in R. Krause's *Enzyklopädie der mikroskopischen Technik*, vol. 2. Berlin (Urban & Schwarzenberg).
501. Thannhauser, S. J., Boncoddo, N. F., & Schmidt, G., 1951*a*. *J. biol. Chem.*, **188**, 423.
502. —— —— —— 1951*b*. Ibid. **188**, 427.
503. Thorpe, J. F., 1907. *Journ. chem. Soc.* (Trans.), **91**, 324.
504. Thorpe, J. F., & Whiteley, M. A., 1937–1954. *Thorpe's dictionary of applied chemistry*. 11 vols. London (Longmans, Green).
505. Tolstoouhov, A. V., 1928. *Stain Tech.*, **3**, 49.
506. Trembley, A., 1744. *Mémoires, pour servir à l'histoire d'un genre de polypes d'eau douce, à bras en forme de cornes.* Leide (Verbeek).
507. Tuan, H.-C., 1930. *Stain Tech.*, **5**, 135.
508. Turchini, J., 1919. *C. r. Soc. Biol.*, **82**, 1134.
509. —— 1922. *Arch. Morph. gén. exp.*, **11**, 3.
510. Underhill, B. M. J., 1932. *Journ. roy. micr. Soc.*, **52**, 113.
511. Unna, P. G., 1887. *Dermat. Stud.*, **4**, 1.
512. —— 1890. *Monat. prakt. Dermatol.*, **11**, 366.
513. —— 1891. *Zeit. wiss. Mikr.*, **8**, 475.
514. —— 1895. *Monatsch. prakt. Dermatol.*, **20**, 365.
515. —— 1911. *Arch. mikr. Anat.*, **78**, 1.
516. —— 1913. *Virchows Arch.*, **214**, 320.
517. —— 1915. *Arch. mikr. Anat.*, **87**, 96.
518. —— 1922. *Centralbl. f. Bakt.*, 1 Abt., **88**, 159.
519. Unna, P. G., & Tielemann, E. T., 1918. Ibid., 1 Abt., **80**, 66.
520. Vakaet, L., 1952. *Biol. Jaarbook*, **19**, 192.
521. Vendrely, C., 1950. *Arch. d'anat. d'hist. d'emb.*, **33**, 113.
522. Vendrely-Randavel, C., 1949. *Acta anat.*, **7**, 225.
523. Venkataraman, K., 1952. 'The chemistry of synthetic dyes'. 2 vols. New York (Academic Press).

524. Vickerstaff, T., 1950. *The physical chemistry of dyeing*. London (Oliver & Boyd).
525. Vigelius, W. J., 1886. *Mitt. zool. Stat. Neapel*, **6**, 499.
526. Virchow, H., 1885. *Arch. mikr. Anat.*, **24**, 117.
527. Wahrman, J., & Zahavi, A., 1953. *Bull. res. Council Israel*, **3**, 265.
528. —— —— 1955. *Nature*, **175**, 600.
529. Watson, J. M., 1943. *J. roy. micr. Soc.*, **63**, 20.
530. Weigert, C., 1884. *Z. wiss. Mikr.*, **1**, 290.
531. Weigl, R., 1910. *Bull. internat. Acad. Sci. Cracovie*, Series B (no vol. number), 691.
532. Weil, A., 1929. *J. biol. Chem.*, **83**, 601.
533. Werner, A., 1908. *Ber. deut. chem. Ges.*, **41**, 1062.
534. Westphal, E., 1880. Chapter in P. Ehrlich's *Farbenanalytische Untersuchungen zur Histologie und Klinik des Blutes*, p. 17.
535. Wetzel, G., 1920. *Arch. mikr. Anat.*, **94**, 568.
536. Wheeler, H. L., & Jamieson, G. S., 1905. *Amer. chem. J.*, **33**, 365.
537. Whitman, C. O., 1885. *Methods of research in microscopical anatomy and embryology*. Boston (Cassino).
538. Wiame, J. M., 1947. *Biochem. Biophys. Acta*, **1**, 234.
539. Wieland, H., & Scheuing, G., 1921. *Ber. deut. chem. Ges.*, **54**B, 2527.
540. Wigglesworth, V. B., 1952. *Quart. J. micr. Sci.*, **93**, 105.
540a. Wilson, E. B., 1925. *The cell in development and inheritance*. New York (Macmillan).
541. Wolman, M., 1955*a*. *Internat. Rev. Cytol.*, **4**, 79.
542. —— 1955*b*. *Quart. J. micr. Sci.*, **96**, 337.
543. Wolman, M., & Greco, J., 1952. *Stain Tech.*, **27**, 317.
544. Woodroffe, D., 1941. *Fundamentals of leather science*. Waddon (Harvey).
545. Woollard, H. H., Weddell, G., & Harpman J. A., 1940. *Journ., Anat.*, **74**, 413.
546. Wormell, R. L., & Kaye, M. A. G., 1944. *Nature*, **153**, 525.
547. Wright, J. H., 1902. *J. med. Res.*, **7**, 138.
548. Wüstenfeld, E., 1956. *Z. wiss. Mikr.*, **63**, 86.
549. Yost, D. M., & White, R. J., 1928. *J. Amer. chem. Soc.*, **50**, 81.
550. Young, J. Z., 1935. *Nature*, **135**, 823.
551. Young, R. T., 1928. *Anat. Rec.*, **40**, 351.
552. Zeiger, K., 1930. *Z. wiss. Mikr.*, **47**, 273.
553. —— 1938. *Physikochemische Grundlagen der histologischen Methodik*. Dresden (Steinkopff).
554. Zenker, K., 1894. *Münch. med. Woch.*, **41**, 532.
555. Zimmermann, A., 1892. *Die botanische Mikrotechnik*. Tübingen (Laupp).
556. Zirkle, C., 1928*a*. *Protoplasma*, **4**, 201.
557. —— 1928*b*. Ibid., **5**, 511.
558. —— 1933. Ibid., **18**, 90.
559. Zulkowsky, K., & Peters, K., 1891. *Sitz. kais. Akad. Wiss. Wien, math. naturwiss. Classe*, **99**, Abt. 2B, 251.

Index

Where two or more consecutive pages are entirely devoted to the same subject, only the first is mentioned in the Index.

Each reference to a particular subject that is distinctly more important than other references to the same subject, is distinguished by the printing of the page-number in heavy type.